Contemporary Issues in Technology Education

Contemporary Issues in Technology Education - About this series

Technology education is a developing field, new issues keep arising and timely, relevant research is continually being conducted. The aim of this series is to draw on the latest research to focus on contemporary issues, create debate and push the boundaries in order to expand the field of technology education and explore new paradigms. Maybe more than any other subject, technology education has strong links with other learning areas, including the humanities and the sciences, and exploring these boundaries and the gaps between them will be a focus of this series. Much of the literature from other disciplines has applicability to technology education, and harnessing this diversity of research and ideas with a focus on technology will strengthen the field.

More information about this series at http://www.springer.com/series/13336

P John Williams • Kay Stables
Editors

Critique in Design and Technology Education

Springer

Editors
P John Williams
Science and Maths Education Centre (SMEC)
Curtin University
Perth, WA, Australia

Kay Stables
Goldsmiths, University of London
London, UK

ISSN 2510-0327 ISSN 2510-0335 (electronic)
Contemporary Issues in Technology Education
ISBN 978-981-10-3104-5 ISBN 978-981-10-3106-9 (eBook)
DOI 10.1007/978-981-10-3106-9

Library of Congress Control Number: 2017930627

Printed on acid-free paper

This Springer imprint is published by Springer Nature
The registered company is Springer Nature Singapore Pte Ltd.
The registered company address is: 152 Beach Road, #21-01/04 Gateway East, Singapore 189721, Singapore

This book is dedicated to the memory of Professor Chitra Natarajan.

In 2014, Chitra was a professor at the Homi Bhabha Centre for Science Education, when she committed to contributing a chapter for this book. She developed a rough draft for the chapter before realising late in 2014 that she was too ill to continue. She passed away in April 2015, while all the authors were meeting in France at a workshop to critique and develop the chapters for the book.

Chitra personified what is important in a scholar. While her scholarship was of the highest standard, she was always approachable and accessible, and respectful of other people's positions. Her contribution to Technology Education and, in particular, to this book is greatly missed. But the contribution she made in her lifetime lives on to inspire and inform us all, as is reflected in "Embracing lives, chasing passions: Memoirs of Chitra Natarajan" by R. Rajagopal. His words capture the wonderful essence of Chitra as "a staunch practitioner of collaborative and multidisciplinary learning-teaching model, she left an indelible footprint in research at the interface of science, technology and society; design and technology education; project based learning; role of diversity in science education; and environmental education. ... She wrote extensively and reflectively on scientific, technological, educational and societal issues; and left behind a rich legacy of invaluable visions and expressions."

Contents

Contributors

Cecilia Axell Department of Social and Welfare Studies, Linköping University, Norrköping, Sweden

David Barlex Independent Consultant, Leicestershire, UK

Marc J. de Vries Delft University of Technology, Delft, The Netherlands

Jacques Ginestié Aix-Marseille Université, Marseille Cedex 4, France

Mishack T. Gumbo University of South Africa, Pretoria, South Africa

Steve Keirl Goldsmiths, University of London, London, UK

Susan V. McLaren University of Edinburgh, Edinburgh, UK

Bill Nicholl University of Cambridge, Cambridge, UK

Stephen Petrina The University of British Columbia, Vancouver, BC, Canada

Niall Seery Department of Design and Manufacturing Technology, University of Limerick, Limerick, Ireland

David Spendlove The University of Manchester, Manchester, UK

Kay Stables Goldsmiths, University of London, London, UK

Belinda von Mengersen National School of Arts, Australian Catholic University, Sydney, NSW, Australia

Terry Wilkinson York University, Toronto, ON, Canada

P John Williams Science and Maths Education Centre (SMEC), Curtin University, Perth, WA, Australia

About the Authors

Cecilia Axell is a postdoctoral researcher in technology education at the Department of Social and Welfare Studies, Linköping University, Sweden. She holds a PhD in technology education. Her research focuses on the pedagogic content of fictional children's books which can link human relationships to technology. In her PhD thesis, she examined how technology and technological development in society are discussed in relation to views of nature and views about the future in a selection of Swedish children's books written in the last century.

David Barlex is an acknowledged leader in design and technology education, curriculum design and curriculum material development. He taught in comprehensive schools for 15 years before taking university positions in teacher education. He directed the Nuffield Design and Technology Project and was educational manager for Young Foresight. David is well known for his interest and expertise in developing curriculum materials that support pupil learning from a constructivist perspective. He uses this approach to develop young peoples' ability to understand and critique the design decisions made by professional designers and those they make themselves in design and technology lessons.

Dr. Marc J. de Vries is professor of science and technology education at Delft University of Technology in the Netherlands and affiliate professor of Christian philosophy of technology at the same institute. He is also assistant professor of philosophy and ethics of technology at Eindhoven University of Technology, the Netherlands. Currently, he is the editor-in-chief of the *International Journal of Technology and Design Education* (Springer) and series editor of the *International Technology Education Studies* book series (Sense Publishers). He is the chairperson of the board of the PATT Foundation that organises the international Pupils' Attitudes Towards Technology conferences. He wrote a monograph on philosophy of technology for technology educators (*Teaching About Technology*; Springer) and a history of 80 years of Philips Research (Amsterdam University Press).

Prof. Jacques Ginestié is the current director of the School of Education (ESPE) at Aix-Marseille University, France's largest university. He manages the research unit ADEF and the federation SFERE-Provence, which coordinates 16 research units on the theme 'Learning and Education'. He has researched in technology education for many years, notably about the efficiency of teaching-learning processes with a particular interest in the role of technical languages and for the interactions between teacher, student and knowledge. He has focused on teacher education and impacts on educational policies. He is a member of several editorial and scientific committees and academic societies.

Mishack T. Gumbo is a professor of technology education (Indigenous Knowledge Systems). He is attached to the Department of Science and Technology Education in the College of Education at the University of South Africa. His duties include teaching undergraduate technology education students and supervision and facilitation of workshops for master's and doctoral students. He has published papers in scholarly journals, contributed chapters in books, wrote a few books and presented papers at national and international conferences. His areas of interest are IKS, technology education teachers' PCK and distance education and e-learning. He is currently leading a strategic intervention project in MSTE in Mpumalanga Province. He is coeditor of two books and a lead editor of one book, all of which are nearing their completion. He serves on ITEEA and Africa Update Boards and is a member of SAARMSTE. He is external examiner of universities nationally and internationally.

Steve Keirl is a curriculum theorist whose research advocates technological and design literacy that is critically, ethically and democratically focused. As such, he strongly resists STEM as a curriculum construct. Steve's research helped inform the 2001 (Birth-Age 18) South Australian Design and Technology Curriculum for which he was chair of the Technology Expert Working Group and subsequently principal author. Critiquing, as one of three strands running throughout that curriculum, was then a 'world first' and has subsequently been recognised internationally for its value to design and technology education. Steve is currently reader in design education at Goldsmiths, University of London.

Susan V. McLaren is a senior lecturer (design and technology education) at the University of Edinburgh, Scotland. Her research interests include pedagogies and principles of design for sustainability and circular economy thinking, discriminators of creativity and progression, real-time multimodal formative assessment, curriculum planning and professional development, interdisciplinary learning and partnership working.

Bill Nicholl After studying design, Bill taught design and technology (D&T) for 9 years in London high schools. In 2000, he set up the course for training teachers of D&T at the University of Cambridge. He teaches across a number of courses including PGCE, MEd and MPhil level, as well as supervises PhD students. His research interests are focused around sociocultural approaches to creativity and in particular the role of the teacher in teaching for [design] creativity. He is on the advisory board for the Center of Excellence for Technology Education (CETE), as well as the advisory boards for two international journals. He has published in peer-reviewed journals and has worked in China, Poland, Ireland and India.

Stephen Petrina specialises in science and technology studies (STS), STEM education and curriculum studies. He is in the Department of Curriculum and Pedagogy at the University of British Columbia. He manages the How We Learn (Media and Technology Across the Lifespan) Lab, funded by the Social Sciences and Humanities Research Council of Canada. Stephen is co-founder and codirector of the Institute for Critical Education Studies. He has recently published in *Hacking Education in a Digital Age*, *British Journal of Educational Technology*, *New Frontiers in Technological Literacy*, *Workplace: A Journal for Academic Labor* and *History of Education Quarterly*.

Dr. Niall Seery is a lecturer in initial teacher education at the University of Limerick and is course director for the Technology Teacher Education programme. Niall also serves as an associate professor at the KTH Royal Institute of Technology in the area of technology education. In 2010, he established and is currently the director of the Technology Education Research Group (www.TERG.ie) at the University of Limerick, which aims to advance technology education and support the continuous development of practitioners, initial teacher education students and second level pupils. He has conducted a number of research projects focusing on pupil learning and assessment and actively supervises doctoral level students. Niall has also worked on a number of funded research projects.

David Spendlove is head of initial teacher education at the University of Manchester. In addition to writing books on assessment and emotional literacy, David has written extensively on design and technology education for almost 20 years. He is a coeditor of *Design and Technology Education: An International Journal*; a member of the editorial board for the *International Journal of Innovation, Creativity and Change*; and a member of DRS/Cumulus/Design-Ed International Scientific Review Committee and Review Board. In 2013, David was awarded an 'Outstanding Contribution' award by the Design and Technology Association.

Kay Stables Kay started her career as a textiles teacher, joining Goldsmiths, University of London, in 1986 as a researcher on the UK Government's Assessment of Performance Unit's Design and Technology project. A founder member of the Technology Education Research Unit (TERU), she has directed and contributed to projects in primary and secondary education in the UK and overseas. With Richard Kimbell, she authored the TERU retrospective, Research Design Learning (2007). More recently, together with Steve Keirl, she published *Environment, Ethics and Cultures: Design and Technology Education's Contribution to Sustainable Global Futures (2015)*, an edited collection of theoretical and practice-based approaches. Recent research has focused on design, creativity and sustainable development, the use of digital tools in assessment (the e-scape project) and designerly well-being. Current research includes creating dialogic frameworks for supporting the development of D&T capability, including in digital environments. Kay has a keen interest in international issues in design and technology education and has been a visiting scholar at institutions in Australia, Canada, New Zealand, Sweden and the USA. She is currently coeditor of *Design and Technology Education: An International Journal* and is an editorial board member of the *Journal of Technology Education*.

Belinda von Mengersen is a lecturer in technology and course coordinator for the Bachelor of Teaching/Bachelor of Arts (Technology), National School of Arts at Australian Catholic University. She specialises in textiles technology and textiles and design pre-service education. A background in visual art and design education and research informs her teaching and research in design and technology education. Research interests include reflective writing, creative thinking, conceptual thinking, critical thinking, design thinking, practice-led research, e-portfolios, comparisons between pedagogical practice in visual art and design and design and technology and the inherently interdisciplinary and dynamic nature of design-related fields like textiles, design and design and technology.

P John Williams is professor and director of the Science and Mathematics Education Centre (SMEC) at Curtin University in Perth, Western Australia, where he teaches and supervises research students in technology education. Apart from Australia, he has worked and studied in a number of African and Indian Ocean countries and in New Zealand and the USA. He directed the nationally funded *Investigation into the Status of Technology Education in Australian Schools*. His current research interests include mentoring beginning teachers, PCK and electronic assessment of performance. He regularly presents at international and national conferences, consults on technology education in a number of countries and is a long-standing member of eight professional associations. He is the editor of the *Australasian Journal of Technology Education*, advisory editor of the *International Journal of Technology and Design Education* and series editor of the Springer *Contemporary Issues in Technology Education* and is on the editorial board of five other professional journals. He has authored or contributed to over 220 publications, and in 2011, he was elected to the International Technology and Engineering Education Association's Academy of Fellows for prominence in the profession.

Terry Wilkinson (BFA, BEd, MEd) is currently a doctoral candidate in the Faculty of Education at York University (Canada). Over her career, she has worked with children with cognitive and behavioural exceptionalities in formal and informal settings and enjoyed teaching design and technology and science and technology in middle school classrooms across the Toronto District School Board. Research interests in critical and visual literacies, design thinking and problem-solving have also informed Terry's practice as an instructor and sessional lecturer for the initial teacher and continuing education programmes at the Ontario Institute for Studies in Education, University of Toronto.

Critique in Design and Technology Education: About the Book

P John Williams and Kay Stables

The goal of this book is to set a historical, philosophical and pragmatic context for critique in design and technology education and provide a framework for critique and discussion about the integration of critique into the teaching and learning of design and technology in schools. The wonderfully diverse discussion and application of notions of critique attest to the diversity of the eminent design and technology education researchers who have contributed chapters to this book.

The term design and technology (D&T) education is used throughout the book to designate the curriculum area, although different terms such as technology, technologies, technology and enterprise and technology and engineering may be used in different countries. Authors may use other related terms when referring to specific examples.

In some quarters, there is resistance to the term *critique* as pretentious jargon, suggesting alternative terms such as review, reflect or analyse are adequate substitutions. However, the history of notions of critique, its philosophical roots and its implications go well beyond reviews, reflections or analyses. In fact in the chapter "Critiquing Teaching: Developing Critique Through Critical Reflection and Reflexive Practice", Susan McLaren suggests it is time to reconceptualise notions of reflection to develop something more meaningful.

Many of the authors of chapters in this book define critique in a way that suits the context of their discussion. This may seem initially confusing, and it may have made the book seem more unified if there was just one definition. However, the variety of definitions is indicative of the rich discourses that pervade the design and

PJ. Williams (✉)
Science and Maths Education Centre (SMEC), Curtin University, Perth, WA, Australia
e-mail: pjohn.williams@curtin.edu.au

K. Stables
Goldsmiths, University of London, London, UK
e-mail: k.stables@gold.ac.uk

PJ. Williams, K. Stables (eds.), *Critique in Design and Technology Education*,
Contemporary Issues in Technology Education, DOI 10.1007/978-981-10-3106-9_1

technology education profession and provide opportunity for debate and, indeed, critique. A critique in technology is a systematic analytical assessment of an element of technology – it may be a technology itself, a product, a process or a material. In a holistic sense, it is an element of a person's technological literacy, a fundamentally critical disposition that is brought to bear on all things technological. In the chapter "Critiquing Teaching: Developing Critique Through Critical Reflection and Reflexive Practice", Susan McLaren applies critique to the art of teaching and in that context describes it as effortful, uncomfortable and disruptive. In the chapter "The Identification and Location of Critical Thinking and Critiquing in Design and Technology Education", David Spendlove provides a fairly straightforward definition that considers critical thinking as *reflective thinking focused on deciding what to believe or do*. He goes on to address the semantics of the terminology in that the 'critical' can be considered as the process, whilst the 'critique' can be considered the outcome. Steve Keirl in the chapter "Critiquing as Design and Technology Curriculum Journey: History, Theory, Politics and Potential" proposes critique is both a noun and a verb, and for John Williams in the chapter "Critique as a Disposition", critique as a disposition is a verb.

The first group of five chapters in this book deals with a range of bases for critique: philosophical, historical, designerly, thinking and knowledge systems.

In the chapter "Philosophy as Critique", Marc de Vries presents the base of philosophy as a fundamental and enduring way of critiquing reality (including design, engineering, technology and everything related), and the strength of philosophical critique is that it is based on philosophy's primary method, namely, argumentation. He distinguishes two different ways in which philosophy does its critical work. The first critique refers to what so-called 'analytical philosophers' do: they analyse terminology. In the chapter "Alternative Knowledge Systems", Mishack Gumbo takes up this point and incorporates it into his critique of Western knowledge systems, a world view which restricts the consideration of alternative world views. The second type of critique for which philosophy can be a useful tool is related to debate itself. This is what people will intuitively associate with the term 'philosophy': dealing with the big questions.

In the chapter "Critique of Technology", Stephen Petrina provides a history of the critique of technology, which in this chapter begins with the spiritual critique of technology and proceeds historically through cultural criticism and social, psychic, ontic and identic critiques. Differentiated from the spiritual critique that preceded it, cultural criticism of technology emerged in the fifteenth and sixteenth centuries as a mode of describing and depicting the mechanical arts. In the eighteenth and nineteenth centuries, spiritual critique was displaced through a rejection of religion and theology as sources of modern authority. With the spiritual ground undermined, social, psychic, ontic and identic critics of media and technology compete for defensible ground for leverage of their particular critique. Stephen proposes that critique of technology is currently in historical decline and suggests the need for a renewal of spiritual critique to come full circle in this history of critique. He does, however, somewhat pessimistically conclude with Latour (2005) that critical discourse has of late become impotent. It has no leverage point left; critics of technology have no reliable or stable ground for their critiques.

The position Kay Stables takes in the chapter "Critiquing Design: Perspectives and World Views on Design and Design and Technology Education, for the Common Good" is that the nature of design provides a fundamental basis for critique. As a way of broadening the perceptions of design and designing within the subject of design and technology, Kay provides a critique from historic, social, cultural and sustainable perspectives. This focus is derived from a concern that many learning experiences provided in design and technology education in schools are not consistently as challenging, engaging and meaningful as they could be and that it is often a limited understanding of design's importance and potential that underpins a restricted approach.

Kay outlines the shifting thinking and understanding of design, and a designer's role, that has significantly changed the territory and scope of design, particularly within the last 50 years. Changes have emerged through designers' questioning and challenging, shifting discourses by positioning design not so much as a thing to be critiqued but as a tool for critique in its own right. Critical and speculative designs are proposed as ways of operating outside the tight constraints of design that simply service industry and so reinforce global capitalist values. This form of design offers a wide range of opportunities for design to pose questions, provoke and inspire. It provides a useful and invigorating stance on critique as 'not necessarily negative; it can be a gentle refusal, a turning away from what exists, a longing, wishful thinking, a desire, and even a dream. Critical designs are testimonials to what could be, but at the same time, they offer alternatives that highlight weaknesses within existing normality' (Dunne and Raby 2013, pp. 34–35).

The goal in this discussion is to open up new possibilities for thinking about what design could be in the context of design and technology education in mainstream schooling. The forms of design that currently exist within this area of schooling vary from country to country and in many ways are driven by the history and culture of design, technology and education within local and national settings. However, there are some common threads highlighted in critiquing design that transcend national contexts, such as the need to place design and technology activities in socially and culturally relevant contexts, to recognise plurality in approaches to designing and to embed an ecological critique that respects a broader canvas of world views – ideas that are picked up later in the book by Mishack Gumbo in the chapter "Alternative Knowledge Systems" and Terry Wilkinson in the chapter "Politicizing the Discourse of Consumerism: Reflections on the Story of Stuff".

David Spendlove takes a more personal approach in the chapter "The Identification and Location of Critical Thinking and Critiquing in Design and Technology Education", developing an argument that engaging in a process of critical thinking leading to critique facilitates agency and self-understanding when engaged in design activities. He bases this on a conceptual framework of three main theories: critical theory, critical pedagogy and critical design, the intersection of which is the location of design and technology critical thinking. In the chapter "Critiquing as Design and

Technology Curriculum Journey: History, Theory, Politics and Potential", Steve Keirl also uses critical theory and its educational forms of critical literacy and critical pedagogy as a source of critiquing in design and technology.

David proposes that agency, within a critical design-thinking framework, is the intentional ability to exercise some control over one's thinking, environment and subsequent existence and action. This notion of intentional ability is aligned with John Williams' discussion (chapter "Critique as a Disposition"), which focuses on dispositions and how design and technology teachers can support the formation of specific dispositions. David's argument provides a counterpoint for the somewhat pessimistic conclusions of Stephen Petrina in the chapter "Critique of Technology", by exploring and developing an 'escape hatch' from the reproduction of orthodox practices and demonstrating how thinking can be opened through engaging in critical and metacognitive decision-making processes. In doing so, David provides further insight into the proposals made by Kay Stables in the chapter "Critique as a Disposition" for developing agency through more critical, speculative approaches.

Mishack Gumbo in the chapter "Alternative Knowledge Systems" critiques the knowledge system of design and technology, reminding readers that there are a range of different knowledge forms, each informed by culture and context. Mishack uses indigenous knowledge systems (IKS) as a counterpoint to Western knowledge systems (WKS) as a way to critique design and technology education. Taking as his starting point that the popular version of universal Western knowledge is currently but wrongly promoted as global knowledge, the technological knowledge that is being taught to students and how it is taught has generally not been critical of the domination of Western approaches. Rather, it has complied with dominant discourses because of their control function through which they determine what can be said and thought, and who can speak, when and with what authority.

Mishack posits that this is not just a problem of the knowledge used in education but that in a broader sense, some solutions to the problems facing Africa, for instance, lie in the need to understand the dynamics of indigenous knowledge. Steve Keirl in the chapter "Critiquing as Design and Technology Curriculum Journey: History, Theory, Politics and Potential" explores this issue as a binary approach to knowledge or literacy – 'cultural literacy', based on the Anglo-colonial literary canon for an elite, and 'functional literacy' for everyone else. Mishack's critique invites a culturally inclusive design and technology curriculum, which offers students equal opportunities and multiple perspectives to facilitate and broaden their understanding of technology, whilst at the same time ensuring dignity of all knowledge forms and accommodation of indigenous cultures.

The second group of five chapters applies notions of critique to certain overarching aspects of design and technology education.

In the chapter "Critiquing as Design and Technology Curriculum Journey: History, Theory, Politics and Potential", Steve Keirl elaborates critiquing as a way of thinking, acting and being in design and technology education. He defines critiquing as the purposeful, practical and metaphorical deconstruction and analysis of any product, process or system in order to expose the values and intentions behind designs, the unanticipated applications of technologies and the relationships between people and technologies. As with designing, new meanings and knowledge emerge from critiquing, and new realisations emerge for seeing, judging and living in the designed world.

In this chapter, Steve situates his discussion of critique as a developmental journey towards a design and technology curriculum in which critiquing is fundamental to design and technological literacy. He reminds us of the substantial role that critical theory has played in bringing critique to education: 'Rather than celebrating objectivity and consensus, teachers must place the notions of critique and conflict at the centre of their pedagogical models... Critique must become a vital pedagogical tool' (Giroux 1983, p. 62). Steve usefully analyses the relationship between critiquing and designing and suggests that they both:

- develop socially valuable attributes in students
- develop thinking styles and confidence
- are valid components of D&T curriculum for all students
- are valid components of general democratic education for all students
- reject fact learning or rote learning
- are necessary for arriving at a best defensible compromise
- are undervalued in organisations

Critiquing is a tool that serves the design enterprise. In fact, effective designing demands deep critiquing.

In the chapter "Critique as a Disposition", John Williams proposes that educators should approach critique as a disposition to be developed. Dispositions are concerned with not only what a student can do but what a student is disposed to do, so addressing the often prevalent gap between abilities and actions. The essentiality of action in a disposition aligns with the manifold notions of activity within design and technology education – an education that is not passive – conceptually, it involves the construction of new knowledge, and practically, it involves movement and action and construction. Design and technology education therefore goes beyond the possibly conceptual although activity-based notions of technological literacy and ensures an activity-based end point. This aligns well with Steve's well-argued position in the chapter "Critiquing as Design and Technology Curriculum Journey: History, Theory, Politics and Potential" that critiquing is a fundamental component of design and technological literacy.

The careful structuring of classroom activities in sequences that are designed to elicit dispositions to critique is a fundamental design and technology teaching activity. In the chapter "Empathy as an Aspect of Critical Thought and Action in Design and Technology", Bill Nicholl extends John's discussion to explore the dispositional dimension to critical thinking in relation to empathy and also utilises the notion of critical thinking dispositions as a goal that can be achieved through 'inclusive design', a user-centred approach to designing.

The position Bill takes in the chapter "Empathy as an Aspect of Critical Thought and Action in Design and Technology" is that inclusive design is crucial for the development of critical thinking dispositions and for engendering empathy. To this end, designers have developed a range of 'signature pedagogies' that have been introduced into schools and include user observation, focus groups, simulation and role playing. This introduction into schools is supported by Susan McLaren in the chapter "Critiquing Teaching: Developing Critique Through Critical Reflection and Reflexive Practice" where she alludes to the processes of teaching as designing and teacher as designer.

Bill suggests that direct and indirect empathy tools and, in particular, the use of role-play and perspective taking (Mead 1934) can be used iteratively and make up some of the signature pedagogies that help form habits of the mind as they 'provide the early socialisation into the practices and values of the field' (Shulman 2005, p. 59). In order to get closer to the lives and experiences of users, understanding their situation as well as their feelings requires an empathic approach to designing, which is part of an overall disposition to think critically.

Throughout the book, there is a call to reconceptualise reflection and develop more meaningful critique within the context of design and technology teaching, but this is perhaps most clearly articulated by Susan McLaren in the chapter "Critiquing Teaching: Developing Critique Through Critical Reflection and Reflexive Practice". Active engagement with critique of teaching practice facilitates professional learning and professional development. It is through critique that teachers can ensure they are professional, fluid and informed in their responses as, and when, scenarios and contexts demand and true to their personal ethics. In order to be in a position to engage in critique of teaching, a teacher needs to accept that their personal values can be challenged by others. This frame of mind demands that design and technology teachers develop the skills to critique *what* they are doing and *why* they are doing *what* they are doing.

The value of critical reflection lies in developing informed practice through professional learning, creative growth and critique, which has the potential to transform teaching.

This is an iterative and continuous process, more a frame of mind or ongoing habit (a disposition maybe, according to John in the chapter "Critique as a Disposition"), reappraised as their career progresses, as society and the needs of learners change and as understanding of design and technology, and design and technology education, develops.

Like Stephen Petrina in the chapter "Critique of Technology", Jacques Ginestié also uses a historical approach as the basis for chapter "A Critique of Technology Education for All in a Social and Cultural Environment", not to trace the history of technology education but to develop understandings about why the original ambitions for the area have not been achieved. To accomplish this, he uses an epistemological and sociocultural critique to develop a broad understanding of the world in which humans are living and developing – recognising that this is ultimately educational whenever societies organise schools for the transmission and development of knowledge. This includes the interrelations between school subjects and the efficiency of the teaching-learning process.

Jacques concludes that there is no simple answer to the current disaffection of pupils and lack of learning progress. A very old elitist tradition, reinforced by the strong logic of academic subjects, contradicts the democratic principles of free education for every child, whatever his or her social origins, for ensuring equal opportunities. In France, the introduction of 'technology education for all' was proposed as an answer to the vocational system that helped France move from a rural to an industrial economy, but this approach has since been overtaken by social changes. Design and technology was proposed as a place where pupils no longer made things but the place where they understood why.

Two major lines of action emerge from this educational policy. The first axis concerns the curricula relating to design and technology and also how it relates to other subjects. The second is the need for continued professionalisation of design and technology teachers. In France, the integration of these two axes has resulted in a recasting of schools. This has involved changes in curricular structures, including emphasising the interdisciplinary, rethinking the academicism of the learned knowledge and paving the way for educational practices that promote a project-based approach and problem-solving. The generalised message from this chapter is that the individuation-socialisation dialectic is another way to think of the place of design and technology as an essential part of modern education for all.

The final set of chapters extends understandings of critique in design and technology education by considering pedagogic practices that allow such understandings to be used transformatively to build the critiquing capabilities of young people.

In the chapter "Disruptive Technologies", David Barlex takes the topic of disruptive technologies and explores how these can be used as a way to support learners to develop skills in critiquing technology. He focuses explicitly on practices and applications of critique and how these can be used within learning and teaching, moving critique beyond the academic, into a proactive position that deepens learners' understanding of technologies and their ability to challenge, rather than accept, technological developments. By drawing on a categorisation created by the McKinsey Global Institute (Manyika et al. 2013), he suggests a framework of criteria that define what a disruptive technology is and identifies a list of nine technologies he considers appropriate and relevant for learners to critique, such as additive manufacturing, artificial intelligence, big data, the Internet of Things and robotics. Maintaining his position in opening up learning and teaching practices, he introduces approaches that can be used to critically explore new technologies and the disruptive impacts they may have. Scenario building is presented as one such approach that can be used as a way of exploring his framework of criteria, and how this might be used in the context of robotics is explored in considerable detail.

Whilst the chapter provides concrete approaches to learners' developing skills in critiquing, David is also careful to raise cautions, for example, about a need to engage and support learners to help them develop 'a critical frame of mind' to critique complex areas. Within this, he stresses the importance of learners having sufficient knowledge and understanding of what they are critiquing. He also highlights how important it is that the curriculum itself is one that recognises the value of critique and includes it in assessment structures.

This chapter provides an additional basis for the notions John Williams develops in the chapter "Critique as a Disposition" related to dispositions, in that design and technology knowledge and understanding are the vehicles through which critical dispositions are developed. It also illustrates ways of introducing critical speculation, for example, through scenario building, that has resonance with ideas put forward in the chapter "Critiquing Design: Perspectives and World Views on Design and Design and Technology Education, for the Common Good" by Kay Stables.

Cecilia Axell presents a chapter that also provides direct support for learning and teaching approaches that develop skills of critique. She does this in the chapter "Critiquing Literature: Children's Literature as a Learning Tool for Critical Awareness" by drawing on children's literature as a fertile space for learners to think critically about technologies. She illustrates the potential of this through a careful selection of children's fiction that provides opportunities to explore a range of values and attitudes through the 'technology landscapes' introduced within the narratives. Through the range of stories chosen, she illustrates how children's fiction can open up technological understandings in terms of artefacts themselves and also the historical, cultural and social contexts in which they are used. This allows her to show the richness of issues and values that can be explored, such as old versus new technology, the ways technologies have been used in colonisation, the conflicts that arise when new technologies threaten environmental sustainability and the enduring nature of some technologies.

Cecilia's aim is not to just present the potential of the stories but also to provide suggestions for using the literature to help learners develop their critical thinking skills, including by engaging in writing stories themselves. She highlights the value of fictional narrative as a way to 'problematise' technology in a distinctly different way than a textbook might that, in doing so, makes conflicting ideas, issues and value positions more understandable and accessible.

Through this chapter, Cecilia makes links with chapter "Critiquing Design: Perspectives and World Views on Design and Design and Technology Education, for the Common Good" by Kay Stables, using stories to illustrate how the user of an artefact defines its purpose and value. She also opens up perspectives from different world views, developing threads in both Kay Stables and Mishack Gumbo's chapters. The stories chosen, and the unpacking of the narratives within, also provide vivid illustrations of the notions of the subject-object relationship and sociocultural inscription discussed by Jacques Ginestié in the chapter "A Critique of Technology Education for All in a Social and Cultural Environment".

Niall Seery shifts the focus of critique to design processes in the chapter "Modelling as a Form of Critique", discussing the importance of seeing critique as a significant element of modelling in design and technology education. Here, he includes modelling as a practice to support critical inquiry and speculation – not just understanding 'the world "*as it is*" but also "*as it could be*"'. He identifies the starting point for this as the imaginative activities that small children engage in, seeing exploration and speculation that leads to insight, and the learning that results from it, as being at the core of such activities. Niall provides a rich account of different meanings and uses of the terms 'model' and 'modelling' and how these are manifested in both cognitive and physical, external contexts. Through this, he discusses ways in which each supports critical thinking, highlighting the capacity for modelling to enable us to 'navigate the unknown'. Communication, with self and others, is pinpointed as a key value of modelling that focuses thoughts and ideas in ways that enable them to be critiqued as a basis for further generative modelling.

With small children, it is often a parent who acts as the mediator and mentor for encouraging critical reflection on a child's explorative or speculative 'modelling'. Niall provides examples of this, such as helping a child separate 'good' and 'bad' ideas after the child has explored whether an iPhone floats in the bath. In school design and technology, the teacher takes this role, and through the chapter, Niall places emphasis on issues and practices of learning and teaching through modelling. He draws attention to the special place this area of learning provides for understanding the value of modelling as a form of critique and of opportunities for direct application of the understandings that result. Identifying a danger of modelling that supports transmissional teaching – modelling to arrive at a 'correct' answer – he highlights the importance of learning in design and technology that exploits the 'wicked problem' nature of designing, the absence of one 'correct' answer and the need to learn to manage uncertainty. He recognises the complexity of a pedagogic approach based on this idea but provides a detailed and insightful discussion of aspects of this complexity, such as the challenges and opportunities of using heuristics and the place and role of knowledge. His discussion is populated with examples from research in design and technology, and consideration is given to implications for classroom practice. He does not deny the challenge of operating in the 'messiness' of modelling as a form of critique but makes clear its potential for effective learning.

In this chapter, Niall extends further the speculative dimension of critique introduced in the chapter "Critique as a Disposition" by Kay Stables and supports the critiquing-design relationship advanced by Steve Keirl in the chapter "Critiquing as Design and Technology Curriculum Journey: History, Theory, Politics and Potential". There is also resonance with Susan McLaren who, in the chapter "Critiquing Teaching: Developing Critique Through Critical Reflection and Reflexive Practice", promotes designerly approaches to self-critique in the professional practices of teachers, including its generative potential in opening up new possibilities for learning and teaching.

In the chapter "Hyper Design Thinking: Critique, Praxis and Reflection", Terry Wilkinson continues the thread of considering critique as a practice within classrooms. The question she explores relates to the nature of a critical pedagogy of consumption in design and technology education and ways in which this can positively politicise learning experiences. Terry's chapter takes the reader into a consideration of policy issues and pedagogic practices that deal with critique in the contexts of ethics, ecology, technology and consumerism. She begins by providing certain key insights into critical pedagogy and then builds this into a context of consumption. Illustrating ways in which promotion of the consumption of goods and services through advertising has become deeply rooted and embedded in societal learning, she raises the potential of education as a transformative space for changing perspectives. She explores this in more detail through critiquing education policy, taking an example from her own experience as a teacher in Ontario, Canada. In critiquing Ontario's science and technology curriculum, she highlights a policy that has drawn inspiration from the science, technology, society and environment (STSE) education movement and has taken account of UNESCO's Decade of Education for Sustainable Development. Whilst commending its intentions, she illustrates ways in which a policy with good intentions can still become rooted in technocratic perspectives.

Terry counters this by raising the possibilities of design activism and presents an alternative view through providing a cameo of the approach taken through Annie Leonard's video *The Story of Stuff*. She illustrates ways in which the video provides an engaging and at times amusing critique of hyperconsumerism and design's role in creating product obsolescence. Drawing on her own teaching experience of using this resource with 12-year-olds, she exemplifies ways in which the video provoked a critical response in the learners as they considered issues of fairness both for the consumers who purchased such products and the workers who were being exploited in the process of manufacturing the products. Whilst presenting a critique of the resource and an analysis of implications of its use, she highlights the possibilities for learning and teaching practices that it affords. Like Niall Seery in the chapter "Modelling as a Form of Critique", Terry recognises the challenges of taking a more critical approach to learning and teaching in design and technology education and balances these against the educational value of a more critical, politicised stance.

This chapter provides valuable illustration of ideas developed in the chapter "The Identification and Location of Critical Thinking and Critiquing in Design and Technology Education" by David Spendlove in relation to the contribution of critical pedagogy to critical thinking in design and technology education, particularly in the territories of civil liberties, political and economic power and consumption, and of an ethical dimension explored in the chapter "Critiquing as Design and Technology Curriculum Journey: History, Theory, Politics and Potential" by Steve Keirl.

There are also some parallels with Susan McLaren who, whilst not focusing on learners in the chapter "Critiquing Teaching: Developing Critique Through Critical Reflection and Reflexive Practice", highlights the value of teachers using critique to make visible the assumptions embedded in a teaching context, such as underlying policy. There are clear links between the ecological concerns that are considered, with those expressed in the chapter "Critiquing Design: Perspectives and World Views on Design and Design and Technology Education, for the Common Good" by Kay Stables.

The final chapter in the book explores ways in which critique that is embedded in process in design and technology education can be a way of deepening design thinking and design development processes. As Niall Seery has done, in the chapter "Hyper Design Thinking: Critique, Praxis and Reflection", Belinda von Mengersen focuses on processes of designing that are relevant to all and any design and technology education practices. In her case, the focus is on the value of critique as part of integrated, continuous 'praxis' (the integration of theory and practice). Drawing from established practices in art and design education, she discusses the potential of a shift towards more critical behaviours and dispositions that could deepen design thinking and that can be developed through existing design and technology pedagogics. A key aspect of this is a focus on metacognition, in which critique acts as the nexus between making (practice) and knowing (theory), supporting the articulation of reflection, either orally or through written language. She further underscores the importance of reflection within an iterative process of design and technology, again supporting the concept of praxis.

Drawing from theories of critical thinking and 'signature' pedagogies from art and design such as the design 'crit', the importance of language-based methods and dialogue is stressed, along with the need for vocabularies that allow learners to make meaning from reflection. Citing Sullivan (2010), Belinda identifies three valuable domains of practice – discursive, dialectical and deconstructive – that can support a rounded approach to critique. She also identifies parallels between the practice of writing and the practice of designing and the importance of learners seeing writing as valuable by linking it directly to what they are trying to achieve. A range of different processes are provided that suggest alternative forms of writing, for example, John Wood's (2004) tetrahedron approach that supports non-linear modes of writing and free flow 'stream of conscious' writing that can be used as a 'warm up' to get writing and ideas flowing. Finally, Belinda makes the points that timely formative feedback, development of a vocabulary of critique, and the use of creative processes of writing are all significant in supporting learners to develop a critical frame of mind and apply this in design development.

Belinda's chapter underscores the significant role of disposition presented by John Williams in the chapter "Critique as a Disposition" and Bill Nicholl in the chapter "Empathy as an Aspect of Critical Thought and Action in Design and Technology" and also echoes Susan McLaren's position in the chapter "Critiquing Teaching: Developing Critique Through Critical Reflection and Reflexive Practice" for a need to shift reflection to a more active, critical stance. There are also clear links to the history and significance of critique embedded in the South Australian Curriculum, chronicled by Steve Keirl in the chapter "Critiquing as Design and Technology Curriculum Journey: History, Theory, Politics and Potential".

As can be seen, the logic of the organisation of this book is explicit, beginning with philosophical, conceptual and historical contexts for critique, through the application of critique to design and technology education, and concluding with discussions of exemplars of critique as a way of deducing pedagogical practices which are conducive to critique. This logic indicates the usefulness of the book for a range of purposes: developing deeper understandings of the foundations of critique, integrating critique into current practice and developing new ideas about how to develop a critical disposition in students.

In searching for the place of critique in design and technology education, the approach taken in many of the chapters aligns with Bourdieu's (1991) notion of the 'heretic break', away from the:

- Vocational and academic spectrum
- Privileged knowledge that proliferates and that is manifested through reproduction of existing knowledge and practices
- Ubiquitous, unspoken and sometimes unconscious beliefs about the nature of knowledge and learning

and towards:

- The critiquing of epistemological beliefs and the engagement in wider socially responsible contexts
- Meaningful substantive critical reflection about the significant contribution of design and technology to a broader education provision
- The development of student capability to articulate a critical frame to their relationship with technology

This book is the first to be dedicated to critique as a practice within the field of design and technology education and offers educators and tertiary students a source of ideas and techniques for critiquing design and technological processes, products and pedagogies. As expressed by Steve Keirl in the chapter "Critiquing as Design and Technology Curriculum Journey: History, Theory, Politics and Potential", it is the hope of all the authors that this book will contribute to the recognition of critique as being a valid, vibrant component of the best design and technology education practice addressing inequity, injustice, sustainability and other ethical issues that arise in any of our realms of co-existence.

References

Bourdieu, P. (1991). *Language and symbolic power*. (G. Raymond & M. Adamson, Trans.). Cambridge, MA: Harvard University.

Dunne, A., & Raby, F. (2013). *Speculative everything: Design, fiction and social dreaming*. Cambridge, MA: MIT.

Giroux, H. A. (1983). *Theory and resistance in education: A pedagogy for the opposition*. London: Heinemann.

Latour, B. (2005). *Critical distance or critical proximity? A dialogue in honor of Donna Haraway*. Unpublished manuscript. Retrieved from http://www.bruno-latour.fr/sites/default/files/P-113-HARAWAY.pdf

Manyika, J., Chui, M., Bughin, J., Dobbs, R., Bisson, P., & Marrs, A. (2013). *Disruptive technologies: Advances that will transform life, business, and the global economy*. McKinsey Global Institute. Retrieved from http://www.mckinsey.com/insights/business_technology/disruptive_technologies

Mead, G. H. (1934). *Mind, self and society*. Chicago, IL: University of Chicago Press.

Shulman, L. (2005). Pedagogies. *Liberal Education, 91*(2), 18–25.

Sullivan, G. (2010). *Art practice as research: Inquiry in visual arts* (2nd ed.). Thousand Oaks: Sage.

Wood, J. (2004). The tetrahedron can encourage designers to formalize more responsible strategies. *Art, Design & Communication In Higher Education, 3*(3), 175–192. doi:10.1386/adch.3.3.175/1.

Part I
The Basis of Critique

Philosophy as Critique

Marc J. de Vries

Abstract Philosophy of technology is a discipline that provides insights into the critical reflection on technology. There are two types of philosophical critique in that respect: an analysis of what we mean by technology and related terms like technological knowledge, technological design and technological literacy and an analysis of the way technology and culture interact. The first type of critique aims at providing a language and terminology for having a proper discussion on technology, and the second type deals with that discussion itself. In this chapter a survey of what has been developed in both types of reflections on technology is presented.

Keywords Analytical philosophy • Continental philosophy • Ontology of technology • Epistemology of technology • Ethics of technology

1 Introduction

1.1 Philosophy of Technology for Critiquing Technology

Philosophy of technology is relevant for technology education in several ways (De Vries 2005a). It can help to develop a sound conceptual framework for curricula and materials. It can also become part of the content of teaching about technology to provide students with material for reflection on the nature of technology as it relates to humans, nature and society. In this book, design and technology education as a means for making critical evaluations of technology is an important focus. Here, too, the philosophy of technology can be an important source of inspiration. Critical analysis of technology, both in terms of what we mean by it and the way it interacts with humans, nature and society, is at the very heart of this discipline. This chapter serves as an overture to the book in that it provides a survey of what philosophy of technology has to offer in that respect. I start by distinguishing two different ways in which philosophy does its critical work. Then I move on to show how each of

M.J. de Vries (✉)
Delft University of Technology, Delft, The Netherlands
e-mail: m.j.devries@tudelft.nl

PJ. Williams, K. Stables (eds.), *Critique in Design and Technology Education*,
Contemporary Issues in Technology Education, DOI 10.1007/978-981-10-3106-9_2

these two modes of critique played out in the (short) course of history of philosophy of technology. I then briefly sketch what this means for design and technology education. But the real elaboration of this will be in the remaining chapters. I want to emphasise that for the sake of introducing philosophy as critique in a concise and accessible way, I will simplify the complexity of the philosophical approaches I will discuss. The reader should see this chapter primarily as an invitation to start (or continue) reading philosophy of technology and design literature and become more acquainted with all the nuances in the debates, more than can be presented in this chapter.

1.2 Two Types of Philosophical Critique

Philosophy is the systematic reflection on reality. This suggests that nothing can escape philosophy's attention. This is the case indeed. Perhaps this is what irritates people about philosophy. It does not seem to have any boundaries. Natural sciences are limited to studying natural phenomena that have a non-intentional character. Psychology is limited to studying humans and intentional phenomena. Linguistics study the way people develop and use languages, and in a similar way all scientific disciplines have their boundaries, within which scientists in that discipline are supposed to stay. This holds even though we encounter an increasing number of academic fields that deal with a variety of aspects of technology. For these, terms like multidisciplinary, interdisciplinary or transdisciplinary are used. But even for those, it holds that they do not have the breadth in scope that philosophy has. A field like environmental studies encompasses many aspects such as the physical, the chemical, the biological, the psychological, the social and the economical, but still it focuses on one particular topic, namely, the natural and technological environment of humans. For philosophers there seems to be no boundaries at all in terms of objects and phenomena that are studied. Another annoying aspect of philosophy is that it does not seem to have any concrete methods. A stereotype is that it is just 'reflection', without any specific indication of how this should be done. Is it just a matter of sitting and thinking? A third stumbling block for an appreciation of philosophy is that it does not seem to have a clear set of criteria for what can be said to be valid and reliable outcomes and what is merely a hypothesis, suggestion or speculation. What is scientific about philosophy? Or does anything go? Without proper answers to these questions, design, engineering and technology educators may feel well justified in regarding philosophy of design and technology as of little value.

There are, however, answers to these questions, although one may not like or agree with them. Let me start with the last two questions: yes, philosophy does have methods and, yes, there are criteria for judging what is scientific and what is not. The most important method in philosophy is argumentation. When I come up with a claim about what we mean when we say 'I know that screwdrivers are for driving screws', we have to defend this claim against objections. The claim might be: it means that I myself believe that screwdrivers are for driving screws and that I have found some sort of justification for that belief (e.g. I have seen most people using them for this purpose), and it is true. That seems quite reasonable. Someone, however, may object as follows: but what does it mean for this claim to be 'true'? Screwdrivers may well be used for other purposes. Can merely the name be sufficient for judging an improper use of the device? It functions very well for opening coconuts or tin can lids. For claims to 'knowledge', then more is needed than belief. We need some sort of agreement between all of us to allow one to claim to 'know' that this is a screwdriver. Philosophers who made the claim may come up with a next claim and say: 'Ok, I will adjust my definition to make it contain a social agreement about screwdrivers'. Then the opponent may come up with an argument against the new definition, and this goes on for as long as there is a disagreement about the definition of knowledge. This agreement will probably never be reached. That is what also annoys people: there seems to be no end to philosophical debates. That, however, is an unfair critique, as it holds for other humanities and sciences as well. When John Horgan's book *The End of Science* was published in 1996, immediately everyone started protesting: how can there be an end to the natural sciences? We have thought this many times, but always new phenomena were discovered that begged an explanation, and often these phenomena posed a challenge to existing explanations for other, already known phenomena. So why should philosophical inquiry have an end? Besides that, philosophical inquiry does not turn in circles ad infinitum, as people may think. In past centuries of philosophical exploration, we made progress in our effort to define what we mean by the claim 'I know that . . . '. We still may not have found a watertight definition, but we do know a lot about where definitions can lead. Other methods used in philosophy are dialectics (present the solution and arguments for criticism by other philosophers, and help them judge their own) and hermeneutics (a set of normatively binding rules by means of which the process of interpretation should proceed; it is good to note that the term also has a wider meaning in which it serves as 'an interrogation into the deepest conditions for symbolic interaction and culture in general' as the Stanford Encyclopedia of Philosophy phrases it). These last-mentioned methods are necessary to include as they take away the idea that philosophy is all about providing watertight definitions and conceptualisations. Philosophy is also, and perhaps primarily, more about raising doubts (sometimes 'systematic doubt' is even seen as the primary method in philosophy) and questions than providing answers.

What remains the case is that philosophy has no boundaries when it comes to objects of study. That does not mean that philosophy is an unstructured and fuzzy field of reflections. There are some sub-disciplines that do have a sharp focus (Morris 1999). The example above features epistemology, a sub-discipline of philosophy that aims at defining what we mean by 'knowledge', what types of knowledge can be distinguished and how they differ from each other. For design, engineering and technology educators, this can get exciting when we ask what makes technological knowledge different from natural science knowledge. Often it is said that design and technology is not a knowledge domain of its own right, but merely 'applies' knowledge from natural sciences. What if philosophers have shown that this is incorrect and there are features and examples of technological knowledge that clearly cannot have been derived from natural sciences? Ontology deals with different ways in which things can 'exist'. 'Things' should then be taken broadly to encompass not only objects but also events. But events exist in a different way than objects. I cannot touch my birthday as I can touch the birthday cake. Yet, my birthday can be said to exist, no less than the birthday cake. For design, engineering and technology educators, this seemingly vague and abstract question may become more interesting when we ask if perhaps technical artefacts 'exist' in a different way than 'natural objects'. Metaphysics concerns world views. 'Beyond physics' is the literal meaning, and it indicates that physics only deals with what can be observed, but that there is more than this. We all hold beliefs about the nature of reality. We have, for instance, beliefs about the mind: is it just a side effect of matter or is it something that may be related to matter, but also has non-material properties? A sub-discipline is the philosophy of mind and it becomes a matter of interest for design, engineering and technology education when it is used to reflect on the nature of computers and in general, artificial intelligence (AI). AI makes us wonder about ourselves: are we just sophisticated computers, because all this talk about mind and soul is just metaphorical, or do we differ fundamentally from those devices because we have something that they do not have, in spite of the fact that the likeliness between them and us becomes more and more impressive. Finally, there is the domain of ethics, probably the only one in which people intuitively see some usefulness. Ethics includes moral philosophy and questions of morality: what do we believe is morally proper and what is not? What do we mean, anyway, by 'morally proper'? Obviously, this is a sub-discipline in philosophy that design, engineering and technology educators often find themselves interested in, although at the same time they often hesitate to bring it into classrooms for fear of being accused of indoctrination. This fear illustrates a fundamental misunderstanding about what ethics is. And that, in turn, illustrates that our scepticism about philosophy is often related to a lack of knowledge of what it really is. Another value-related sub-discipline of philosophy is aesthetics, which obviously is also quite relevant to technology. The term 'axiology' is often used, either as an umbrella term for ethics and aesthetics or to indicate the search for principles underlying both.

Philosophy can also limit itself by focusing on a specific domain of human activities. Thus, we have philosophy of science, philosophy of religion, philosophy of law and philosophy of design, engineering and technology. Each of these may cover questions similar to the ones we identified when discussing epistemology, ontology, metaphysics and ethics, but deals with them in a specific way. To use the term 'applying' them is a bit dangerous here, too, because sometimes, discussions in these specific domains give rise to new questions in the general sub-disciplines. Epistemology of technology, for instance, has shown that the traditional approaches of trying to repair the 'justified true belief' account of knowledge (that was also the basis of my example) does not address fundamental features of technological knowledge such as its normative dimension (later I will return to this).

In this chapter I want to show that philosophy is a fundamental way of critiquing reality (including design, engineering, technology and everything related). The strength of philosophical critique is that it is based on philosophy's primary method, namely: argumentation. It is not just buckshot scattering various ideas and questions, but a well-founded form of critique that cannot be dismissed by shrugging shoulders. Philosophy can be used as critique in two distinct ways, each of which has led to a certain tradition in philosophy (Verkerk et al. 2015). These are indicated by the terms continental philosophy and analytical philosophy. Immediately it is clear that these are non-philosophical terms, as they are incomparable. Continental is a geography-related term and indicates that philosophers in this tradition generally live(d) and work(ed) on the European Continent. Analytical indicates a way of thinking. It is conceivable that people thinking analytically live and work on the continent, as well as in other places. So the terms are in fact inappropriate for dividing all philosophers into two separate groups. That is why nowadays the terms are often abandoned, although a good alternative has not yet been found. But whatever the names, two distinct ways of critiquing can be identified.

The first critique refers to what so-called analytical philosophers do: they analyse terminology. This is often seen as the more 'exact' part of philosophy, where argumentation plays the most vital role. My example from epistemology derives from that realm. Before starting a discussion of dangers or advantages of having knowledge of some socially sensitive issue, we have to agree on what we mean by knowledge to prevent complete chaos in the debate. We may think that our conventional, daily-life use of terms suffices, but that is often an overestimation of our intuitive analytical capabilities. We do have a lot of inconsistencies in the way we use terms, but this often goes unnoticed. Let us take the debate about technology: is it applied science or is it not? Some people say 'Yes'. They come up with examples such as transistors, lasers and the atomic bomb and show how much these were the result of research in physical sciences. Others, though, say 'No' and come up with the steam engine, glass lenses and early airplanes and show that their inventors had hardly any solid knowledge of the natural phenomena on which their inventions were based. For the philosophy of design, engineering and technology, the analytical approach in philosophy leads to questions like: what do we mean by 'design', by 'technology', by 'engineering', by 'technical artefacts', by 'technological knowledge', by 'requirement', by 'optimisation', etc.? Is design a knowledge-producing or knowledge-using process? That depends on our answers to these questions. So first we have to agree on what we want them to mean (a matter of choice indeed!) and only then we can have a philosophical discussion.

The second type of critique for which philosophy can be used is related to debate itself. This is what people will intuitively associate with the term 'philosophy': dealing with the big questions 'Why are we on earth, why do we love or hate each other, why do we feel the need for developing knowledge and why do we have technology? Or why should we, perhaps, sometimes not have technology or have it only in certain ways or under certain conditions? Let us call this, for lack of a better alternative, the human/social approach in philosophy.

The analytical and the human/social approach clearly need each other. For a proper debate, proper terminology is needed. Vice versa, a proper terminology is useless without a debate in which this terminology is used. In design, engineering and technology education, too, we need both forms of critique. Our terminology needs critiquing because we want to teach students proper terminology, and we first need to be consistent ourselves. The ideas that have been developed in the analytical (sometimes called the Anglo-Saxon) tradition can be valuable. Also we want to make students aware of social debates related to technological developments, and for this the ideas that have been developed in the continental tradition can be useful. As philosophy always tries to go to the very basics in critiquing and characterising, this also makes philosophy relevant for education. In education we want to start simple and only then move on to the complexities. If, for instance, we want to teach about all sorts of technical devices, it makes sense first to start by showing what the basic characteristics of any technical artefact are. The philosophy of design, engineering and technology has developed ideas about that (I will come back to those later); these insights can help to start simple and only then move on to details.

1.3 The Early History of Philosophy of Design, Engineering and Technology

It is remarkable how late the philosophy of technology emerged in the field of philosophy. Given the importance of technology in society and its impact on human life, one would expect that this would be one of the first phenomena to arouse the interest of philosophers, but it was not. Philosophical reflection on science preceded reflection on design, engineering and technology, although it is not hard to show that technology is older than science and has at least as much influence on the way we think and live. Perhaps it was the strong association with material reality that held back philosophers from developing a systematic reflection on design, engineering and technology (or designers, engineers and technologists systematically reflecting on their discipline). Philosophers were perhaps so much focused on the mind and so much less on matter that they did not mind about matter. In other words: matter did not matter so much to them. Whatever the reason may have been, the philosophy of technology is probably not older than approximately 50 years or so. Following interest in mechanical philosophy, one of the first publications with 'philosophy of technology' in the title was *Philosophie der Technik* by Ernst Kapp, published in 1877 (Mitcham 1994). But for a long time, this book remained a stray stone. Perhaps the first publication that can be called a philosophy of technology after Kapp's book was the essay *Die Frage nach der Technik* (The Question Concerning Technology) by the German philosopher Martin Heidegger in 1954. In these two publications, we can already recognise the difference between the analytical and the continental approach. Kapp's intention was to identify the nature of technology. Accordingly, technology was a human projection of body organs. A biface is a projection of our fist, made because our fist is not strong enough to split stones. Pots and pans are projections of our hands, made because we each have only two and yet need to hold a lot of food and liquids in a household. Glass lenses are a projection of the lens in our eye, made because it does not produce a sharp image. And so on. It was much later that this origin of the philosophy of technology was picked up again in the extended mind idea, developed by Andy Clark and David Chalmers in their 1998 article with the same name: the idea that notepads, electronic memories and the like are in fact extensions of our human mind that make the boundary between mind and matter become fuzzy. So Kapp's early work in the philosophy of technology was primarily analytical. His way of critiquing technology was by revealing its basic intentions: to extend the human body. Heidegger's work, on the contrary, should be seen in the human/social realm. The idea developed in his essay was that technology has taught humans an instrumentalist way of looking at reality. We hardly enjoy anymore the beauty of a tree; rather, the first thought that comes into our minds is: how many planks can I make out of that tree? He called this the 'Gestell' of technology, a sort of framing in our minds. In *Die Technik und die Kehre* (The Turn) (1949/1962), he continues this critique by claiming that technology forces nature to provide resources for humans. Still today, this view of technology seems to be popular, and this was stimulated particularly, of course, since we became aware of how much we have 'tortured' nature so that it has become seriously distorted. Different as they are, both Kapp's and Heidegger's approaches have relevance for today.

For a period of time, not much was done in the analytical realm. Perhaps the two most noticeable exceptions are the French philosopher Gilbert Simondon (see Ginestié in this volume) and the Dutch philosopher Hendrik van Riessen. Simondon worked on the meaning of the term 'technical object' (we would nowadays say 'technical artefact'; De Vries 2008, 2010). According to him, what we mean by that is not just the tangible object as it stands before us, but also the whole line of development in which it stands. The object is also a process of becoming, or as Simondon calls it, the 'individuation' of the device. He uses the same term for humans, and it should be read in the context of an evolutionary perspective. In this process, the artefact more and more integrates different functions, and this is what Simondon calls 'concretisation'. Simondon did not have an engineering education, but he had carefully investigated numerous devices in their technical details. This, too, made his book *Du mode d'existence des objets techniques* (1958) an exception in the philosophy of design, engineering and technology of that time. Van Riessen also focused on the nature of technical artefacts and saw them as entities that functioned in different aspects of reality: a bridge, for instance, is not just a physical thing (obeying the conditions set by the laws of the physical aspect of reality) but also a spatial thing, an economical thing, an aesthetic thing and even a 'trust' thing (referring to an aspect of reality that entails that all things can be the subject or the object of trust). Both Simondon and van Riessen used their analyses to show that there is nothing inhuman about technology, but that rather technology is an integral part of our human existence. This illustrates that the analytical approach, generally speaking, was more positive about technology than the continental approach. The critique of technology in the analytical approach was primarily focused on making us aware how rich the existence of technical artefacts is: it is not just a single object, but stands as a constellation in a long line of development and functions in many different aspects of reality and therefore can be studied from many different points of view.

1.4 The Continental Approach

As stated before, Kapp, Simondon and van Riessen for a long time were exceptions in that they represented the analytical approach, whereas most philosophers of technology in the second half of the twentieth century took the continental road. It was not until the end of the twentieth century that the interest for the analytical approach was revived. Unfortunately, the continental approach was largely negative about technology, as we already saw with Heidegger. Another philosopher who wrote about technology in a very gloomy mood was Jacques Ellul. His view of technology was that it had become almost entirely autonomous (Ellul 1990). As it played such an important role in culture and society, it was a system with endless feedback. We did not choose computerisation and atomisation, they just overtook us.

The exceptions to the rule that continental philosophers were very critical of technology were Karl Marx and Friedrich Engels. They held a deterministic view of history inspired by another German philosopher, Georg Hegel, and believed that the disappearance of capitalism and the liberation of the labour class would come as an historical necessity and technology would be the lever to cause that. It was the labouring class in particular that had technology in their hands because they were the ones that were responsible for the production of goods. This would enable them to overthrow the dominance of the capitalists. For that reason, Marx and Engels were very positive about the role of technology in society. This is also why communist countries had a high respect for technology education. In the former Eastern European countries, 'polytechnic' education, as it was often called, had a prominent place in the curriculum, long before other Western countries had it as a compulsory subject. An excellent conceptual development for technology was done in the context of polytechnic education, among others by Dietrich Blandow (1992). Unfortunately much of that got lost when polytechnic education was abandoned almost immediately after the breakdown of the communist system in the late 1980s.

Both the existentialist and the neo-Marxist approach were continued after the death of their respective founding fathers. Still today they form two major forms of critique concerning the position of technology in human lives, nature, society and culture. Perhaps the most prominent current philosopher of technology who writes in the traces of Heidegger is Albert Borgmann, an American (note how inappropriate the term continental philosophy is). In line with Heidegger, he criticises technology for having caused a situation in which we are entirely focused on commodities that are provided by the devices that surround us. We forget to ask the basic question of the purpose of these devices. Why do we want these commodities? Is there no deeper purpose of life? Those are the sort of questions that Borgmann poses. His term for the role of devices as purely providing for commodities is the 'device paradigm' (Borgmann 1984). Accordingly, our personal experience of reality has become much poorer than before because of this paradigm. In the past, when we wanted to bring our home to a comfortable temperature, we went into the forest, cut wood for our stove, brought it home with great effort, and then made a fire, which often was also quite a challenge. Now we just programme the thermostat a bit and up goes the temperature. Not a very rich experiencing of reality, according to Borgmann. Our experience is poorer also in the sense that the device paradigm makes many experiences uniform. This is necessary to make the commodity affordable in terms of economies of scale. We make a meal by putting a ready-to-eat meal in the microwave oven, push a few buttons and after a couple of minutes we take out the meal and eat it. It tastes exactly the same each time we use it. Compare that with the cooking of a meal from basic ingredients: a bit of this and a bit of that and each time a bit different. Isn't there much more pride and satisfaction in the meal that is prepared that way than in the case of the microwave bite? What is the proper treatment for this diagnosis? This is what Borgmann calls 'focal activities': activities that focus our attention on reality so that we get a more intense and rich experience. For instance: go by bike instead of by car, cook your own meal, attend concerts or play an instrument instead of listening to 'canned' CD or mp3 music, etc. Borgmann realises that we have already gone too far to make this the basis for our lives, but he makes a plea for at least giving it some room in our routines, so that we still feel the difference between those rich experiences and what the device paradigm offers.

Also in Heidegger's vein, but more balanced in his critique, is Don Ihde. He calls himself a post-phenomenologist to indicate that he takes a step further in the line of the phenomenologists, which is closely related to the existentialist line. His particular interest is the way in which technology intervenes in our experience of reality. We seldom 'see' reality directly, but mostly via technological instruments. Ihde distinguishes different modes in which this happens (Ihde 1990). One he calls the hermeneutic relation. Here the awareness of the role of technology is most certainly important. A domain that Ihde often uses to illustrate this is that of medical imaging. What we see is an image that represents a reality 'behind' it. The MRI scan picture represents a body, but in a way that needs interpretation (hence the term 'hermeneutic'). It is even the case that one and the same scan can be represented in different ways, depending on the need of the doctor (the oncologist gets a different image of the same scan than the neurologist). An example where we confuse things if we do not realise this role of technology is when we admire an image of the stars produced by the Hubble telescope. We see many colours and mistakenly think that stars have different colours, while actually the colours in the image represent temperatures. Overlooking the hermeneutic relation in that case leads to misunderstandings. That can also happen in the case of the other mode, the 'alterity relation'. In that case, we see a reality that is created by the technology itself. The most obvious example is games, including multi-user games like Second

Life and other virtual worlds. We can create our own avatar and observe that world, but what we see is not 'real'. As long we are aware of the alterity role of technology, that is not a problem, but when we confuse virtual reality and 'real' reality (oops, how confusing can it get!), then we make a mistake. The more advanced the games and virtual worlds become, the more they look like the real world and the easier the confusion emerges.

We have seen how the existentialist and post-phenomenologist approach critique technology: by pointing out how technology can take our attention away from the richness of reality and the deeper questions of life by providing us means for simple (but uniform) commodities and how technologies create our images of technology, which can be enriching as long as we recognise the role of technology. This becomes misleading when we overlook this role. Let us now see how the neo-Marxist line has also continued after its founding fathers had died.

The most prominent here is Andrew Feenberg. His primary concern is the role of technology not only in the lives of individuals but in society as a whole. He takes up a point that Herbert Marcuse brought forward, namely, that the rationality that reigns in technology is a choice, not a necessity. People may think that there is one optimal solution to any socio-technological problem, but what is optimal depends on one's rationality. The rationality that determines current technological development is mostly that of capitalism and economic growth. Feenberg challenges that rationality and shows that it is well possible to open up this rationality and allow for other rationalities to 'hack' technologies and use them for different goals. His empirical evidence for that is the Minitel system in France. Originally designed for dissemination of government information in libraries and shopping malls, hackers literally hacked the system and started using it for exchanging information (Feenberg 1999). Feenberg applauds this and developed a view of technological developments in which there is an explicit place for this democratisation of technology. He writes about primary and secondary instrumentalisation. In primary instrumentalisation a socio-technological problem is decontextualised so that engineers can design a solution to this problem. In secondary instrumentalisation, the solution is recontextualised, and in Feenberg's view, this is not just a matter of passively accepting what the engineers have come up with, but actively adapting it to the needs of (specific groups in) society. Thus, technology can become a vehicle by which all social classes can have a say in the way technology is used in society. This would then be an alternative for technologies being used to consolidate the power of the ruling class. Feenberg's secondary instrumentalisation would find ways of turning that upside down and adapting the bridges and parkways to make them play a totally different role. In a way Feenberg resembles Ihde in that he continues a line of continental philosophy of technology, but turns the rather gloomy image of technology into one in which technology can also be seen in a more positive light. Information technology is perhaps the most prominent example that shows how technology can be a means for democratisation. At the same time, Winner would argue that it can be a means for controlling society and economy and raises important questions concerning privacy and reductionist tendencies by treating people as 'data' (Winner 1986).

Another philosophy, neither really analytical nor really continental, in which technology was appreciated in a positive way is pragmatism. As with the next approach that will be mentioned here, the religion-inspired one, pragmatism, is discussed in a section on continental philosophy although it does not fit there quite naturally. Pragmatism emerged mainly in the USA, which is clearly not continental ('continent' meaning the European continent). On the other hand, pragmatism is not primarily concerned with conceptualising, as in analytical philosophy. What pragmatism has in common with what is called continental philosophy in this chapter is that it provides answers to very basic questions about what life is about. That communality seems to be more important than the geographical mismatch with the name 'continental philosophy'. As it is very common in philosophical literature to use only two basic categories in surveys of philosophical 'schools', in this chapter I stick to the continental versus analytical dichotomy and put pragmatism and religion-inspired approaches under continental. In the philosophy of technology, Larry Hickman is a well-known representative of pragmatism. Hickman follows the traces of John Dewey whose conviction was that no a priori values should determine our decisions, but we should always try out different options and see what works. Hickman sees engineers as the best examples for how to do that. Engineers develop prototypes and test them, and eventually practice determines what is the best solution (Hickman 2001). What works is good. Or in pragmatic epistemology, what works is true. Education was seen already by Dewey as the context in which we should learn this way of thinking by doing. Education is not the transfer of a priori truths, but the development of insights by trying out what works and what does not. Thus, in pragmatism technology becomes the model for all social decision making. By all means, a positive critique.

There are also philosophers of technology who have religion as a starting point. These, too, do not sit comfortably in the continental 'stream'; as among these, we also find philosophers that are definitely analytical (the epistemologist Alvin Plantinga being a well-known example). In this section the focus will be, however, on those religion-inspired approaches that are continental in the sense that they are not primarily involved in conceptualising but in answering basic questions of life. One of such 'schools' is the 'reformational philosophy', inspired primarily by the Dutch philosopher Herman Dooyeweerd, who also developed a view on technology. In this view, technology is put into a double perspective. On the one hand, technology is seen as a God-given opportunity to work in His creation, and at the same time it is done by humans who also have evil tendencies and may use their technological capacities for greedy and selfish purposes, thereby neglecting the intrinsic value of God's creation and creatures (Schuurman 1997). A similar view is often held by Islamic philosophers of technology, but they do not recognise an original 'paradise' state in which humans were not yet fallen in sin. Both Christian and Islamic philosophers of technology expect the end of history to come when God intervenes and brings (again) a state of perfection (Jochemsen and Van der Stoep 2010). Here, too, is a difference between a Christian and an Islamic perspective in that in Islam the place of humans in this new state is decided purely on the basis of their own acts, while in Christianity the notion of sins being forgiven is crucial for determining whether someone ends up in the new paradise or in hell. I mention this religion-inspired notion of paradise because it plays an important and often overlooked role in current technological developments, particularly in the rhetoric that accompanies it (Noble 1999).

In this context, the notion of utopia should be mentioned. A utopia (literally a non-place, to indicate that it does not really exist) is a place where everything is perfect and ideal. We see this particularly in the promises that accompany nanotechnology. It is claimed that this technology is so fundamental because one day it will allow us to manipulate individual atoms; hence, we can make everything we want and also exactly the way we want it to be (De Vries 2005b; Peterson and De Vries 2012). That includes biological 'devices'. There is a discipline of 'synthetic biology', but it has not yet perfected the manipulation of individual atoms (rather biologists work with larger 'chunks' of living matter). But it promises that one day we will be able to build life, and then build perfect life, without vulnerability to disease and death. We see a similar promise in virtual worlds: in such worlds all sorts of boundaries in reality are non-existent. In a virtual world we can create ourselves as an avatar that can have any shape we want ourselves to have. The 'no boundaries' ideal can also be found in advertisements. 'Endless fun', 'endless shrimp', 'unlimited broadband' and 'unlimited data' show the ideal of doing away with all limitations that we see as barriers to perfect happiness. Religion-inspired philosophy critiques these types of promises by pointing out that we humans cannot and should not bring about this state of perfection. Although, we do have a duty to seek to try to improve the world in which we live. It is not only religion-inspired philosophers that warn against this utopian thinking in technology. Dutch philosopher Hans Achterhuis points out the main danger of utopian thinking: at a certain stage we are willing to pay any price to get to the utopia realised (De Vries 2012).

In summary, the continental critique of technology was mostly critical, particularly in the early years. Technology was blamed for gaining autonomy and control, for depriving us from a feeling of freedom, from taking away attention from the important questions in life, and for creating a distorted or confused view of reality. Later the picture became more balanced, not just negative. Technology was acknowledged also to be a means for enriching our view of reality and for changing or democratising society.

1.5 The Analytical Approach

We will be shorter about the analytical approach as much less has been written in this realm than in the continental approach. As we have seen, the analytical approach was perhaps the beginning of the philosophy of design and technology, but for a long time remained the exception rather than the rule. It was not until the 1980s that the analytical approach became more popular. It was a different type of philosopher who became interested in the analytical approach. Usually, continental philosophers are *philosophes 'pur sang'*: they have a philosophical background only. Those who became interested in the analytical approach in the philosophy of design, engineering and technology usually had both a philosophical and an engineering (or natural or physical science) background. Perhaps it was annoyance about the naivety with which continental philosophers often wrote about technology, a naivety that revealed their lack of internal understanding of technology and engineering that renewed analytical philosophy's reflection on technology. Whatever may have been the real motive, the fact is that these analytical philosophers started not by writing in very general and abstract terms about technology, but by examining empirical studies made by historians and sociologists and also by examining literature from within the engineering profession. This was sometimes called the 'empirical turn' in the philosophy of design, engineering and technology (Kroes and Meijers 2000), and it mirrored a similar turn in the philosophy of science (created by such philosophers as Bruno Latour, John Law, Michel Callon and others).

The analytical approach critiques technology in a different way than the continental approach. The latter is more concerned with how technology has an impact on individuals and society. The analytical approach deals with critique of our understanding of technology itself, of our understanding of the nature of technology. Three domains were particularly seen as interesting: the ontology of technical artefacts, the epistemology of technology and engineering sciences and the methodology of design. Let us take a closer look at each of those and see how they critique naïve understandings of design, engineering and technology.

Artefacts play an important role in our intuitive notion of technology. But what makes an artefact an artefact? Our first hunch is of course: it has been made by humans. That distinguishes it from natural objects like beehives, beaver dams or stones and trees. But it is not always easy to tell whether an object is human-made or natural. Can you tell the difference between a flower that grew without any human intervention and one that was cultivated in a greenhouse? And does it matter anyway when you want the flower to bring some beauty into your house? Is the origin enough to make the distinction between a natural object and a technical artefact? In the analytical philosophy of technology, the idea has been developed that what makes technical artefacts distinct from natural objects is that they have a relational nature, apart from the physical nature that natural objects also have. The stone in the woods that no one notices has a physical nature (it can be described in terms of its shape, weight, colour, etc.). But the screwdriver I use has more than that. It is a screwdriver because we as humans regard it to be an object with which we can drive a screw. Even stronger: we may also decide to regard it is a means for opening tin cans. Evidently the physical nature is intrinsic, and it is not up to us to decide what its weight, colour, etc. is. But for the screwdriver to be a screwdriver, it takes a human to decide that it is a screwdriver (or a paperweight, a can opener, etc.). This relational nature of technical artefacts is something that we easily overlook, which makes us think that artefacts can only do what they are designed for. That stimulates a deterministic view of technology. The dual nature view liberates us from a deterministic view of technology (or at least provides an important underpinning for a nondeterministic view).

In the epistemology of technology, a main focus has been to show that technology is not just applied science, as has been suggested since the 1950s. This, too, is a critique of conventional ideas about technology that may hamper us to develop a view of technology in which things are not just as they are, but are a matter of decision making. The applied science approach suggests that scientific knowledge can be translated into products in a deductive way (because it suggests that no new knowledge is developed in the process). This again suggests determinism in technological developments. But in the philosophy of technology, we have become aware that most technological knowledge cannot be derived from science, and this is the normative dimension (De Vries et al. 2013). Engineers have knowledge of functions. But function does not tell us what the artefact does. It does not describe the behaviour of the artefact. It rather tells us what the artefact should enable us to do. The screwdriver is a screwdriver not because it actually drives screws, but because I want it to drive screws. If it does so, and it does it well, I call it a 'good' screwdriver, and if it does not do it well, or if it is broken, I call it a 'bad' screwdriver or a 'broken' screwdriver, but still a screwdriver anyway. This is different in physical science. If an electron does what electrons normally do, the scientist will not call it a 'well-functioning electron', but just 'an electron'. If it does not do what an electron normally does, the scientist will not call it a 'bad' electron or a 'broken' electron, but he then decides that it is a different particle. The physical scientist does not judge reality, but only describes it. Engineers do have an opinion about reality, and in fact what they do is talk about a desired reality rather than the actual reality. This normative dimension can also be seen in other categories of engineering knowledge, such as knowledge of norms and standards, knowledge of rules of thumb, knowledge of good practice and so on (Meijers and De Vries 2009). Natural and physical scientists have that kind of knowledge also, but then it is part only of their knowledge of the methodology of their discipline, never part of what they see as the content of their discipline itself (they do not see it at the same level as the laws of nature they discover). The immediate consequence of the normativity in technological knowledge is that values play an inherent role in technology. These can be purely functional, but often ethical values also come in. When do we regard a car as 'good'? Only when it brings us from A to B? Most people would not be content with that. A 'good' car is also one that is environmentally friendly, safe (preferably both for the driver and for other people that may get hit by the car in an accident), user-friendly, etc. The notion of normativity in technological knowledge stimulates a view of technology in which values are an integral element, contrary to much of the past in which we thought that technology is instrumental and neutral. Again we see that analytical philosophy, although it 'only' deals with conceptualising, does contribute to critiquing technology and our views of technology.

We also see this in the domain of methodology. Here we do not mean that term as a list of prescriptive phases or steps that can be taught to new designers. Methodology literally means: the study of ways in which . . . . (And then some activity should be added, like 'science' of 'design', which then leads to 'methodology of science', especially 'design methodology'.) In the early days of design methodology, it was suggested that ideal design flowcharts could be developed, irrespective of the field of design (Cross 1984). The overall structure of such frameworks was usually: analyse the problem – synthesise possible solution – evaluate against the criteria and choose for further elaboration. Now we know that (1) design is too domain-specific to make such general schemes work and (2) even within one domain, design problems and designerly ways of thinking differ so much that such schemes are too abstract and idealised to work in concrete and complex reality. Our intuitive ideas about design appeared to be too naïve and simplistic. We now recognise the complexity and variation of design processes much more than before. The same, by the way, holds for science (we also dropped the belief that something like 'the' scientific method exists). Prescriptive views of design stimulate deterministic views of technology. Again we see how analytical philosophy of technology points out that such a view is not supported by systematic reflection on the nature of design, engineering and technology.

1.6 Implications for Design and Technology Education

This chapter is meant to be a sort of 'overture' to following chapters in which lines will be drawn to education. I will, therefore, just make a brief transition here to those chapters and leave the more concrete applications to the authors of those chapters. We have seen that there are two modes of philosophical critique of technology: the continental and the analytical. Both have shown their value in the past. How far removed is all this philosophising from the practice of design, engineering and technology education? Unfortunately, often quite far. Educational practice still today often suggests a deterministic view of technology, in spite of the fact that we do a lot more design work than before. But as long as in that design work, we think that design processes can be prescribed according to a scheme that represents the 'logical and optimal' design process, we make students think that 'it just has to be this way'. Similarly, we may hold to a purely descriptive view of technological knowledge. And again, similarly misleading may be a strictly non-relational view of artefacts as neutral instruments. Add to this the ideas we have seen in continental philosophy of technology in which values play a very important role. Those can be personal values (the quality of our individual experience of reality) or environmental, social and cultural values (relations between sub-cultures or social classes). Using both analytical and continental philosophy of technology will enable critique of an educational practice that implicitly or explicitly presents a deterministic perspective on technology. Such a perspective is not fruitful in the context of technological literacy (unless we define technological literacy in a very narrow and instrumental mode). Technological literacy implies that (future) citizens are not just able to use the technological devices around them properly, but also make sophisticated judgments about various aspects and phases of technological developments. In subsequent chapters of this book, this will be elaborated for various aspects of design, engineering and technology education.

References

Blandow, D. (1992). *The elements of technology for education*. Eindhoven: Eindhoven University of Technology.

Borgmann, A. (1984). *Technology and the character of contemporary life: a philosophical inquiry*. Chicago: University of Chicago Press.

Cross, N. (1984). *Developments in design methodology*. Chichester: Wiley.

Ellul, Jacques. (1990). *The technological bluff* (Geoffrey W. Bromiley, Trans.). Grand Rapids: Eerdmans.

Feenberg, A. (1999). *Questioning technology*. London: Routledge.

Hickman, L. (2001). *Philosophical tools for technological culture*. Bloomington/Indianapolis: Indiana University Press.

Horgan, J. (1996). *The end of science: Facing the limits of knowledge in the twilight of the scientific age*. New York: Broadway Books.

Ihde, D. (1990). *Technology and the lifeworld: From garden to earth*. Bloomington: Indiana University Press.

Jochemsen, H., & van der Stoep, J. (Eds.). (2010). *Different cultures, one world. Dialogue between Christians and Muslims about globalizing technology*. Amsterdam: Rozenberg Publishers.

Kroes, P., & Meijers, A. (Eds.). (2000). *The empirical turn in the philosophy of technology*. Oxford: Elsevier Science.

Meijers, A. W. M., & de Vries, M. J. (2009). Technological knowledge. In J. K. Berg Olson, S. A. Pedersen, & V. F. Hendricks (Eds.), *A companion to the philosophy of technology* (pp. 70–74). Wiley-Blackwell: Chichester.

Mitcham, C. (1994). *Thinking through technology*. Chicago: University of Chicago Press.

Morris, T. (1999). *Philosophy for dummies*. New York: Wiley Publishing, Inc.

Noble, D. (1999). *The religion of technology: The divinity of man and the spirit of invention*. New York: Penguin Books.

Peterson, M., & de Vries, M. J. (2012). Do new technologies give rise to new ethical issues? Some reflections on nanotechnology. In C. Kermisch & M.-G. Pinsart (Eds.), *Nanotechnologies: Towards a shift in the scale of ethics?* (pp. 87–100). Brussels/Paris: EME/CEI.

Schuurman, E. (1997). *Perspectives on technology and culture*. Potchefstroom: Institute for Reformational Studies.

Simondon, G. (1958). *Du mode d'existence des objets techniques*. Paris: Aubier.

Verkerk, M., Hoogland, J., van der Stoep, J., & de Vries, M. J. (2015). *Philosophy of technology: An introduction for technology and business students*. New York: Routledge.

de Vries, M. J. (2005a). *Teaching about technology. An introduction to the philosophy of technology for non-philosophers*. Dordrecht: Springer.

de Vries, M. J. (2005b). Analyzing the complexity of nanotechnology. *Techné, 8*(3), 62–75.

de Vries, M. J. (2008). Gilbert Simondon and the dual nature of technical artifacts. *Techné, 12*(1), 23–35.

de Vries, M. J. (2010). Introducing Van Riessen's work in the philosophy of technology. *Philosophia Reformata, 75*(1), 2–9.

de Vries, M. J. (2012). Utopian thinking in contemporary technology versus responsible technology for an imperfect world. *Perspectives on Science and Christian Faith, 64*(1), 11–19.

de Vries, M. J., Hansson, S. O., & Meijers, A. W. M. (Eds.). (2013). *Norms in technology*. Dordrecht: Springer.

Winner, L. (1986). *The whale and the reactor: A search for limits in an age of high technology*. London: University of Chicago Press.

Critique of Technology

Stephen Petrina

Abstract Diverse relationships to technology and media are expressed or emerge over time, including hopeful enthusiasm and critical resistance. For all the enthusiastic and critical analyses, there are few extensive histories of the critique of technology. This chapter historicizes critical relationships to technology, which range from detachment and skepticism to implicit resistance and explicit opposition or rejection. Relationships to technology and media have immediate implications for culture, economics, and education, but the focus here is on long-term historical implications. This begins with the spiritual critique of technology and proceeds historically through cultural, social, psychic, ontic, and identity critiques. In the final analysis, questions are raised for educators and researchers: If critique barely changes a thing, including youth consciousness, what is its utility? If it has been enough for criticism and critique to offer a counter to progress narratives, then how effective has this been?

Keywords History • Critique • Cultural criticism • Spirituality • STS

This chapter is modest history in expanding "The Frustration of Technology" (Stern 1937) and big history in conceptual granularity, chronological scale, and geographic scope. This is a history of the critique of technology, giving historical depth to Mitcham's (1994) "ways of being-with technology" (pp. 275–299). Diverse relationships to technology and media are expressed or emerge over time, and this chapter specifically historicizes critical relationships, which range from detachment and skepticism to implicit resistance, such as featherbedding, and explicit opposition or rejection. Relationships to technology and media have immediate implications for culture, economics, and education, but the focus here is on long-term historical implications (Fox 2002; Kozinets 2008; Rybcynski 1983). This chapter provides a backdrop for key themes in the book, including how grounds for various philosophies emerge and the practice of critiquing technology.

S. Petrina (✉)
The University of British Columbia, Vancouver, BC, Canada
e-mail: stephen.petrina@ubc.ca

PJ. Williams, K. Stables (eds.), *Critique in Design and Technology Education*, Contemporary Issues in Technology Education, DOI 10.1007/978-981-10-3106-9_3

This chapter begins with the spiritual critique of technology and proceeds historically through cultural criticism and social, psychic, ontic, and identic critiques. Differentiated from the spiritual critique that precedes, cultural criticism of technology emerges in the fifteenth and sixteenth centuries as a mode of describing and depicting the mechanical arts. In the eighteenth and nineteenth centuries, spiritual critique is displaced through a rejection of religion and theology as sources of modern authority. With spiritual ground undermined, social, psychic, ontic, and identic critics of media and technology compete for defensible ground for leverage. The history of critique is a search for ground. This chapter historicizes the critique of technology as well as critique as a practice that has run out of steam (Latour 2004). Disinterestedness, "critical distance" from, or "free relation" to media and technology— seductive, Romantic orientations revived in the 1940s—have been instrumental in critique's gradual decline. The critique of critique has quickened the decline. The conclusion questions the short-term future of machinic critique and long-term renewal of spiritual critique. A story of how we have critically related to technology over time has potential to change these relationships.

Although media and technology were not common in English until the seventeenth century, both are loanwords from ancient Greek and Latin. Homer's (ca. 760 s BCE) μέσος, διὰ, and τέχνη in *the Iliad and Odyssey* are eventually Latinized as *media*, *dia*, and *technê* and generally Anglicized as media, means, and technology. Demonstrating the interrelationships, Aristotle's (ca. 330 BCE/2006, 1447a) *Poetics* analyzes διὰ τέχνης, which can be translated as media, means, *or* technology. Plato and Aristotle developed διαλεκτικῇ (dialectic), or means of incisive speech and reason, and judge forms of language and writing as *media* or means and *technê* (Cuomo 2007). Both are judgmental of τέχνας as technique books or textbooks. When Chaucer (ca. 1395, 219–220) intimates that impious and pious alike are deployed as "goddes instrumentz and meenes" for good or ill, three centuries later he might have said "God's technology and media."

1 Spiritual Critique of Media and Technology, 550 BCE–1400 CE

Accounting for prehistories of cultural evolution, where two hominids debate the merits of a stone implement or pictograph, the first and best critique of technology is Genesis. Predating the Bible, Homer does not offer a critique inasmuch as a demonstration of the design and use of technology via divine intervention. Indeed, for over 2500 years and to this moment, the Garden, Tree of Knowledge, Cain and Abel, Noah's Ark, Tower of Babel, and Babylon are commonly raised to illustrate the fourfold of spirituality, nature, humanity, and technology and so accentuate critique. Although the Bible is ambivalent and enthusiastic toward technology in places, it is quite judgmental and critical in others. In Genesis 4:1–16, for instance, YHWH is extremely judgmental of Cain's culture and technology in that it upsets the course of nature, but in Genesis 6–10 inspires Noah to construct a sophisticated technological fix to escape the wrath of torrential waters. The course of nature is restored and technology is a saving grace, while at the same time the flood destroys a corrupted generation, culture, and technology. Critique of technology is given dimension in Revelation 18 and takes a form of uncompromising judgment (Greek κρίσις, *krísis*, derived from χρίνω or κρίνω, *krínō, krinein, krisis*, Latin *iudicium, discrimen*) of merchants and luxuries as God (YHWH, Θεὸς, *Deus*) proceeds to level Babylon.

Renewed to describe the tower of servers projecting an aura of information through the cloud for an effortless scale to the luminosity of the heavens and how predictably this latest conveyance atrophies and stifles dialogue, the Tower of Babel is history's longest-standing critique of technology. If in Genesis 6–10, YHWH is determined to disperse nature, including human nature and in Genesis 11:1–9 is a diffusion of culture and technology. If an ark or boat is a vehicle for conveying or saving souls, a tower, effectively a stairway to Heaven, is not. In Genesis 11:1–9, Yahwist (J) gives this storied critique of building materials, plans, purpose, and architectural structure. Kafka (1917) adds that the plans were deferred to future generations, which "recognized the senselessness of building a heavenly-reaching tower; but by that time everybody was too deeply involved to leave the city" (p. 39). Generations find themselves too implicated or over-committed to abandon the tower's extension and Babylon's expansion. Although a story of the concentration of power and apotheosis, the more focused critique of idolatry comes later in Isaiah, where the final fall of Babylon materializes the "day of the Lord" or judgment (13:1, 6). Isaiah 44:9–20 is satiric iconoclasm, as vanity is exposed in idolizing artifice and the nature from which it derives. Babel and Babylon archetypes or prototypes are reiterated over time as an opposition or tension between the production, consumption, and mediation of things versus development of higher values or more simply between vice technology and noble deeds.

Eastern traditions also remind humans of the obligations of choosing daily or discerning between good works and bad. From the wisdom of Confucius, Mencius warns of overdirecting one's energies and desires toward worldly things at the expense of cultivating virtue. Following Lao Tzu, a contemporary and rival philosopher of Confucius, Chuang Tzu (2001) cautions against attachment to worldly things, as they come from and return to nothing (Kieschnick 2003; Mitukuni 1979). In a story of "The Goose that Cackled," he comments paradoxically that troubles arise one-sided, "treating things as things but not letting them treat you as a thing" (p. 121). Through the first century BCE, Buddhism was spread from India along the Silk Road west through Arabia and east through China. Somewhat like Hebraic, Christian, Confucian, and Taoist teachings, early Buddhists juxtapose the humble spiritual quest of the monk against the ambitious material concerns of the privileged and rich. In the ancient biography, the Bodhisattva Siddhartha "renounced power and worldly pleasures, gave up his kingdom, severed all ties, and went into homelessness" (Aśvaghoṣa, ca. 110/ Aśvaghoṣa 1894, p. 25). With distinctions in practice between India and China, Buddhists attacked material trappings with rigor, suggesting that humans are subjected to miserable delusions (suffering) of the senses in seeking pleasure and meeting desire in objects or technologies, which invariably in their hollow promise leave one empty. Of course like Christians, few lived like hermits, monks, or nuns, and for the common Buddhist, there were degrees of detachment from the material world. With disagreement on various aspects, Buddhism shares with Hinduism wariness toward material things and attachment or desire generated.

In revelation, Muhammad acknowledges Hebraic law, but the Qur'an (2:83–84, 7:142–145) (Translated by A. Y. Ali 1934) does not repeat the commandments verbatim. The first is iterated as divine law ("Worship none but Allah") (2.83), while the lesson on the fabrication and veneration of idols is given as a demonstration of Abraham's faith and iconoclasm (21.51–70). In addition, the Qur'an (95:4–5) makes it clear that worldly goods yield to good works, all of which yield to the spiritual: "We have indeed created man in the best of moulds," the spiritual, "Then do We abase him (to be) the lowest of the low," the material. "Except such as believe and do righteous deeds," or good works, "For they shall have a reward unfailing." With a generally positive view of crafts as a source of livelihood (Ghabin 2009, pp. 124–132), the Qur'an (59:24) cautions against imitating by icon or competing with Allah's creations, and Muhammad's traditions (*hadiths*) reinforce an injunction against the fabrication and veneration of idols (e.g., statues of Buddha). Muhammad (ca. 624/Muhammad 1932) paid close attention to what Arab artisans made and how they conducted themselves in commerce and trade (Ghabin 2009, pp. 31–39, 133–148). *Hadith*, along with a system of *hisba*, a moral and religious duty to righteousness or in the Qur'an (22:41) given as "enjoin the right and forbid wrong," worked to subject the market and its artifacts to a higher spiritual calling (Ghabin 2009). By 650, a venerated theme of spiritual judgment on design, production, and representation, suggesting or leading to action, regulation, and reform, characterized the critique of technology.

In the *Cratylus*, Plato (ca. 380 BCE) recalls Socrates discerning between two types of wisdom. The first, Socrates notes, is spouted by Heraclitus, who says "'everything gives way and nothing stands fast,' and, likening the things that are to the flowing of a river, he says that 'you cannot step into the same river twice'" (402a). Wisdom (*phronêsis*, *sophia*) in this sense, Socrates clarifies, "signifies the grasping (*epaphê*) of this motion" (412b). In response to Heraclitus, he continues, "but some are moving quickly, others slowly. So what moves quickly is not all there is" (412c). Socrates ends the dialogue disagreeing with Heraclitus' doctrine that things are "flowings or motions," suggesting that wisdom instead also means grasping what is slow and enduring (440b). Throughout the dialogues, and quite vividly in the *Cratylus*, Plato demonstrates reverence toward distant ancestors (*prógonos*) and the ancients (*palaiós*), standard bearers of both might and right, common, although not uniform, across antiquity.

Although Plato does not use the word *ματαιοτη*, often translated as vanity (*vanitas*), the implication is that some knowledge and technologies are vain. Are all designers or technologies susceptible to vanity? For instance, Paul says in Romans 8:20: "For the creature was made subject to vanity"; humans and nature are corruptible. In *Institutes of Oratory* (90–95 CE), Quintilian speaks of *ματαιοτεχνιά*, *mataeotechnia*, a charge of emptiness or judgment of vanity passed on specific crafts and technologies or more generally qualifying technology as inherently empty or vain (Book II, 20:3). The Tower of Babel, its plans, and knowledge of its construction are *mataeotechnologia*, are they not? Similarly and somewhat like Plato, Pyrrhonists developed skepticism (*σκεπτικός*, *zetesis*) as a mode of questioning or criticizing technical expertise (e.g., Sextus Empiricus, 200 CE). To be sure, the means through which ancients critiqued technology were varied and not limited to "skepticism" (McLeish 2014, pp. 213–267; Mitcham 1994, pp. 277–283; Whitney 1990).

In some wisdom traditions, ancestors and the ancients are, rather than suspended in a distant past or immemorial antiquity, retired to another land or kingdom over the mountains or waters and may return some day to restore what was lost or stolen. Following the Spanish conquest in the 1520s, chroniclers conveniently exaggerated Aztecah (Aztecs) confusion of their god Quetzalcoatl with the sudden appearance of invader Cortés (Lockhart 1993). As Sahagún (ca. 1545), a Franciscan monk, documented Aztec spirituality he also refuted it, belief for belief, as so much idolatry. The Aztecs conveyed practices of honoring gods of "the fire, the water, the wind, the sun," as "by means of them we live; they guide us, they protect us. They support, they carry" (p. 56). Despite Sahagún's dismissal, Aztec crafts and inventions were subordinate to the higher powers.

2 Cultural Criticism and Critique of Media and Technology, 1450–1820

As Augustine (ca. 413/1909) documented the gradual decline of the Roman Empire, he stayed centuries of invasion and war by recounting the daily prayer and ritual necessary to momentary solace from burdens and temptation of life in the "earthly city" (Book 19:28). This city is divided, "one in worldly possession, the other in heavenly hope"; the latter anticipates dedication to good works on Earth to serve the city of God (Book 15:21) (Ellul 1951). When Gibbon (1802) summarized the decline, he too stayed centuries of the diffusion of religion and war with the daily turn of the wheel and technologies for food and livelihood. Fortunately for humans, despite a declining Empire, he wrote, "the more useful, or, at least, more necessary arts, can be performed without superior talents, or national subordination… the scythe, the invention or emblem of Saturn, still continued annually to mow the harvests" (pp. 581–582). By December 800 or the day Pope Leo crowned Charlemagne Emperor of Rome, which marked the beginning of what 400 years later was called the Holy Roman Empire, the spiritual or metaphysical ground for the critique of technology was well established. Evident in Chaucer by the fourteenth century, discontinuity or opposition between ancient and modern, east and west, and orient and occident emerged. Over the next 150 years, as ships sailed for colonization, exploration, and mission, creating slave and trade routes, a division between old and new world, and civil and civilized or cultured and cultivated versus primitive was established. Along with aggressive expansion and iconoclasm, within theology was long-standing concern for the edification of humans, including education in the crafts for commerce (Bynum 1973). Customary processes of writing, copying, and printing helped spread critique, but by this time, it was becoming evident that religion and spirituality were creative or inspiring forces of technology (Noble 1997; Ovitt 1986; White 1967, 1975).

Gutenberg's innovation on the traditional screw press with moveable type and printing of the Bible (Latin Vulgate) from 1450 to 1455 best represents new forms of the spiritual promotion of technology. In the *Catholicon of 1460*, Gutenberg acknowledges in his only colophon that "this excellent book" was completed, "not by reed, stylus, or quill, but with the miraculous and harmonious concurrence of punches and types cast in moulds" (quoted in Stillwell 1936, p. 11). Hence, he gives praise to God, the book, and technology. After Gutenberg's death in 1468, Schoeffer follows the tradition in 1473 but now expands the praise to engravers and printers: "Not without the aid of native artisans did Moses achieve the design of the tabernacle and Solomon that of the temple" (quoted in Stillwell 1936, p. 17). Pope Innocent VIII in 1487 decreed praise on the new technology and condemnation for printing anything that challenged Christian faith: "the evil influence of badly conducted arts of printing constitutes to-day the greatest danger to society." The result was prepublication regulation by censors. Pope Leo X decreed in 1515, again with a mix of praise, caution, and regulation: "Wherefore, that that invention, so advantageous to extending the glory of God, to the increase of the faith, and the diffusion of the arts and sciences, may not have the contrary result and become an obstacle to the salvation of souls, we have deemed it advisable to direct our attention to the printing of books" (quoted in Schroeder 1937, p. 504). A decree that the press was neither good nor bad, reserving judgment on how it is used, characterized cultural criticism of technology.

Literary style took advantage of the presses to describe *and* depict technology. In 1470, the *Orthographia* included the now famous Fichet Letter, a commentary on the new presses that printed "with speed, elegance and beauty." Gutenberg deserved praise for inventing "divine and praiseworthy things, in so much as he cut letters of a such a sort that whatever can be said, or thought, can be immediately written or copied" (quoted in Stillwell 1972, p. 92). The first books integrating typeset and engraved block prints were religious, such as Pfister's *Biblia Pauperum* (1460), but printers also established a market for depicting and describing mechanics and the mechanical arts: *Verona Valturius* (1472), *De Architectura* (1521), *Pirotechnia* (1540), *Humani Corporis* (1543), *De re Metallica* (1556), *Das Ständebuch* (*Book of Trades*) (1568), *Livre des Instruments Mathematiques et Mechaniques* (1569), *Diverse et Artificiose Machine* (1588), *Teatro Nuovo di Machine et Edificii* (1607), *The English Improver* (1649), *Mysteries of Nature and Art* (1654), *Orbis Pictus* (1657), *Humane Industry* (1661), *Theatrum Machinarum Novum* (1662), *Mechanick Exercises* (1683–1703), and *Lexicon Technicum* (1704). From Biblical descriptions, Bruegel's depictions of the Tower of Babel (1563) are ominous. Spiritual critique, such as the *Speculum Vitae Humanae* (1570), and contempt for handiwork were common but looked antiquated against the cultural criticism of media and technology found in the new illustrated books (Knoespel 1992; Long 1997).

In the *Preliminary Discourse to the Encyclopedia*, d'Alembert (d'Alembert 1751) draws a bridge to the past, but the expanse is excessive. Progress appears as just a chain of small steps on the surface as "the imagination of the moderns was reborn little by little from that of the ancients" (p. 66). One continuity spanning the divide, he continues, is that "the mechanical arts, which are dependent upon manual operation and are [still] subjugated ... Subsequently it became a reason for holding them in contempt—so much does poverty harm everything that accompanies it" (pp. 41, 42). However, Diderot's (1751) "Prospectus" puts the expanse in stark terms by contrasting moderns with ancients or more recently with sixteenth-century inventories of knowledge: "Think of the progress that has been made since their time in the sciences and the arts! Think of the many truths that are unveiled today which were not dreamed of then!" And "laws of sound criticism were entirely unknown" (pp. 108, 109). Chambers' *Cyclopaedia* tried to account for changes circa 1728, says Diderot, but "everything was lacking on the subject of the mechanical arts. Chambers read books, but he saw scarcely any artisans" (pp. 110–111). "It is thus that we have become convinced of men's [and women's] ignorance concerning most of the objects in this life and of the difficulty of overcoming that ignorance" (p. 124).

In *Frankenstein; or, the Modern Prometheus*, Shelley (1818/1831) picks up the pace of the story with the closing of a lecture on modern progress and Victor becoming a disciple of science. Modern scientists' eyes seem to merely "pore over the microscope or crucible," the lecturer concludes, but they "have indeed performed miracles... and even mock the invisible world with its own shadows." "So much has been done," Victor resolves afterward, but "more, far more, will I achieve: treading in the steps already marked, I will pioneer a new way, explore unknown powers, and unfold to the world the deepest mysteries of creation" (p. 34). After 2 years of studies in chemistry, anatomy, and mechanics, he became "capable of bestowing animation upon lifeless matter" and commenced on "the creation of a human being" (pp. 38, 40). On a November night, he succeeds. Galvanized or given life, the new creature is a thing of beauty but overnight becomes a "miserable monster" through the pains of Victor's fear and guilt. At the moment of creation, he confides, "the beauty of the dream vanished, and breathless horror and disgust filled my heart" (pp. 44, 43). Hence, he leaves the creature to fend for itself. From then on, Victor is a "shadow of a man" tormented by the "monstrous Image" he "endued with the mockery of a soul still more monstrous" (p. 163). Through the end, the creature realizes and avenges its existential condition, "wretched, helpless, and alone" and wanting to destroy "every vestige of cultivation in the garden" and more pointedly destroy Victor Frankenstein and "all thou lovedst" (pp. 112, 120, 198): "I, the miserable and the abandoned, am an abortion, to be spurned at, and kicked, and trampled on" (p. 200). The Tower of Babel stands for the spiritual critique of technology, while *Frankenstein* became a prototype for cultural criticism.

3 Social Critique of Media and Technology, 1840–1900

In the *Critique of Pure Reason*, published in 1781, about the time Boulton and Watt were designing a rotary steam engine to drive cotton spinners in mills, Kant developed critique as a method to set limits on what can be known beyond the ground of experience and how to firm this up as the empirical and objective ground of reason (p. xix). Rather than "criticism of books and systems," Kant founded critique to "expose the groundless nature of the pretensions of" two faculties, reason and understanding (p. 54). He proceeded with the *Critique of Practical Reason* (1785) and *Critique of Judgment* (1790) and a critique of *Religion within the Limits of Mere Reason* (1793) in the face of a Censorship Edict and charges of insubordination by the King after the second edition in 1794. Critics adjusted the method (i.e., simply put, "expose the groundless nature of the pretensions of...") or "critical theory" to an increasingly open field of objects. Kant (1781) proclaimed: "Our age is the age of criticism, to which everything must be subjected," including the "sacredness of religion," "authority of legislation," "monopoly of the schools," power of steam, and natural liberty of capital (p. xix).

Carlyle (1829) took this for granted and proposed that the modern age was "not an Heroical, Devotional, Philosophical, or Moral Age, but above all others, the Mechanical Age... It is the Age of Machinery." "Philosophy, Science, Art, Literature, all depend on machinery," he added (pp. 442, 443). More profound was the interpenetration of culture, humanity, and machinery: "Not the external and physical alone is now managed by machinery, but the internal and spiritual also.... [Humans] are grown mechanical in head and in heart, as well as in hand" (p. 444). Protests against machinery in Britain were common, including the Luddites' efforts in breaking shearing and spinning frames and looms from 1811 to 1813, but government protected the industrialists' capital through police or military force and legislation. A few decades later, in her research for *Mary Barton*, Gaskell (1848) documented Manchester workers, living from hand to mouth, lamenting: "There's never been good times sin' spinning-jennies came up." "Machines is th' ruin of poor folk," they complained (p. 133). The Age of Machinery was prolonged with humans reduced to appendages and inventors and machines becoming Promethean, as Shelley suggested. Machines became the new heroes.

Engels and Marx faced stories that forces of production were now a force of history or, more specifically, the steam engine was an engine of progress. In the wake of this progress was the displacement of labor by capital (Stern 1937). Following his "Outlines of a Critique of Political Economy", Engels (1845) begins *The Condition of the Working-Class in England* with an observation that the agricultural and industrial proletariat in Great Britain begins "with the invention of the steam-engine and of machinery for working cotton. These inventions gave rise, as is well known, to an industrial revolution, a revolution which altered the whole civil society" (p. 1). The "victory of machine-work over hand-work in the chief branches of English industry was won," says Engels, "and the history of the latter from that time forward simply relates how the hand-workers have been driven by machinery" (p. 7).

In short order, a self-contained machine became a mega-machine; England became the "workshop of the world."

In 1807, England abolished their African slave trade but not slavery. In 1808, the Northern United States did the same. The number of slaves in the USA nonetheless increased from 697,879 in 1790 to 3,179,589 in 1850 (Ballantyne 1857, p. 427). Following the cotton gin patent in 1794 and diffusion across southern states in the early 1800s, the number doubled between 1820 and 1850. Slaves were in turn driven toward torturous production rates, with increases in "per hand" yields measured at 500 or 600 %. A Presbyterian preacher only half jokingly commented how before the cotton gin a slave was worth \$300–\$400, and afterward was worth \$600, and soon \$900 "and then there was no such thing as moral law... then 1,000 or 1,200 dollars, and slavery became one of the beatitudes" (Beecher 1863, p. 15). If machines were made to make history could the adversity experienced by those who worked for the machines power an alternative force of history?

When in 1844 Marx declares that "the critique of religion is essentially completed" and "the critique of heaven is transformed into the critique of the earth," he suggests that the spiritual ground of critique had been sufficiently undermined (Marx 1844a, pp. 131, 132). While moderns removed Heaven as a fulcrum for a lever that could move the Earth, Marx recognized the futility of replacing an ancient cosmos with the quicksand of modern culture. Countering the relativity of criticism of one text to the next, or one art (e.g., fine, liberal) to another (e.g., mechanical, menial), he and Engels discover social bedrock in materialism.

From Kant, it was apparent that neither criticism nor critique alone was solvent action, which is to say the idea of culture or philosophy elevated above everyday life or handiwork became a modern convenience (e.g., "a free attitude to its object"). From then on, impartial criticism and indifferent critique were recognized as baseless, groundless, or powerless. In the *Manifesto*, Marx and Engels (1848) acknowledge that "steam and machinery revolutionised industrial production" and modern industry is now led by the bourgeoisie, but deny these developments as social progress (p. 8). Bourgeois "class culture" is but "a mere training to act as a machine" with "children transformed into simple articles of commerce and instruments of labor" (pp. 19, 20). Famously, Marx and Engels conclude: "Let the ruling classes tremble at a Communistic revolution. The proletarians have nothing to lose but their chains" (p. 31).

The "new geography" was formed in the 1860s to make sense of the alarmingly observable influence of humans on nature. Signifying changes, *Man and Nature* represents a comprehensive attempt to describe the "modes in which human action has been or may be most injurious or most beneficial in its influence upon the physical conditions of the earth" (Marsh 1867, p. 10). Cultivating the garden had taken on new dimensions. Although canals, such as the Suez with excavation beginning April 1859, dams, locks, sluices, sewers, mines, and tunnels were immediately noticeable in scale, the other extremes were easily overlooked or taken for granted as inexhaustible. For example, the abundant numbers of birds and their migratory habits seemingly protect them from numerical reduction or extinction. The wild bird population suffered through disorientation and destruction of habitat and scientific specimen production, but in the first issue of *Audubon Magazine*, Thaxter (1887) questions the consumption of birds for hats. Birds such as egrets and plovers were plucked for plumes and feathers or stuffed for design and decoration. Was exploitation of labor and nature a side effect or inherent in "economic growth"?

Through the nineteenth century, social and immanent critique of media and technology were given form beginning with the work of Engels and Marx. They also gave the critique of morals form, which was made more scathing by Nietzsche in *Beyond Good and Evil* (1886) and *Genealogy of Morals* (1887). Cultural criticism, despite its misgivings, was moved to represent nature within "machine in the garden" criticism and moral critique (Marx 1956, 1964). Arts and crafts criticism and agrarian and environmental critique of machines offered new ground to humanize or naturalize some practices by revolutionary and regulatory action counter to capitalism. Arts and crafts, for instance, agitated for dignity against so-called "labor-saving" machines, which ultimately "reduce the skilled labourer to the ranks of the unskilled" and "intensify the labour of those who serve the machines" (Morris 1885, pp. 36–37). Against the subordination or subservience critique, capitalists and managers eventually argued the opposite: the new system elevates the unskilled worker to the dignity of machine hand (Taylor 1911, pp. 146–148). Artisans and trade unions fought against mechanization throughout the century but by its turn resignation set in: "resistance to machinery is futile" (Barnett 1926, p. 211; Hays 1905, p. 27).

4 Psychic and Ontic Critique of Media and Technology, 1910–1970

As the expanse of da Vinci's works was published, most from the Paris Codici and Codice Atlantico between 1891 and 1894, he was established as "a pioneer of the modern spirit." One historian wrote that da Vinci culminated a 200-year process, which began with "Dante, the first modern man," in distinguishing the "Modern from the Middle Age... His attitude towards life was, in a word, thoroughly modern" (Thayer 1894, pp. 514, 510, 508, 515). Freud (1910) acknowledges that da Vinci "became the first modern natural philosopher," as "he learned to depreciate authority and to reject the imitation of the 'ancients' and constantly pointed to the study of nature as the source of all wisdom." But, Freud reasons, "we would say that the 'ancients' and authority only corresponded to the father, and nature again became the tender mother who nourished him" (p. 102). Da Vinci's flying machines, the "human bird," were notoriously reduced to childhood sexual fantasies. Freud was certain that "Leonardo bears out what we must assume from our investigation of children of our times, namely, that his childhood investigation [into machines] was directed to sexual matters" (p. 109). As Klein (1923) began her analysis of children, she asked of da Vinci's (1505, p. 420, folio Sul Volo cover 2 r) infamous prophesy (i.e., "The great [human] bird will take its first flight from the back of its great swan;

it will fill the universe with amazement and all literature will tell of its fame."): "Does not this mean winning the mother's recognition of his genital achievements?" "Leonardo's genital activity, which played so small a part as far as actual instinctual gratification was concerned, was wholly merged in his sublimations" (p. 99).

If in recapitulation theory, says Freud (1910), "psychic development of the individual is a short repetition of the course of development of the race," culture, or society, then the individual unconscious undergirds the social (p. 60). Social ground, bedrock granite on the surface of Engels and Marx, was substrate and layered, readily liquefied or made molten by the psychic magma beneath. Marx's (1844b) "real, corporeal *man*, man with his feet firmly on the solid ground, man exhaling and inhaling all the forces of nature," collapsed (p. 153). The cultural, social, and immanent were fluid while interiorized, "psychic impressions" of the self provided seemingly indestructible ground (p. 60). A frozen Siberia of ethical or social ground could be melted by one hot, steamy psychic molecule. If Marx (1859, p. 11) and Engels needed the sum total of social "relations of production" or economic structure for ground and trusted an entire proletariat to materialize as a fulcrum, Freud needed only the minutiae of psychic content to ground a small lever and demonstrate the insubstantial gravity of supposed heavyweights such as da Vinci and modern machines.

Heidegger was elected on 21 April 1933 as Rector of the University of Freiburg and commenced to enforce the Nazis' Civil Service Act of 7 April, which dismissed Jews from university positions. On 1 May he joined the Nazi Party and eventually removed his dedication in *Being and Time* (Heidegger 1927) to Husserl, his mentor and friend. *Being and Time* finds ontological ground for a phenomenology of the near and commonplace, following Husserl's maxim, "'to the things themselves'" (p. 24). After the war, in July 1945, Heidegger faced a denazification committee and with Freiburg under French occupation was forced to give up his position in January 1946. The prohibition on lecturing was lifted in September 1949 and he resumed teaching at Freiburg in 1951. After giving a lecture titled "The Enframing" in December 1949, Heidegger (1954a, b) retitled the essay "The Question Concerning Technology" (QCT) for a lecture in Munich in 1953. Heidegger (1954b) proposes that in a process of questioning we establish a "free relationship" to technology to discern its essence (p. 3). Following this "way of thinking" promised a chance of deliverance from technology: "Everywhere we remain unfree and chained to technology, whether we passionately affirm or deny it" (p. 4). His insight that we establish a free relation to technology reiterates Hegel's (1807) premise that organic beings essentially exist for themselves and assume "a universal and free relation to inorganic nature" (p. 271). In 1955 Heidegger rephrases this as "releasement toward things" (*Gelassenheit zu den Dingen*) (p. 54) and in a 1957 lecture clarifies the "free relation" as a "step back": "the step out of technology and technological description and interpretation of the age, into the *essence* of modern technology which is still to be thought" (p. 52).

The further one drilled down into the ground of being and essence, the more one pulled core samples of the unconscious, psyche and soul, or subjectivity. For the Frankfurt School, which was closed and displaced from its material foundations in the spring of 1933 while its faculty left in exile from the Nazis' Civil Service Act, finding social, moral, psychic, or ontic ground for a critique of media and technology was crucial. One critique deconstructed, eroded, or undermined the ground of the other and, where ethical critique prepared ground, Freud (1930, pp. 75–91) found narcissistic quicksand. On Marx and Engels' *Kapitalkritik*, critique of ideology and commodity fetishism, and Freud's *Kulturkritik*, the Frankfurt School built an unmoored critique of instrumental reason and rationality (Feenberg 1991, 2002; Habermas 1981a, b; Horkheimer 1947; Leiss 1972). Horkheimer and Adorno's (1947) *Dialectic of Enlightenment*, or "critique of enlightenment" with its own inherent instrumentality (p. xviii), was a helpful addition to cultural criticism and the critical historiography of media and technology, e.g., *Instinct of Workmanship* (1914), *Metropolis* (1926), *Man and Technics* (1931), *Technics and Civilization* (1934), *Technology and Society: The Influence of Machines* (1941), *Mechanization Takes Command* (1948), *Empire and Communications* (1950), *A History of Mechanical Inventions* (1954), *The Gutenberg Galaxy* (1962), and *The Machine in the Garden* (1964). In *The Technological Society*, Ellul (1954) demonstrated that postwar efforts to subordinate technology to ethics or morals reduced to "technical humanism." Technology freely absorbed these humanist critiques: "A few printed pages out of the deluge of printed matter will never make the butterfly a revolutionary" (pp. 337, 424).

The ground of critique was made material once again in Carson's *Silent Spring* in June 1962 in the *New Yorker*. A "grim spectre has crept upon us almost unnoticed," she begins, but this time it is not communism, as that was being snuffed out by both oppression and despondence. The spectre now was the relatively rapid 25-year change in magnitude of humans and their technologies in an assault on the environment: "the contamination of the air, earth, rivers and seas with dangerous, and even lethal, materials" (p. 35). "The pollution of our environment has many sources—radioactive wastes, fallout from nuclear explosions, domestic wastes from cities and towns, and chemical wastes from factories as well as the new fallout from chemical sprays," she continues, dangerously affect our natural resources (p. 64). Birds, with their universal appeal and primordial power, had fallen silent she reports in part fable and part fact as DDT and other insecticides entered food chains and upset ecologies: "In the mornings, which had once throbbed with the dawn chorus of robins, catbirds, doves, jays, and wrens, and scores of other bird voices, there was now no sound; only silence lay over the fields and woods and marshes" (p. 35).

Carson passed away in 1964, while *Silent Spring*, adding activism to cultural criticism, was catalytic for the environmental critique of media and technology in the late 1960s and 1970s. *Silent Spring* along with *Unsafe at Any Speed* (1965), *The Greening of America* (1970), *Nuclear Power and its Critics* (1971), *Limits to Growth* (1972), and *Small is Beautiful* (1973), which defined appropriate technology, characterized newfound links among environmental activism, criticism, and critique. Technology assessment, technological forecasting, and environmental impact assessment signaled policy responses to anticipate risk or industrial accidents, crises, disasters, diseases, and hazards (Bauer 1995). Curricular reforms, such as science, technology and society (STS), science and technology studies (STS), and design and technology (D&T), offered educators and students in the late 1960s and 1970s experiences and insights for making activism and ecocriticism integral to technological literacy (Kimbell and Stables 2008; Petrina 1992, 2000, 2014).

5 Identic Critique of Media and Technology, 1975–2001

Social and psychic critiques were running out of steam and suspended in token groundbreaking. As Lyotard (1979) observes in *The Postmodern Condition*, "we cannot conceal the fact that the critical model in the end lost its theoretical standing and was reduced to the status of a 'utopia' or 'hope,' a token protest." The "social foundation" with its narrative of emancipation had "blurred to the point of losing all of its radicality" (p. 13). Moreover, he notes, in "contemporary society and culture—postindustrial society, postmodern culture—the grand narrative has lost its credibility" (p. 37). Despite post-foundational talk and dwarfing of narratives or traditions, ontic or categorical identity seemingly offered insurmountable, solidified, protected ground, enshrined in the *Universal Declaration of Human Rights* on 10 December 1948 and civil rights in most countries to buttress against discrimination and inequality.

Indigenous activists in the 1970s and 1980s reclaimed some of the ground or land taken through colonial expansion. For example, in Australia, Aborigine northern lands taken in 1863 were reclaimed through traditional territory rights in 1976. Upon first contact of the British Empire in Cook 1770, Cook concluded that the indigenous "seem to have no fixed habitation" yet the Aborigines had been there for 60,000 years (p. 320). In the 1970s in Australia, this point was explicit: land defines what it means to be Aborigine. O'Shane argued in 1979 that the Australian government's management of unceded land "threatens the continued exploitation of our people and our natural resources by multinational mining corporations" (quoted in Burgmann 2003, p. 69). In 1981, she became the first woman and Aborigine to head up a government department in Australia. In America on 27 April 1763, Chief Pontiac addressed a council of tribes gathered on the banks of the Ecorse River and appealed for unity in sieging Fort Detroit and halting British incursions on traditional lands. Neolin, of the Lenni Lenape (of the Wolf), which joined with Pontiac and his Ottawas for various sieges, had journeyed to paradise to see the Master of Life (Keesh-she'-la-mil'-lang-up) (Hunter 1971). The spirit messenger brought wisdom, which Pontiac (1763) articulated: "This land where ye dwell I have made for you and not for others. Whence comes it that ye permit the Whites upon your lands? Can ye not live without them" (pp. 28–30)?

Fifty-one African countries or republics gained independence during the postwar era and through the 1970s; nearly the entire continent became postcolonial and again vulnerable to western growth and development or "dependency domination" (Okolo 1983, p. 237). The question of whether precolonial or indigenous technologies were as effective or more sustainable than colonial and postcolonial or capitalist technologies was reiterated as the concept of the "third world" was exploited during this time (Mytelka 1989). This was no trivial question given 17,500 years of agricultural technologies and 2500 of iron production in Africa before first contact in Guinea in 1450 (Bokoum 2004). In Nigeria, for example, colonial scrap iron policies "succeeded in strangling the local industry and arresting the further development of indigenous technology" despite reports from the smelters at Sukur that their implements lasted twice as long as those made from the cheaper, imported scrap (Emeagwali & Abubakar 1999, p. 29). The last of the indigenous iron smelting furnaces was taken out of production in the early 1980s. Traditions and continuity with the past were severed, disrupting obligations of the Sukur to their ancestors (Childs and Killick 1993).

Feminists in the 1960s and 1970s documented the gendering of media and technology and directly challenged an entrenched doctrine of separate spheres. Rich (1972) summarizes an alarm that sounded in the 1960s as a realization that among "the devastating effects of technological capitalism" there was now an inability to "envision new human and communal relationships." The supremacist postmodern man was discredited in the return of the repressed modern woman. A century of feminist critique of media and technology was validated as the scale and scope of failures of patriarchy increased. As women reclaimed control over oral contraceptive technologies and policies through the 1960s and 1970s they recouped key elements of control over their bodies and careers. Woman's bodies and subjectivities constituted sacrosanct ground while expanding livelihoods meant undermining essentialisms and reclaiming culture. "While we socialize our men to aspire to feats of mastery," Cowan (1979) observes, "we socialize our women to aspire to feats of submission." "Boys play with blocks; girls play with dolls. Men build; women inhabit. Men are active; women are passive. Men are good at mathematics; women are good at literature." "We trained our women to opt out of the technological order," she concludes, "as much as we have trained our men to opt into it.... women who might wish to become engineers or inventors or mechanics or jackhammer operators would have to suppress some deeply engrained notions about their own sexual identity" (p. 62). Through the late 1970s and early 1980s, women changed practices through critique and direct action and corrected the record of cultural criticism to demonstrate inherent biases of media and technology, e.g., *Dynamos and Virgins Revisited* (1979), *Killing Us Softly* (1979), *Women and the Mass Media* (1980), *Machina ex Dea* (1983), *More Work for Mother* (1983), and *Machinery of Dominance* (1983).

This criticism was well placed, yet experiences of women in kibbutzs of Israel or the Soviet Union, mestizas and Latinas in the Americas, Muslim women in the Middle East, Chinese women growing up as communists, or African diaspora and Bengali women tested limits of identity politics. Identifying as differently abled, gay or lesbian, aboriginal, indigenous, postcolonial, or religious provided claims to cultural distinction as well as protected ground. All claimed rightful parity with the women's and civil rights movements and anti-capitalism. "Standpoint epistemology" was developed in the mid-1980s to account for how or why identities are differently grounded, located, positioned, and oriented toward objects, human, material, and spiritual (Ahmed 2006; Haraway 1985; Harding 1986, p. 660; Romany

1991). With the "ontology grounding 'Western' epistemology" "undermined, probably fatally," identity is, at best, something fractured and partial (Haraway 1985, pp. 152–153, 155).

Human identities and protected ground are needlessly irrefutable or weighty next to cyborgs, agile and strategic as they are, which revel in fragmentation and the loss of identity, needing only transitory signals to ground critique and go. Haraway (1985) describes a cyborg as "a kind of disassembled and reassembled, postmodern collective and personal self" (p. 163). Cyborgs are the beings of recrafted bodies and minds reliant on communications and biology's intent on "the translation of the world into a problem of coding" (p. 164). "From one perspective, a cyborg world is about the final imposition of a grid of control on the planet," she advances, "the final appropriation of women's bodies in a masculinist orgy of war." "From another perspective, a cyborg world might be about lived social and bodily realities in which people are not afraid of their joint kinship with animals and machines" (p. 154). "The machine is us, our processes, an aspect of our embodiment," Haraway advises. On shaky ground, "we can be responsible for machines; they do not dominate or threaten us" (p. 180). Banking on random access memory, the "cyborg would not recognize the Garden of Eden" (p. 151). Inessentials, "'god' is dead; so is the 'goddess'" (p. 162).

6 Conclusion

Has the search for ground necessarily come full circle and back to spiritual critique of media and technology? After all, the Tower of Babel offers a venerated theme of judgment on design, production, and representation, suggesting or leading to action, regulation, and reform. Or is this the era of cyborgenic and robotic critique? The machinic critique of media and technology, the most immanent of all, is epitomized in the illustrious Mac versus PC ads that ran from 2006 through 2009 and Cortana versus Siri ads beginning 2014. The long-standing process of company criticizing company, pots calling kettles black, is resolving in the machinic critique of machines (Petrina 2014). All is not yet lost for humans; cultural criticism, describing and depicting media and technology, has merits. The mere performance of critique has value, however groundless. Critics, whether of architecture, art, design, media, engineering, or technology, invariably face a "gulf between the creative and critical" (Forster 1947, p. 17). For practitioners, accusations remain that criticism is "grotesquely remote from the state responsible for the works it affects to expound" (p. 13). At the same time, abstraction, asceticism, renunciation and spiritual detachment have nothing over the "liquid relationship to possessions" that global capitalism and critique machines necessitate (Bardhi et al. 2012). Distantiation is adrift.

In addition to charges of futility is a realization that critique has run out of steam (Latour 2004; Marek 1986). The realization is reflexive and not merely more post-critical commentary, another critique of critique (Petrina 2012). The problem is not so much a cooptation of critique by the other side, conspiracists, corporations, deniers, machines, etc., as Latour (2004) suggests. “What has critique become when a French general, no, a marshal of critique,” asks Latour, claims “that the Twin Towers destroyed themselves under their own weight, so to speak, undermined by the utter nihilism inherent in capitalism itself” (p. 228) (i.e., Baudrillard 2002, p. 8)? Nor is it that critique is easy, if not cheap; action is comparatively costly (Latour 1998, pp. 94–95). The problem is that critics of media and technology have no reliable or stable ground for their critiques. This chapter historically traces what Latour (2005) infers: “Critical discourse has of late become impotent. It has no leverage point left” (p. 4). Machinic critique forms its own short-lived ground, while social, psychic, ontic, and identic critiques are uninformed and removed from a world transfixed by religious and spiritual devotion.

References

Ahmed, S. (2006). *Queer phenomenology*. Durham: Duke University Press.

Ali, A. Y. (1934). *The Holy Qur-ān: English translation and commentary* (1st ed.). Lahore, PK: Shaik Muhammad Ashraf.

Aristotle. (ca. 330 BCE/2006). *Poetics* (J. Sachs, Trans.). Newburyport: R. Pullins.

Aśvaghoṣa. (ca. 110 CE/1894). Prince Siddhattha becomes Buddha (T. W. R. Davids, Trans.). In P. Carus (Ed.), *The gospel of Buddha* (pp. 7–46) Chicago: Open Court.

Augustine. (ca. 413/1909). *The city of God (de civitate Dei)* (vol. 2) (J. Healey, Trans.). Edinburgh: John Grant.

Ballantyne, T. (1857). The cotton crisis; and how to avert it. *British Quarterly Review, 26*, 416–448.

Bardhi, F., Eckhardt, G. M., & Arnould, E. J. (2012). Liquid relationship to possessions. *Journal of Consumer Research, 39*(3), 510–529.

Barnett, G. E. (1926). Chapters on machinery and labor. *Quarterly Journal of Economics, 40*(2), 209–231.

Baudrillard, J. (2002). *The spirit of terrorism and Requiem for the twin towers* (C. Turner, Trans.). New York: Verso.

Bauer, M. (Ed.). (1995). *Resistance to technology: Nuclear power, information technology and biotechnology*. Cambridge, MA: Cambridge University Press.

Beecher, H. W. (1863). American rebellion: Speech. *Cowen Tracts*, 1–25.

Bokoum, H. (Ed.). (2004). *The origins of iron metallurgy in Africa: New light on its antiquity: West and Central Africa*. Paris: UNESCO.

Burgmann, V. (2003). *Power profit and protest: Australian social movements and globalisation*. Crows Nest: Allen & Unwyn.

Bynum, C. W. (1973). The spirituality of regular canons in the twelfth century: A new approach. *Mediaevalia et Humanistica, 4*, 3–24.

Carlyle, T. (1829, June). Signs of the times. *Edinburgh Review, 49*, 439–459.

Carson, R. (1962, June 16 23, 30). *Silent spring*. New Yorker, 35–99, 31–89, 35–67.

Chaucer, J. (1850). The frerers tale. In *The works of Geoffrey Chaucer and others being a reproduction in facsimile of the first collected edition 1532* (pp. 112–116). London: Alexander Moring.

Childs, S. T., & Killick, D. (1993). Indigenous African metallurgy: Nature and culture. *Annual Review of Anthropology, 22*, 317–337.

Chuang, T. (2001). *Chuang-tzu: The inner* chapters (A. C. Graham, Trans.). Cambridge, MA: Hacket.

Confucius. (1883). *Analects* (J. Legge, Trans.). London: Trubner.
Cook, J. (1770/1893). *Captain Cook's journal during his first voyage round the world made in H.M. Bark "Endeavour" 1768–71*. London: Elliot Stock.
Cowan, R. S. (1979). From Virginia Dare to Virginia Slims: Women and technology in American life. *Technology and Culture, 20*(1), 51–63.
Cuomo, S. (2007). *Technology and culture in Greek and Roman antiquity*. Cambridge, MA: Cambridge University Press.
d'Alembert, J. R. (1751/1963). *Preliminary discourse to the Encyclopedia of Diderot* (R. N. Schwab, Trans.). Indianapolis: Bobbs-Merrill.
Diderot, D. (1751/1963). Prospectus. In *Preliminary discourse to the Encyclopedia of Diderot* (R. N. Schwab, Trans.). Indianapolis: Bobbs-Merrill.
Ellul, J. (1951/1970). *The meaning of the city* (D. Pardee, Trans.). Grand Rapids: Eerdmans.
Ellul, J. (1954/1964). *The technological society* (J. Wilkinson, Trans.). New York: Vintage.
Emeagwali, G. T., & Abubakar, N. (1999). Colonialism and African indigenous technology. *African Technology Forum, 7*(2), 27–30.
Engels, F. (1845/1892). *The condition of the working-class in England in 1844* (F. K. Wischnewetzky, Trans.). London: Allen Unwin.
Feenberg, A. (1991). *Critical theory of technology*. New York: Oxford University Press.
Feenberg, A. (2002). *Transforming technology: A critical theory revisited*. New York: Oxford University Press.
Forster, E. M. (1947, July 1). On criticism in the arts, especially music. *Harper's Magazine*, 9–17.
Fox, N. (2002). *Against the machine: The hidden Luddite tradition in literature, art, and individual lives*. London: Island Press.
Freud, S. (1910/1916). *Leonardo da Vinci: A psychosexual study of an infantile reminiscence* (A. A. Brill, Trans.). New York: Moffat, Yard & Company.
Freud, S. (1930). *Civilization and its discontents* (J. Strachey, Trans.). New York: W. W. Norton.
Gaskell, E. (1848). *Mary Barton: Tale of Manchester life*. London: Chapman and Hall.
Ghabin, A. (2009). *Hisba, arts and craft in Islam*. Wiesbaden: Harrassowitz.
Gibbon, E. (1781/1802). *The history of the decline and fall of the Roman Empire* (Vol. II). London: Frederick Warne.
Habermas, J. (1981a/1984). *The theory of communicative action: Reason and the rationalization of society*. Vol. 1 (T. McCarthy, Trans.). Boston: Beacon.
Habermas, J. (1981b/1987). *The theory of communicative action: Lifeworld and system: A critique of functionalist reason*. Vol. 2 (T. McCarthy, Trans.). Boston: Beacon.
Haraway, D. J. (1985). A cyborg manifesto: Science, technology and socialist-feminism in the late twentieth century. In *Simians, cyborgs, and women: The reinvention of nature* (pp. 149–181). New York: Routledge.
Harding, S. (1986). The instability of the analytical categories of feminist theory. *Signs, 11*(4), 645–664.
Hays, D. A. (1905). Automatic machinery. In *Proceedings of the glass bottle blowers' association of the United States and Canada* (pp. 27–33). Camden: C. S. McGrath.
Hegel, G. (1807/1910). *The phenomenology of mind [spirit]*, Vol. 1 (J. B. Baillie, Trans.). New York: MacMillan.
Heidegger, M. (1927/1953/1996). *Being and time* (J. Stambaugh, Trans.). New York: SUNY Press.
Heidegger, M. (1954a). Die frage nach der technik. In C. G. Podewils (Ed.), *Die künste im technischen zeitalter. Gestalt und Gedanke; Jahrbuch, 4*, 70–108.
Heidegger, M. (1954b/1977). The question concerning technology. In *The question concerning technology and other essays* (W. Lovitt, Ed. & Trans.). New York: Garland.
Heidegger, M. (1955/1966). *Discourse on thinking* (J. M. Anderson & E. H. Freund, Trans.). New York: Harper & Row.
Heidegger, M. (1957/2002). The onto-theo-logical constitution of metaphysics (J. Stambaugh, Trans.). In *Identity and difference* (pp. 42–74). Chicago: University of Chicago Press.
Horkheimer, M. (1947). *Eclipse of reason*. New York: Oxford University Press.
Horkheimer, M. & Adorno, T. W., (1947). *The dialectic of enlightenment*. London: Allen and Lane.

Hunter, C. E. (1971). The Delaware nativist revival of the mid-eighteenth century. *Ethnohistory, 18*(1), 39–49.
Kafka, F. (1917/1975). The city coat of arms (W. Muir & E. Muir, Trans.). In *Parables and paradoxes* (pp. 37–39). New York: Schocken.
Kant, I. (1781). *Critique of pure reason*. New York: Harcourt Brace and Co..
Kieschnick, J. (2003). *The impact of Buddhism on Chinese material culture*. Princeton: Princeton University Press.
Kimbell, R., & Stables, K. (2008). *Researching design learning: Issues and findings from two decades of research and development*. Dordrecht: Springer.
Klein, M. (1923). Infant analysis. In E. Jones (Ed.), *Contributions to psycho-analysis, 1921–1945* (pp. 87–116). London: Hogarth.
Knoespel, K. J. (1992). Gazing on technology: *Theatrum Mechanorum* and the assimilation of Renaissance machinery. In M. Greenberg & L. Schachterle (Eds.), *Literature and technology* (p. 66). Bethlehem: Lehigh University Press.
Kozinets, R. V. (2008). Technology/ideology: How ideological fields influence consumers' technology narratives. *Consumer Research, 34*(6), 865–881.
Latour, B. with Hermant, E. (1998/2006). *Paris: Invisible city* (L. Carey-Libbrecht, Trans.). Retrieved http://www.bruno-latour.fr/sites/default/files/downloads/viii_paris-city-gb.pdf
Latour, B. (2004). Why has critique run out of steam? From matters of fact to matters of concern. *Critical Inquiry, 30*(2), 225–248.
Latour, B. (2005). *Critical distance or critical proximity? A dialogue in honor of Donna Haraway*. Unpublished manuscript. Retrieved http://www.bruno-latour.fr/sites/default/files/P-113-HARAWAY.pdf
Leiss, W. (1972). *The domination of nature*. Boston: Beacon.
Lockhart, J. (1993). *We people here: Nahuatl accounts of the conquest of Mexico*. Berkeley: University of California Press.
Long, P. O. (1997). Power, patronage, and the authorship of ars: From mechanical know-how to mechanical knowledge in the last scribal age. *Isis, 88*(1), 1–41.
Lyotard, J.-F. (1979/1984). *The postmodern condition: A report on knowledge* (G. Bennington & B. Massumi, Trans.). Manchester: Manchester University Press.
Marek, J. (1986). Marxism as product of the age of the steam engine. *Studies in Soviet Thought, 32*(2), 155–161.
Marsh, G. P. (1867). *Man and nature; or, physical geography as modified by human action*. New York: Charles Scribner.
Marx, K. (1844a/1970). *Critique of Hegel's 'Philosophy of right'* (A. Jolin & J. O'Malley, Trans.). Cambridge: Cambridge University Press.
Marx, K. (1844b/1988). *Economic and philosophic manuscripts of 1844* (M. Milligan, Trans.). Amherst: Prometheus.
Marx, K. (1859/1904). *A contribution to the critique of political economy* (N. I. Stone, Trans.). Chicago: Charles H. Kerr & Company.
Marx, L. (1956). The machine in the garden. *New England Quarterly, 29*(1), 27–42.
Marx, L. (1964). *The machine in the garden*. New York: Oxford University Press.
Marx, K., & Engels, F. (1848/1888). *Manifesto of the communist party* (S. Moore, Trans.). London: William Reeves.
McLeish, T. (2014). *Faith and wisdom in science*. Oxford: Oxford University Press.
Mencius. (ca. 350 BCE/1861). *The works of Mencius* (J. Legge, Trans.). London: Trubner.
Mitcham, C. (1994). *Thinking through technology. The path between engineering and philosophy*. Chicago: University of Chicago Press.
Mitukuni, Y. (1979). The Chinese concept of technology: A historical approach. *Acta Asiatica, 36*, 49–66.
Morris, W. (1885). *Useful work v. useless toil*. London: Socialist League.
Muhammad. (ca. 624/1932). *The English translation of the Holy Traditions (Hadith) of the Holy Prophet Muhammad* (Part 1) (M. M. Ilahi, Trans.). Lahore: Ripon Press.

Mytelka, L. K. (1989). The unfulfilled promise of African industrialization. *African Studies Review, 32*(3), 77–137.
Noble, D. (1997). *The religion of technology: The divinity of man and the spirit of invention*. New York: Knopf.
Okolo, A. (1983). Dependency in Africa: Stages of African political economy. *Alternatives, 9*, 229–247.
Ovitt, G. (1986). The cultural context of western technology: Early Christian attitudes toward manual labor. *Technology and Culture, 27*(3), 477–500.
Petrina, S. (1992). Questioning the language that we use: A reaction to Pannabecker's critique of the technological impact metaphor. *Journal of Technology Education, 4*(1), 54–61.
Petrina, S. (2000). The politics of technological literacy. *International Journal of Technology and Design Education, 10*(2), 181–206.
Petrina, S. (2012). The new critiquette and old scholactivism: On academic manners, managers, matters, and freedom. *Workplace: A Journal for Academic Labor, 20*, 17–63.
Petrina, S. (2014). Postliterate machineries. In J. Dakers (Ed.), *New frontiers in technological literacy: Breaking with the past* (pp. 29–43). New York: Palgrave Macmillan.
Plato. (ca. 380 BCE/1996). *Cratylus* (C. D. C. Reeve, Trans.). In J. M. Cooper (Ed.), *Plato: Complete works* (pp.101–156). Cambridge, MA: Hackett.
Pontiac. (1763/1912). Pontiac manuscript (R. C. Ford, Trans.). In C. M. Burton (Ed.), *Journal or narrative of a conspiracy*. Detroit: Speaker-Hines.
Rich, A. (1972, November 30). The anti-feminist woman: Review of *The New Chastity and Other Arguments Against Women's Liberation*. *New York Review of Books*, 34–40.
Romany, C. (1991). Ain't I a feminist? *Yale Journal of Law and Feminism, 4*(1), 23–33.
Rybcynski, W. (1983). *Taming the tiger: The struggle to control technology*. New York: Viking Press.
Sahagún, de B. (ca. 1545/1970). *Florentine codex: General history of the things of new Spain. Book 1—The gods* (A. J. O. Anderson & C. E. Dribble, Trans.). Santa Fe: School of American Research.
Schroeder, H. J. (1937). *Disciplinary decrees of the general councils: Text, translation, and commentary*. St. Louis: B. Herder.
Shelley, M. W. (1818). *Frankenstein; or, the modern Prometheus*. London: Colburn and Bentley.
Stern, B. J. (1937). The frustration of technology. *Science & Society, 2*(1), 3–28.
Stillwell, M. B. (1936). *Gutenberg and the Catholicon of 1450: A bibliographical essay*. New York: E. B. Hackett.
Stillwell, M. B. (1972). *The beginning world of books, 1450–1470*. New York: Bibliographical Society of America.
Taylor, F. W. (1911). *Shop management*. New York: Harper.
Thaxter, C. (1887). Woman's heartlessness. *Audubon Magazine, 1*(1), 13–14.
Thayer, W. R. (1894). Leonardo Da Vinci as a pioneer in science. *The Monist, 4*(4), 507–532.
White, L. (1967, March 10). The historical roots of our ecologic crisis. *Science, 155*, 1203–1207.
White, L. (1975). Medieval engineering and the sociology of knowledge. *Pacific Historical Review, 44*(1), 1–21.
Whitney, E. (1990). Paradise restored: The mechanical arts from antiquity through the thirteenth century. *Transactions of the American Philosophical Society, 80*(1), 1–169.

Critiquing Design: Perspectives and World Views on Design and Design and Technology Education, for the Common Good

Kay Stables

Abstract This chapter critiques design and design practices from historical, social, cultural and sustainable perspectives as a basis for opening up a broader perspective on the ways design and designing are seen within mainstream design and technology education in schools. This chapter is divided into four broad sections. The first section explores the ways that design practitioners, theorists and historians critique past and present practices of design from within the profession. This is followed by an outlining of approaches that some designers have taken in using design itself as a way of critiquing society and culture. The focus then turns to design and technology education and highlights concerns that have been identified both at school and higher education level. Finally, consideration is given to examples that illustrate positive approaches to bringing broader and more critical approaches to design and technology in classrooms, including ways that are developed in detail in further chapters in this book.

Keywords Design practices • World views • Critical design • Design activism • Utopian and dystopian design

1 Introduction

The aim of this chapter is to discuss, expand critique and disrupt some mainstream ideas about the nature and potential of design and then explore the relevance of this discussion in the context of school design and technology projects. This aim is premised on a concern that many learning experiences provided in design and technology education in schools are not consistently as challenging, engaging and meaningful as they could be and that it is often a limited understanding of design's importance and potential that underpins a restricted approach. With this in mind, my intention is to illuminate a far broader canvas of what design practices can be and

K. Stables (✉)
Goldsmiths, University of London, London, UK
e-mail: k.stables@gold.ac.uk

P.J. Williams, K. Stables (eds.), *Critique in Design and Technology Education*,
Contemporary Issues in Technology Education, DOI 10.1007/978-981-10-3106-9_4

what they can achieve as a means to enrich, broaden and deepen critical and creative practices within mainstream schooling.

This chapter will take a critical look at design practices more generally and then focus on a critique of design practices within schools. The phrase 'design practices' is used explicitly in this chapter not as a way of creating a dichotomy between practical and theoretical approaches but to indicate a more holistic perspective of *all* that *designing* constitutes, drawing on Lucy Kimbell's application of Reckwitz's theory of social practices (2002) in which she describes practices as 'a nexus of minds, bodies, objects, discourses, knowledge, structures/processes and agency, that together constitute practices which are carried out by individuals' (Kimbell 2009, p. 4).

2 Critiquing Past and Present Design Practices

2.1 How Do Designers Define and Critique Themselves?

While its roots go deep into human activity, the idea of design as it is viewed today is relatively young. For many its birthplace is seen as the industrial revolution, meaning that, as a discipline and a profession, it has only been in existence for about 150 years. The practices of design have evolved partly in relation to meeting the needs of industrial societies, and these practices have created many valuable innovations. However, in the fullness of time, some design innovations have come to be seen as of questionable value. Historically, design practices have had their critics, not least from within the field of design itself. The first section of this chapter explores historical and current practices of design that arguably have created as many problems in the world as they have solved. The perspectives in this first section are largely those expressed by designers themselves, challenging and critiquing from within.

Design historians and those operating in the area of 'design studies' raise issues about design criticism itself. Is critique too embedded within normative views such that designing is critiqued against what might be seen as the design 'cannon'? Should critique step outside of this in order to provide less insular perspectives (Whitely 1997)? Huygen (1997) takes a historical view of design critique and identifies three forms of criticism: one that focuses on instruction on how to design; one that deals with norms and criteria such as functionality, utility, durability, universal aesthetic and so on; and one that he refers to as cultural criticism, critiquing 'the context of design and its impact on society, and on the ideology and the way it functions' (p. 41).

It is this third form of critique that I see as the most productive to use to frame this chapter, drawing on examples where this is the stance that designers themselves have used to critique design practices from within.

2.2 Utopian and Dystopian Views of Design

Design has been a major force in increasing standards of living and economic prosperity for large sections of the world's population. But in parallel, it has also been a major player in the development of a consumer culture that threatens to massively damage the futures of all living things. Margolin, writing in 1998, links the challenges that have been created in the name of design back to its modern starting point with the statement that 'Since design's beginning, when it was conceived as an art of giving form to products for mass production, it has been firmly embedded in consumer culture' (Margolin 1998, p. 83). He sees this embeddedness as defining the mainstream development of design – citing examples of mass production in the UK, USA and Germany. Many designers see themselves as operating for the common good. But the values of different people, different cultures and different times affect the ways in which individuals and communities see the world, so that one person's 'common good' may be another person's nightmare. Within the post-industrialisation history of design, there have been a series of movements, driven by the vision, ideals and values of groups of designers whose motives have been for progress in society – a utopian aspiration. Dorrestijn and Verbeek (2013) highlight key utopian movements in the modern age – the *Arts and Crafts* movement that sought a utopia that acted as a contrast and alternative to mass production, *modernism* movements that saw industrialisation and technology as forces for social good and *postmodernism* that has been promoted as a utopia of diversity and plurality. They identify these as utopian design movements 'because in these movements designers seem to have been most explicitly concerned with improving peoples' ways of living by means of design' (p. 46). But they also comment that these movements are highly contested in terms of the extent to which their utopian ideals have been achieved.

This contestation has resonance with Margolin's (1998) writings about the negative contribution of design to societies and the lack of critique that was visible during the twentieth century. In making his comments, he points to the small number of visionaries who have taken a critical stance from within – identifying particularly Papanek (1971) as the harshest critic when he wrote:

> Today, industrial design has put murder on a mass-production basis. By designing criminally unsafe automobiles that kill or maim nearly one million people around the world each year, by creating a whole new species of permanent garbage to clutter up the landscape, and by choosing materials and processes that pollute the air we breath, designers have become a dangerous breed. (Papanek 1971, p. ix)

Looking back further in design's history, Margolin cites Buckminster Fuller's critique of an earlier era (1920s) that attacked traditional practices and limitations of industrial design, by focusing on the possibilities of advanced technologies that allowed for economic use of materials – a critique that was evidenced through Fuller's unconventional design thinking as shown in his practice, for example, from his early developments of the 4D 'Dymaxion' house in the 1920s to his geodesic dome of the 1950s and 1960s. Margolin makes the point that, in referring to Papanek and Fuller, he is contrasting one designer who was drawing on wisdom of indigenous peoples (Papanek) and one who was realising his ideas through the latest technological developments (Fuller). In doing so he highlights an enduring dimension of designing as part of human culture and its inextricable link with technology and technological development. He mentions more recent critics such as Gui Bonsiepe, Tomas Maldonado and John Chris Jones but comments that within the twentieth century, with a few exceptions, 'designers have not been able to envision a professional practice outside of the consumer culture. ... most product designers have remained locked into the aims and arguments of their business clients, believing themselves unable to take any initiatives of their own' (Margolin 1998, p. 86).

In exploring perspectives on design for the common good, what quickly becomes apparent are conflicting views, approaches and attitudes, particularly when design is linked closely with technology.

Discussing the contestation of utopian design, Dorrestijn and Verbeek (2013) highlight a reaction against a perception of technology as the 'highway to utopia' that has created a 'dystopian countermovement' (p. 52) preoccupied with the dangers that technology poses. This swing between extremes illustrates the conflict between utopian beliefs and dystopian fears and, for Dorrestijn and Verbeek, an overarching challenge between freedom and constraint. For Walker (2010), writing in the context of sustainable design, the idea that technology will make people happier by providing solutions to social and ecological problems creates a 'technocratic version of societal wellbeing … where our environmental and social problems will be solved through the ingenious application of advanced, super-efficient, non-polluting technologies' that is 'patently false and flies in the face of both logic and the teachings of all the major philosophical and wisdom traditions down the ages' (Walker 2010, p. 104). Feenberg (1999) identifies ideas of technology as the solution to societal problems as being driven by a deterministic concept of technology – one that sees technology as neutral and its increased use as a way of progressing 'civilisation'. In explaining the challenges to technological determinism, he outlines a 'substantive' theory of technology that denies the neutrality of technology. In his view, technology 'embodies specific values. Its spread is therefore not innocent. The tools we use shape our way of life in modern societies where technique has become all-pervasive. In this situation, means and ends cannot be separated' (Feenberg 1999, p. 2). When viewed in this way, a utopian perspective of technological advancement is questionable, in particular when technology starts to impact on humanity and the environment in ways that hadn't been anticipated. In a similar vein to Dorrestijn and Verbeek, questions are raised about the locus of control, for Feenberg whether it is within the technology itself or within the human capacity that created it. This quickly turns the tables on technological utopia, creating images of dystopian proportions – made visible in our imaginations through fictions such as Mary Shelley's Frankenstein, Aldous Huxley's Brave New World and George Orwell's 1984. And so the technological genie is out of the bottle, and no attempts to push it back will be successful. Presenting a more optimistic perspective, Feenberg uses critical theory to explore a further view that:

> recognises the catastrophic consequences of technological development but still sees a promise of greater freedom in a possible future. The problem is not with technology as such but with our failure so far to devise appropriate institutions for exercising human control over it. We could tame technology by submitting it to more a democratic process of design and development. (Feenberg 2006, p. 12)

Through this latter statement, Feenberg makes the link not just to technology but also to the role of design. The two are undeniably and inextricably linked, for example, as portrayed in Archer's straightforward definition of technology as 'knowing how' and design as 'envisaging what' (Archer 1992, p. 8). In this linking, it is important to recognise that design in one way or another is implicated in technological development, important in achieving effective developments but equally culpable when things go wrong.

'Going wrong' can be an unintended consequence of a poorly considered design, possibly a naïve belief that neutrality means that any new concept or product will be used for its original 'good' intent. But Mitcham and Holbrook (2006) point out that the 'going wrong' can also be driven by 'evil' intent, for example, through acts of terrorism. They illustrate this by pointing to examples given by Kemper (2004) such as the use of fertilisers used in car bombs. But even more mundane products have their initial purposes displaced once in the hands of consumers. One only has to look at the number of people wearing highly technically designed running shoes who never run anywhere to see examples of this. Many of us would have opened a tin of paint with a screwdriver! Whether using examples of car bombs or running shoes, it is clear that to anticipate that an object designed for one purpose will not be appropriated (or misappropriated) for some other purpose – what Ihde (2006) calls the Designer Fallacy – is somewhat simplistic. To fail to recognise the many ways in which human beings choose to interact with and utilise the artefacts and systems of the made world, especially from the viewpoint of the professional designer, is a fundamental criticism that has been placed at their door. Tony Fry (2012) takes this debate one step further in exploring the reciprocal relationship between designed objects and humans – the ways in which humans design things that, in turn, affect the ways in which human behaviour is changed. For example, humans have designed mobile phones, and mobile phones in return have 'designed' the ways in which we live our lives, from the ways and speed with which we communicate with each other to the ways in which we access information 'on the go'. This reciprocal relationship he refers to is ontological designing, a facet of the 'designer' in all human beings. He points to the destructive side of this relationship that can be witnessed in the unsustainable aspects of 'western' civilisation – what he calls the nihilistic de-futuring activities of humans, but also to the potential, through human agency, to 'self-recreate' to a positive 'futuring' through ontological designing. Inherent in this idea is a need to recognise the 'future' that de-futuring presents and the need for some significant shifts in thinking including, in Fry's proposition, a shift away from capitalist utopias of modernism. In considering the directions such a shift could take, it is hard to ignore those that prefigured early ideas of utopia as rejecting a life full of abundance and luxury that modernism promised, including the man conceiving the first 'modern Utopia', Thomas More, who, in his fictional work of the same name, written in 1516, promoted the idea of a better life that included an increase in leisure through living a modest lifestyle with a decrease in wants.

Views of utopia are starkly contrasting, and, as those discussed above indicate, one person's utopia is another person's dystopia. But a bigger question, and one that raises issues for how designers can support the idea of 'futuring', is whether, while fictions can create utopias, in reality they are either achievable or desirable. De Vries (2012) makes the point that, from a Christian perspective, only God can create Utopia and that it is more appropriate and valuable for designers to accept that human beings are imperfect and live in an imperfect world and that responsible technological development should take place within this context. In identifying a negative legacy of utopian design, Dorrestijn and Verbeek (2013), in the context of user-influencing design, also challenge the idea of designing for utopia from an ethical standpoint. For them, the aim is to design in a way that finds a balance between coercion and freedom, exploring 'nudge' (a libertarian paternalism approach) and 'persuasive' technology as possible ways forward, recognising that, in the end, 'any design will have unforeseen mediating effects'.

2.3 *Micro and Attainable Utopias*

The utopias critiqued above all fit in some way into what might be called a grand narrative of Utopia – utopia with a capital 'U'. They are characterised by a specific vision, developed by the few in the interests (or not) of the many. In moving away from this, Dorrestijn and Verbeek propose as an alternative a more moderate 'post-utopian social engagement' that would bring 'a conscious and meaningful integration of technology into people's ways of living' (p. 54). They link this proposal to Hannah Arendt's idea (1958) that, in their words, 'discussions of the good life were rooted in plurality. It was not the desire to develop overarching frameworks for one single answer ... but rather inter-action: acting with others' (Dorrestijn and Verbeek 2013, p. 54).

This plurality indicates an increasing attention that designers are paying to more collective approaches – for some, new views of utopia that are driven by a belief in a collective, social interaction. This can be seen in the views expressed by Wood (2007) when he uses the terms 'attainable utopias' or 'micro-utopias' that capture 'a more tentative, temporary, pluralized or truncated version than the ones we may find in the picture books' (p. 3). He contrasts the idea of the conformity and possible fascism of a universal state of Utopia, with what he describes as 'a more interdependent network of "micro-utopias" (i.e. brief, or local utopias) [that] might be both helpful and feasible. ... different types of wisdom that are joined together' (Wood 2007, p. 12–13). Utopias are often characterised as unattainable, but for Wood it is in the space between the thinkable and unthinkable that designing (and in particular collective designing) can operate. He proposes the following approach:

> It is wise to dream beyond what we currently believe to be attainable. Once we have done so, the next step is to co-imagine the dream in a more shareable form. This means exchanging dreams and seeing how they can be conjoined to enhance one another. The third step is to check that we really want what we have dreamed. The fourth step is to see how much of the dream is attainable. The fifth step is to share the task of producing and sharing the dream. (Wood 2007, p. 13)

This small 'u' view of utopias is echoed by Gamez and Rogers (2008) who champion the concept's intrinsic diversity and equity. The views of those speaking for the more micro views of utopia frequently link to democratic views of design that put the designer as part of a team of experts, rather than being what Baynes refers to as the 'hero' designer (2010). A number of newly defined fields of design have emerged from within the profession that recognise the value of collective approaches that bring together expertise from within and beyond the formal disciplines of design and that see the 'user' and other stakeholders as valid contributors. In their own ways, whether defining themselves or their methods as co-design, participatory design, socially responsive design or user-centred design, they are collectively critiquing the idea of the hero designer and recognising the value of more distributed approaches to design, without denying the importance of professional expertise – less the hero, more the team. Manzini (2015) sees these more distributed approaches as key to what he refers to as an 'emerging civilisation' in a world 'when everybody designs' in which social innovation is created by collaboration between design 'experts' (or professionals) and 'nonexperts' – who he refers to as diffuse designers – those who engage with design through their innate human design capability.

The importance of designers working with expertise beyond design also emerges from Fry's critique of the insularity of seeing design as a 'category, discourse, or professional instrumental practice' (2012. p. 91). He argues that to not bring design thinking to other thinking across the arts and sciences is to ignore the complexity of design and its fundamental contribution in human development. Threading through each of the cases for more distributed, collaborative approaches is a motivation for creating 'better' futures – less grand, more attainable.

As more and more designers are finding themselves working in interdisciplinary contexts, there is an emerging concern that maintaining traditional design disciplines (such as graphic, product, fashion, industrial design) creates unnecessary boundaries and that moves towards interdisciplinary or transdisciplinary practices are more realistic and less restrictive. Some go further to talk of post-disciplinary design, drawing from post-disciplinary studies that:

> emerge when scholars forget about disciplines and whether ideas can be identified with any particular one; they identify with learning rather than with disciplines. They follow ideas and connections wherever they lead instead of following them only as far as the border of their discipline. It doesn't mean dilettantism or eclecticism, ending up doing a lot of things badly. It differs from those things precisely because it requires us to follow connections. One can still study a coherent group of phenomena, in fact since one is not dividing it up and selecting out elements appropriate to a particular discipline, it can be more coherent than disciplinary studies. (Sayer 2000, p. 87)

What can be seen from all of the above approaches is the (sometimes implicit) critique of an omnipotent position of a professional designer, an understanding of the importance of collaboration and, within this, a recognition of the potential of the designer in all humans.

2.4 World Views

While designers might work increasingly with broad groups of stakeholders, the vision they bring to a project is critical. Fundamental to ways in which design visions are made manifest is the world view of the designers themselves (and/or those commissioning their work), and there is an increasing divide between designers pushing a consumption model that is premised on a western view of affluence and those who have a commitment to a more inclusive world view that takes account of the design challenges within the context of the geographical east and south and of value positions within different faith and indigenous communities.

David Orr writes about the importance of an ecological world view, which he contrasts with an industrial world view. His critique of design is based on a consideration of these two world views in proposing ways in which design could make a greater contribution to sustainable futures. He is not talking here about a paternalistic view of 'western' designers, designing for communities and cultures that they have only a superficial understanding of, but recognising and learning from the design wisdom in other cultures. 'The starting point for ecological design is not some mythical past, but the heritage of design intelligence evident in many places, times, and cultures prior to our own' (Orr 2002, p. 4). He illustrates this with examples drawn from different cultures such as Amish, Inuit and traditional Balinese agriculture – histories of nature and culture living in harmony. This perspective can also be seen in his definition of what he identifies as ecological design in which:

> The goal is not total mastery but harmony that causes no ugliness, human or ecological, somewhere else or at some later time. And it is not just about making things, but rather remaking the human presence in the world in a way that honors life and protects human dignity. Ecological design is a large concept that joins science and the practical arts with ethics, politics, and economics. (Orr 2002, p. 4)

Orr is presenting an optimistic position. Coward and Fathers, in critiquing design in development contexts, take a more pessimistic view that, despite the potential for otherwise:

> Western or Northern concepts of design have forsaken the discipline's capacity to contribute to the quality of life in favour of its role in adding value and increasing sales and profits. Design in the 'developed' world bears much of the responsibility for peddling visions of a lifestyle that, if not immediately unsustainable for the Western/Northern minority, is certainly unattainable for the majority of the population in the South, which for the most part lives in poverty. (Coward and Fathers 2005, p. 452)

While presenting some recent positive examples of alternative approaches to design methodologies in a range of development contexts, they also identify the negative impacts of imported western views of design.

Balaram (2011), an Indian design academic and industrial designer writing from the perspective of a colonised society, also critiques what he describes as imported design, especially when done in the name of collaboration:

> Apart from the disregard to the difference in culture, climate, social and economic contexts, such a trend eventually kills indigenous creativity and creates dependence. Any form of dependence leads to exploitation in many forms by the foreign country. (p. 97)

He makes a strong plea for a revised view of what a designer should be, placing humanity at the core. He sees this as particularly important because of the growth of technology:

> What is now required more and more is not a skilled designer (by skill I mean knowledge and aesthetic sense included) but a broad-based, socially well-integrated, humane designer with a broad global vision. (p. 102)

Balaram makes a case for a shift from 'object-centred designing' to 'process-centred designing', and in this, it is possible to see the derivation of his description of a designer. He suggests that process-centred designing puts the focus on 'basic human needs rather than materialistic concerns ... The focus here is on people not as consumers but as sensitive human beings. The designer's workplace is not his studio but the site where people live. Much of the designing is not designing for the people but designing with the people' (p. 195).

Fleming, in *Design Education for a Sustainable Future*, provides a powerful yet simple concept through his 'mantra' that 'form follows world view'. Through the book, he returns consistently to the impact that world view has on the ways that any one of us approaches designing, particularly in the context of sustainability, suggesting that the mantra 'asks each of us to examine our intentions, personal values and behaviors' (Fleming 2013, Kindle loc. 1230). His critique is of superficial approaches to sustainable design – 'green design' that he refers to as 'less bad' – and also his critique of the inclusion of sustainability in the design curriculum, to which we will return in the third section of this chapter.

2.5 Design Activism

The increasing priority that designers place on their broader contribution to culture and society has seen the emergence of design activism that acts as a form of critique, both of the practices of designers and of the contexts in which design is operating. Historically, a significant focus for activism has been located in the territory of anti-consumerism – as can be seen through the actions of graphic designers in the 1960s that created the 'First Things First' manifesto, critiquing and taking a stand against graphic design being used as a tool to feed the growth of consumption through advertising. Activism through manifestos can be seen as statements of principle that designers make as a way of going public with intentions that then need to be delivered through their related actions as designers. They can also be statements of practice – as can be seen in Bruce Mau's 1998 'An Incomplete Manifesto for Growth' that actively challenges and disrupts conventional approaches to designing through statements such as:

> **Capture accidents.** The wrong answer is the right answer in search of a different question. Collect wrong answers as part of the process. Ask different questions.
>
> **Ask stupid questions.** Growth is fuelled by desire and innocence. Assess the answer, not the question. Imagine learning throughout your life at the rate of an infant.

And linking back to ideas of post-disciplinarily:

> **Avoid fields. Jump fences.** Disciplinary boundaries and regulatory regimes are attempts to control the wilding of creative life. They are often understandable efforts to order what are manifold, complex, evolutionary processes. Our job is to jump the fences and cross the fields.

Julier (2013) distinguishes between design activism that is aimed solely at changing attitudes (e.g. through posters or manifestos) and that which 'functions in both a utilitarian and politicizing sense ... include[ing] the development of new processes and artifacts, where their starting points are overtly social, environmental, and/or political issues, but where they also intervene functionally in these' (p. 219). Critical in this idea is the importance of agency – the designer taking action in response to their own values and beliefs not the designer as a cog in a wheel doing a job defined and specified by someone else. These views have resonance with those of Fuad-Luke who speaks of design practice that creates 'a counter-narrative aimed at generating and balancing positive social, institutional, environmental and or economic change' (2009, p. 27).

3 Design as Critique

Through the first section of this chapter, a story is told of the shifting thinking and understanding of design and a designer's role that has significantly changed the territory and scope of design, particularly within the last 50 years. Changes have come about partly because of broader changes in the world and in societies. But important changes have also emerged through designers questioning and challenging to shift discourses by positioning design not so much as a thing to be critiqued, but as a tool for critique in its own right. The second section of this chapter will turn to explore design practices where the designer sees critique as a major driver of their work. This will start with unpacking the relatively recent concept of critical and speculative design.

There is a mainstream expectation that designed objects will have been created with a particular purpose in mind – meeting a need, maybe solving a problem and, most likely, ending up as a consumer item to be purchased. The standard assumption around what we mean by purpose is a normative one – to meet individual or society's needs for, say, warmth, protection, transportation and communication. There is a general concern that the designed 'thing' will meet some criteria in terms of technical function, user needs and aesthetic impact. But we have seen from the earlier discussion on design activism that sometimes the need can be to protest, to shift behaviour. It is also clear that one person's purpose or need may be completely at odds with that of another person. Stuart Walker (2006) separates out these two approaches by pointing out that:

> Functional objects do not always have to be all that functional. They do not have to be efficient, effective, economic or even acceptable. Mass-produced products have to be all these things because there is so much capital invested in their production; they have to be profitable. Therefore, the tendency is to play safe and to stay with the tried and true. Understandably, change tends to be incremental and cautious. There are, however, other ways of considering the creation of functional objects, and one of these that is especially useful is 'design as critique'. Design itself can be used as the vehicle of critique and as a means of communication for drawing attention to the inadequacies of current assumptions. Walker (p. 127)

3.1 Critical Practices: Critical Design and Speculative Design

Dunne and Raby are generally acknowledged as the initiators and key proponents of critical design. Showing some resonance with Walker, they give the following rationale for a shift in the use of design:

> The design profession needs to mature and find ways of operating outside the tight constraints of servicing industry. At its worst product design simply reinforces global capitalist values. Design needs to see this for what it is, just one possibility, and to develop alternative roles for itself. It needs to establish an intellectual stance of its own, or the design profession is destined to lose all intellectual credibility and viewed simply as an agent of capitalism. (Dunne and Raby 2001, p. 59)

This alternative and possibly more conceptual approach is one that they have explored extensively. They highlight the value and potential of provocation through speculation, to an extent in the same way that science fiction does. But for them the medium is design practice, not creative writing. In *Speculative Everything: Design Fiction and Social dreaming* (2013), they distinguish approaches to design practices, in a similar way to Walker, between 'industrial production and the marketplace' and a 'parallel channel' of conceptual design that includes 'speculative design, critical design, design fiction, design futures, antidesign, radical design, interrogative design, design for debate, adversarial design, discursive design, futurescaping, and some design art' (Dunne and Raby 2013, p. 11). They see this parallel channel as offering a wide range of opportunities for 'design to pose questions, provoke, and inspire' (ibid) and, most importantly, to critique. They see designers as inherently optimistic and provide a useful and invigorating stance on critique as 'not necessarily negative; it can be a gentle refusal, a turning away from what exists, a longing, wishful thinking, a desire, and even a dream. Critical designs are testimonials to what could be, but at the same time, they offer alternatives that highlight weaknesses within existing normality' (p. 34–35). They see critical design as an activity, not a label, and suggest that 'all good design is critical .... critical thought translated into materiality'.

In reviewing approaches to critical practice, Malpass (2013) suggests that a common link between them is satire, rationality and narrative, but he sees different stances or purposes for approaches. He distinguishes critical design as critiquing the present:

> critical design focuses on present social, cultural, and ethical implications of design objects and practice. It is grounded in critical social theory. ... Through mechanisms of defamiliarization and estrangement, designers such as Dunne and Raby extend the critical distance between the object and the user; in so doing, they make striking comment on current sociotechnical, economic, political, cultural, and psychological concerns. (p. 341)

As an example, Malpass points to Dunne and Raby's 2004 project 'Is the future yours?' in which they 'present a collection of hypothetical products to explore the ethical, cultural and social impact of different energy futures. ... The scenarios included biofuel created from human waste. ... the implication that human beings can or might be transformed from fuel consumers to energy providers' (Malpass 2013, p. 341).

He explains that critical design typically uses fictive scenarios that raise issues through making the person engaging with the designed objects uncomfortable with the concepts presented – in Malpass's word to 'experience a dilemma' – in the example given using child labour to produce energy. Malpass characterises all forms of critical design as involving satire. This example has resonance with satirists from bygone ages – such as Jonathan Swift's *A Modest Proposal*, in which, writing in 1729, he proposed a solution to the Irish 'problem' as breeding children to be cooked like suckling pigs.

Unlike Dunne and Raby, he separates out speculative design as being:

> Situated between emerging scientific discourse and material culture, ... it typically focuses on the domestication of up-and-coming ideas in the sciences and applied technology. It is concerned with the projection of sociotechnical trends, developing scenarios of product roles in new use contexts. It is linked to futures, scenario building and technoscientific research. It is characterized by its inquiry into advancing science and technology. It aims to broaden the contexts and applications of work carried out in laboratories and show them in everyday contexts. (Malpass 2013, p. 338)

Malpass suggests that speculative design operates within Feenberg's (1999) description of a substantive view of technology. His examples include Kerridge's project on biojewellery (Thompson et al. 2006) in which jewellery such as wedding rings were created from the growth of bone cells from each partner, grown to intertwine into the rings. He points to the way that speculative designers 'fast-forward' future scenarios built on current science and technological developments that allow us to see how future use of science and technology might develop, including using future, and often dystopian, fictional contexts.

3.2 *Critiquing Critical Design*

While critical design is critiquing design, there are others who are critiquing critical design. For some, there is concern that these practices come from a stance of privilege, ignoring issues of injustice that make these approaches 'of little value or interest to the majority population of the world who are already living the various negative consequences of past speculations' (Kiem 2014). These views have been countered by Dunne and Raby themselves (2013) and also by others providing examples that exemplify a broader perspective, such as that given by Vu (2011) of design company Droog's project making chairs from the second hand clothes of homeless people.

More in-depth critique comes from Bardzell and Bardzell (2013) who refer to different lenses to view our increasingly technological world such as science technology studies (STS), philosophy of technology and also what they call 'the emerging area of research through design or constructive design' (Bardzell and Bardzell 2013, p. 1), in which they include critical design. Reviewing the value of critical design in the context of human–computer interaction research, they suggest that critical design literature is too underdeveloped to provide practical support and unpack critical design through an analysis using critical theory and metacriticism. Through this they come to the conclusion that 'critical design' as what they see as the narrow definition of Dunne and Raby is one way of looking at it, but that critical design in a more open sense – design that critiques – has much to offer. They exemplify this through Gaver et al.'s 'Prayer Companion' – a design that emerged from research into the potential of new technologies to enhance the lives of the elderly, within which they explored the lives of a cloistered order of predominantly elderly nuns. The Prayer Companion provided this through:

> a resource for the spiritual activity [that] displays a stream of information sourced from RSS news feeds and social networking sites to suggest possible topics for prayers. The nuns have engaged with the device enthusiastically over the first ten months of an ongoing deployment, and ... report that it plays a significant and continuing role in their prayer life. (Gaver et al. 2010, p. 1)

4 Critiquing the Normative Paradigms of Design and Designing Within Mainstream Design and Technology Education

The purpose behind the first two sections of this chapter was to open up new possibilities for thinking about what *design* could be in the context of design and technology education in mainstream schooling. The ways in which design currently exists within this area of schooling varies from country to country and in many ways is driven by the history and culture of design, technology and education within local and national settings. The extent to which design and technology education can be critiqued in a general way across these settings is debatable, but some common threads and themes exist and provide some backcloth against which to begin to explore the potential impact the critique of and by design, as discussed here, could have on future contexts in compulsory schooling.

In England, we have a formalised critique, in the form of evaluation through a national inspection service and in recent years have also found the school subject of design and technology under the spotlight from a range of other stakeholders. Elsewhere (Stables 2012), I have drawn on these recent critiques, identifying a number of concerns including learners being set too many tasks that are formulaic, that are too narrowly focused and that lack challenge. There is too little focus on interdisciplinarity, societal challenges and links with the world beyond schools, resulting in too little emphasis on projects that have genuine social and cultural relevance to young learners. There is also a clear message that more attention needs to be placed on design and for stronger links both with STEM subjects but also with art. More positive critique suggests a consensus that when design and technology is working well in schools, it is taught and learned in ways that are enlightening, inspiring and challenging and that spark enthusiasm and innovation and invoke confidence and pride in learners.

Critiques from elsewhere that have resonance with issues identified earlier in this chapter include concerns for a lack of focus on sustainability issues. Elshof (2006, 2009) highlights a need to move away from a 'product paradigm' that supports a consumerist view of the world. For him, this consumerist view has been the dominant paradigm in design and technology education in which 'productivism as an encompassing belief system offers an uncritical valorization of industry, economic growth, and the consumption of technological products' (Elshof 2006, p. 23).

In a similar vein, Flowers (1998) critiques technology education from an explicitly ecocentric stance, identifying the extent to which an anthropocentric world view dominates design and problem-solving activities, focusing on '"control" over the "human-made and natural environment" to better meet "human needs and wants"' (p. 20).

In a separate article, Flowers (2010) also draws attention to a 'dogmatism ... prevalent in the curriculum, literature, and research in Technology Education' made evident by 'dogmatic uses of a single English word – "the" – to falsely imply uniqueness' (p. 10). He illustrates his point by referring to his self-awareness of his own practice:

> I found myself teaching students about "*the* five families of materials," "*the* six types of material processing," "*the* definition of technology," "*the* rules for brainstorming," "*the* environmental impacts of our obsession with lawns," ... But are there exactly five families of materials, and are these five the five? In each of these instances, I seemed to be attempting to convey to students that one particular model, list, or procedure was the only (or the only important) model, list, or procedure, and they had better learn it. (Flowers 2010, p. 14; my italics)

Of particular importance for the focus of this chapter, he draws attention to the impact that this has had in relation to the unhelpful way in which designing is represented normatively within design and technology education as *the* design process, as if there was just one. This position has been critiqued by others (e.g. Petrina 2000; Lewis 2005) and has dogged both pedagogy and assessment for at least half a century. The tenacity of this representation of designing as a universal method is great – as we are currently witnessing with revisions to the English D&T curriculum where our latest attempt at shifting the dogmatism of the design process has been to refer to 'iterative processes of designing' – only to find that people are now talking about *the* iterative process of design.

Useful critique is also evident in the context of design education in higher education. Margolin (1998) comments on the negative impact on design education of the narrow, consumerist view of design professions and the narrow model of design practice that the students are exposed to. Clune (2008) extends this by highlighting models of design education that, by the way design is defined, support students in designing for unsustainability, giving examples of where a problem is identified through a particular type of product, such as cars, which then focuses the student's mind on redesigning cars, rather than stepping back to look at the broader sociocultural context in which they exist.

As mentioned previously, Fleming (2013) critiques design education that engages students in superficial, 'less bad' approaches to sustainable design. He also comments on the problems created for students by a lack of opportunities for collaborative working by maintaining 'academically reinforced disciplinary silos' (p. 6) and the need for design educators to challenge this position. Questioning a westernised view of design, he states that:

> if form follows world view, and if integration is the new consciousness, then how will that impact design education? The process begins with understanding some core values – inclusion and cooperation – and by pursuing a set of integral core behaviors: beginning with inclusion, the question of "who designs" has new meaning in the age of collaboration, cooperation and integration. (Fleming 2013, p. 4)

5 Achieving a Broader Canvas: Attainable Utopias, Sustainable Futures, Critique and Speculation, Activism and Agency (Insights from Current Practice)

While I have painted a gloomy picture of a reality that is present in in current design and technology education, many of the ideas and understandings from design have already infiltrated the minds and practices of groups of design and technology educators, examples of which are highlighted through the chapters that follow in this book. In this final section, I draw from these examples in order to illustrate positive ways towards enriching design and technology learning and teaching through a broader and deeper understanding of future-facing critical and creative design practices.

Many of the ideas expressed within this chapter about utopian and dystopian views of design when related to new technologies have resonance with David

Barlex's chapter "Disruptive Technologies" on Disruptive Technologies. David provides insights into ways to enable learners to critique disruptive technologies – approaches that allow them to question the validity and impact of the technologies through exploring both constraints and affordances. He suggests a number of approaches, such as drawing on Macnaghten, Davies and Kearnes 'narratives' (2010) to critique (narratives of desire, alienation, the sacred, evil and hope and exploitation) or alternatively through the lenses of 'people', 'market' and 'society'. He also suggests more speculative approaches, such as building future scenarios that in turn can be used for learners to create fictional vignettes that allow learners to explore how technologies might impact on people lives. The value of speculation is further explored in Niall Seery's chapter "Modelling as a Form of Critique" on Modelling as a Form of Critique. In this chapter, Niall presents ways of understanding how speculative modelling can be used as a form of critique that 'explores the world *as it could be* not *as it is*. Future facing speculative scenario building, used to create contexts for design and technology projects also supports the creation of meaningful design briefs which, Fleming reminds us, provide "the consciousness of the project, develops the necessary diverse stakeholders, determines the rules for the co-creative design process, sets the schedule of interactions and clearly illuminates the integrative goals of the project' (2013, p. 6). In my own research, responding to the concerns expressed earlier around formulaic, narrow and unchallenging design projects, learners working in groups who shared an interest in particular societal issues who created their own scenarios and briefs showed maturity, creativity and commitment in their speculations and prototypes for which they felt pride and a sense of achievement (Stables 2013) In a similar vein, Bill Nicholl's chapter "Empathy as an Aspect of Critical Thought and Action in Design and Technology" 'Empathy as an Aspect of Critical Thought and Action in Design and Technology' illustrates how young people can engage in user-centred design methods to develop understanding in a context and build empathy for the users they are designing for. Providing a further example of using scenarios, he shows how imagining you are someone 'suffering capability loss' performing an everyday task can build empathy for the user. He illustrates this further by showing how using user-centred simulation tools such as gloves that simulate arthritis by restricting finger movement or glasses that simulate failing eyesight through blurring lenses allowed young learners to more fully engage with those they were designing for, increasing relevance and resonating with Balaram's (2011) plea for a shift from 'object-centred designing' to process-centred designing that focuses on 'basic human needs rather than materialistic concerns'.

Mishack Gumbo opens up further insight into the importance of keeping design and technology education firmly grounded in societal and cultural contexts. In his chapter "Alternative Knowledge Systems" on 'Alternative Knowledge Systems', he illustrates how an alternative to a traditional western world view shows inclusivity and respect for alternative wisdom and understanding and provides a fresh perspective to enrich approaches to design and technology, both in terms of knowledge and of practices. His chapter echoes those of Orr and Coward and Fathers in questioning the appropriateness of the dominance of a western world view while also showing how much design, technology and education have to learn from indigenous knowledge systems. He illustrates this with examples of indigenous design and technological practices and presents cultural concepts of community and collaborative approaches that resonate with Walker, Fry and Wood and that provide potential for pedagogic approaches, for example, for holism, co-creation and collectiveness and for dealing with values and complexity.

Cecilia Axell, in her chapter "Critiquing Literature: Children's Literature as a Learning Tool for Critical Awareness" on 'Children's Literature as a Learning Tool for Critical Awareness', shows how stories can be used to open up different world-views in relation to design and technology, for example, through her account of Ghanaian author Meshack Asare's *The Canoe Story* that highlights ecological issues and Roald Dahl's *Charlie and the Chocolate Factory* that illustrates technology's role in colonisation. She also introduces pedagogic approaches to bringing criticality and speculation, for example, by presenting children with an incomplete story where they can bring their own creative and speculative design and technological ideas to life through the way they complete the story.

Echoing the concerns of designers (Papanek, Walker, Fry, Margolin, Orr) and design educators (Elshof, Fleming and Flowers), Terry Wilkinson's chapter "Politicizing the Discourse of Consumerism: Reflections on the Story of Stuff" 'Politicizing the Discourse of Consumerism' provides evidence of how both agency and activism can be kindled in young learners whose eyes are opened to the realities of production and consumption of objects in their everyday lives. Using Annie Leonard's *Story of Stuff* as both a model of critique and the context for a project with 12-year-old learners, she illustrates their reactions to gaining insight into realities of globalisation and the critical stance they took as a result. More details of the case study presented from the project can be found elsewhere (Wilkinson and Bencze 2015), but what she makes clear in this book's chapter is the empowerment potential of the use of the resource as a way of enabling 'concrete utopian thinking' that is 'wilful' not 'wishful', 'infused with hope and anticipation' – an idea that shares much with Wood's concept of attainable utopias.

6 Conclusion

Each section in this chapter spotlights areas of interest and concern that are now engaging collections of designers in creating new practices of design and new roles for designers. Many of these practices and roles are currently remote or entirely hidden from experiences of designing that are provided for young people in mainstream schooling. There could be a considerable number of arguments for maintaining this position, but I am not aware that these are currently being made, because such practices are also remote, unseen or outside of the experiences of teachers. Does this mean that a broader, more socially and culturally contextualised experience of designing is not desirable or seen as attainable? If it is the latter, then following some guidance from John Wood (2007) would not be a bad start in exploring ways of opening up possibilities, first by dreaming beyond what we might currently see as attainable, sharing and exchanging dreams and checking whether they are what we really believe to be desirable, how they can become attainable and how to share the task of realising them, however 'micro' they may initially be.

References

Archer, L. B. (1992). The nature of research in design and design education. In B. Archer, K. Baynes, & P. Roberts (Eds.), *The nature of research into design and technology education* (pp. 7–14). Loughborough: Loughborough University of Technology.

Arendt, H. (1958). *The human condition*. Chicago: Chicago University Press.

Balaram, S. (2011). *Thinking design*. New Delhi/Thousand Oaks/London/Singapore: Sage.

Bardzell, J. & Bardzell, S. (2013). *What is "Critical" about critical design?* Paper presented at the CHI '13 Proceedings of the SIGCHI conference on human factors in computing systems.

Baynes, K. (2010). *Models of change: The impact of 'designerly thinking' on people's lives and the environment, seminar 4 modelling and society*. Occasional paper 6. Loughborough: Loughborough University.

Clune, S. (2008). *How you define is how you design: Problematitic definitions in design for sustainability education*. Paper presented at the Changing the change: Design, visions, proposals and tools, Turin.

Coward, T., & Fathers, J. (2005). A critique of design methodologies appropriate to private-sector activity in development. *Development in Practice, 15*(3–4), 451–462.

de Vries, M. J. (2012). Utopian thinking in contemporary technology versus responsible technology for an imperfect world. *Perspectives on Science and Christian Faith, 64*(1), 11–19.

Dorrestijn, S., & Verbeek, P.-P. (2013). Technology, wellbing and freedom: The legacy of utopian design. *International Journal of Design, 7*(3), 45–56.

Dunne, A., & Raby, F. (2001). *Design noir*. Basel: Birkhäuser.

Dunne, A., & Raby, F. (2013). *Speculative everything: Design, fiction and social dreaming*. Cambridge, MA/London: MIT Press.

Elshof, L. (2006). Productivism and the product paradigm in technological education. *Journal of Technology Education, 17*(2), 19–33.

Elshof, L. (2009). Toward sustainable practices in technology education. *International Journal of Technology and Design Education, 19*(2), 133–147.

Feenberg, A. (1999). *Questioning technology*. London/New York: Routledge.

Feenberg, A. (2006). What is philosophy of technology. In J. R. Dakers (Ed.), *Defining technological literacy: Towards an epistemological framework* (pp. 5–16). New York/Basingstoke: Palgrave MacMillan.

Fleming, R. (2013). *Design education for a sustainable future*. London/New York: Earthscan (Routledge).

Flowers, J. (1998). Problem solving in technology education: A Taoist perspective. *Journal of Technology Education, 10*(1), 20–26.

Flowers, J. (2010). The Problem in technology education (A definite article). *Journal of Technology Education, 21*(2), 10–20.

Fry, T. (2012). *Becoming human by design*. London/New York: Berg.

Fuad-Luke, A. (2009). *Design activism: Beautiful strangeness for a sustainable world*. Earthscan from Routledge.

Gámez, J. L. S., & Rogers, S. (2008). Introduction: An architecture of change. In B. Bell & K. Wakeford (Eds.), *Expanding architecture: Design as activism* (pp. 18–25). New York: Metropolis Books.

Gaver, W., Blythe, M., Boucher, A., Jarvis, N., Bowers, J., & Wright, P. (2010). *The prayer companion: Openness and specificity, materiality and spirituality*. Paper presented at the 28th international conference on human factors in computing systems (CHI'10). Atlanta,

Huygen, F. (1997). Report from Holland: Design criticism after postmodernism. *Design Issues, 13*(2, A critical condition: Design and its criticism), 40–43.

Ihde, D. (2006). The designer fallacy and technological imagination. In J. R. Dakers (Ed.), *Defining technological literacy: Towards and epistemological framework* (pp. 121–132). Basingstoke/New York: Palgrave Macmillan.

Julier, G. (2013). From design culture to design activism. *Design and Culture, 5*(2), 215–235.

Kemper, B. (2004). Evil intent and design responsibility. *Science and Engineering Ethics, 10*(2), 303–309.

Kiem, M. (2014). *When the most radical thing you could do is just stop: Or, why the doyens of 'critical' design are the problem with critical design*.

Kimbell, L. (2009). *Design practices in design thinking*. Paper presented at the European Academy of Management Conference, Liverpool.

Lewis, T. (2005). Creativity—A framework for the design/problem solving discourse in technology education. *Journal of Technology Education, 17*(1), 35–52.

Macnaghten, P., Davies, S., & Kearnes, M. (2010). Narrative and public engagement: Some findings from the DEEPEN project. In R. von Schomberg & S. Davies (Eds.), *Understanding public debates on nanotechnologies: Options for framing public policy* (pp. 13–30). Luxemberg: European Union.

Malpass, M. (2013). Between wit and reason: Defining associative, speculative, and critical design in practice. *Design and Culture, 5*(3), 333–356.

Manzini, E. (2015). *Design, when everybody designs: An introduction to design for social innovation*. Cambridge, MA/London: MIT Press.

Margolin, V. (1998). Design for a sustainable world. *Design Issues, 14*(2), 83–92.

Mitcham, C., & Holbrook, J. B. (2006). Understanding technological design. In J. R. Dakers (Ed.), *Defining technological literacy: Towards an epistemological framework* (pp. 105–120). New York/Basingstoke: Palgrave MacMillan.

Orr, D. W. (2002). *The nature of design: Ecology, culture and human intention*. New York: Oxford University Press.

Papanek, V. (1971). *Design for the real world: Human ecology and social change*. London: Thames and Hudson.

Petrina, S. (2000). The political ecology of design and technology education: An inquiry into methods. *International Journal of Technology and Design Education, 10*(3), 207–237.

Reckwitz, A. (2002). Toward a theory of social practices a development in culturalist theorizing. *European Journal of Social Theory, 5*(2), 243–263.

Sayer, A. (2000). For postdisciplinary studies: Sociology and the curse of disciplinary parochialism and imperialsim. In J. A. Eldridge (Ed.), *For sociology: Legacies and prospects* (pp. 83–91). Durham: Sociologypress.

Stables, K. (2012). *Designerly well-being: Can mainstream schooling offer a curriculum that provides a foundation for developing the lifelong design and technological capability of individuals and societies?* Paper presented at the The PATT 26 Conference: Technology education in the 21st Century, KTH, Stockholm, Sweden.

Stables, K. (2013). *Social and cultural relevance in approaches to developing designerly well-being: The potential and challenges when learners call the shots in design and technology projects*. Paper presented at the technology education for the future: A play on sustainability, Christchurch, New Zealand.

Thompson, I., Stott, N., & Kerridge, T. (2006). *Biojewellery: Designing rings with bioengineered bone and tissue*. London: Oral & Maxillofacial Surgery, King's College London.

Vu, T. (2011, February 11). Critical design as constructive provocation. *MIND Design, 36*.

Walker, S. (2006). *Sustainable by design*. London/Stirling: Earthscan.

Walker, S. (2010). Sermons in stones: Argument and artefact for sustainability. *Les Ateliers de l'ethique, 5*(2), 101–116.

Whitely, N. (1997). Introduction. *Design Issues, 13*(2, A critical condition: Design and its criticism), 1–3.

Wilkinson, T., & Bencze, J. L. (2015). With head, hand and heart: Children address ethical issues of design and technology education. In K. Stables & S. Keirl (Eds.), *Environment, ethics and cultures: Design and technology education's contribution to sustainable global futures* (pp. 231–244). Rotterdam: Sense Publishers.

Wood, J. (2007). *Designing for micro-utopias: Thinking beyond the possible*. Aldershot: Gower.

The Identification and Location of Critical Thinking and Critiquing in Design and Technology Education

David Spendlove

Abstract This chapter considers how critical thinking and critique are an essential part of design and technology education. By drawing upon a theoretical framework of three main theories, critical theory, critical pedagogy and critical design, the chapter will explore how engaging in a process of critical thinking leading to critique facilitates agency and self-understanding when engaged in design activities.

Through drawing upon theories of psychology, sociology, politics and philosophy, the formation of epistemological beliefs and robust decision-making will be explored. The chapter concludes by identifying how the 'escape hatch' from reproduction of orthodox practices and thinking can be opened through engaging in critical and metacognitive decision-making processes.

Keywords Critical thinking • Metacognition • Critique • Decision-making

1 Introduction

Critical thinking has been established within educational discourse since the mid-twentieth century, and definitions since this time have continued to evolve. As a starting point within this chapter, I will adopt a fairly straightforward definition that considers critical thinking as 'reflective thinking focused on deciding what to believe or do' (Ennis 1989, p. 4). Within this construct the 'critical' element of critical thinking is etymologically derived from Greek language meaning 'discerning judgement based on standards'; therefore this chapter will consider why such 'discernment in thinking' is important within a design and technology educational context, but equally why such critiquing may be particularly difficult in such an environment. To achieve such critical thinking therefore requires a process of critiquing which can be recognised as a detailed analysis of what Kumar (1996) describes as resulting in 'countering the hypothesis proposed by a peer with an alternative hypothesis' (p. 10). Therefore the 'critical' can be considered

D. Spendlove (✉)
The University of Manchester, Manchester, UK
e-mail: David.Spendlove@Manchester.ac.uk

P.J. Williams, K. Stables (eds.), *Critique in Design and Technology Education*,
Contemporary Issues in Technology Education, DOI 10.1007/978-981-10-3106-9_5

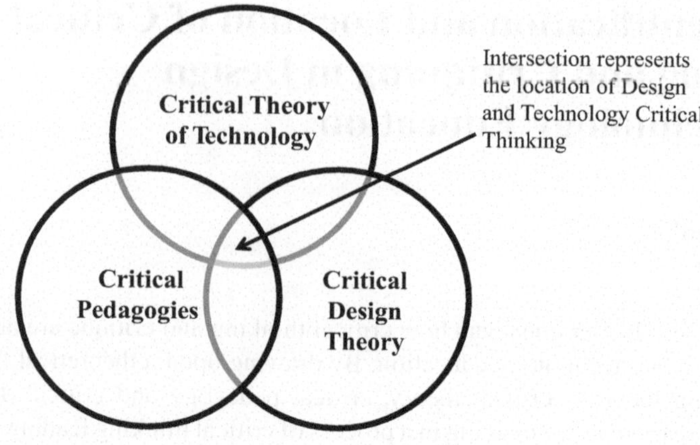

Fig. 1 The location of design and technology critical thinking

as the process, whilst the 'critique' can be considered as the outcome. Within the theoretical framework of the 'critical', it is important to recognise that such critiquing does not exist in a vacuum; consequently this chapter will pose an alternative hypothesis drawing upon three related domains of 'critical', namely:

- Critical design theory (Dunne and Raby 2001)
- Critical theory of technology (Feenberg 2008)
- Critical pedagogies (Apple 1990; Giroux 1994)

The rationale for the above is that each domain draws upon the term 'critical' in the search for 'something better' through challenging accepted norms by offering an 'alternative hypothesis'. The intersection of all three domains (see Fig. 1) represents the design and technology learning environment and recognition that neither the materials, technologies, theories, processes nor procedures we employ nor the educational contexts, assessments and pedagogies utilised are in any sense neutral. Each, the *content*, *context* and *pedagogy*, has a strong cultural and political history such that when we engage in a process of critical thinking, each reveals the lack of neutrality and the often unintended consequences of such limited criticality and associated decision-making.

In this context there has been a plethora of programmes specifically promoting critical thinking in education, often promoting critical thinking as a standalone decontextualised activity, whilst research also has shown that as many as half of experienced teachers, as part of a sample of 10,000 teachers, were rated below 'effective' when attempting to develop students' critical thinking skills (TNTP 2015). The emphasis within this chapter is therefore that critical thinking should be naturally embedded within design and technology education due to the rich and powerful contexts the subject offers for genuine engagement, and as a consequence all teachers need an effective understanding of the underlying principles of critical thinking. To extend this further my view is that a design and technology educational experience without critical thinking embedded within it is therefore benign and impoverished. The challenge for the teacher is subsequently to consider what form the critical thinking will take and what aspects of critical theory and critical pedagogies will inform their practice. In an attempt to unpack this significant challenge, which also represents a unique opportunity, this chapter will focus on:

- The nature of critique in critical thinking
- Critical theory for design and technology
- Critical pedagogies
- The pursuit of agency

The chapter will conclude by identifying ways forward for design and technology education through reconceiving the critical 'thinking' elements of the subject whilst identifying how the pursuit of 'agency' offers unique opportunities for design and technology students.

2 Understanding and Critiquing in Critical Thinking

Previously I have identified critical thinking as reflective thinking focussed upon deciding what to believe or do based upon the application of discerning judgement. Central to this is the challenging and critiquing of often taken for granted views of knowledge, knowledge production and hierarchies of 'privileged knowledge'. Equally, questioning the basis of epistemological beliefs, such as beliefs we have about the certainty of knowledge, the values placed upon knowledge and the control we have over the acquisition, transmission and application of knowledge are all essential features of critical thinking.

Whilst some may therefore consider critical thinking as abstract and something that philosophers may do, critical thinking is also an applied process that results in a critique that has significant value for design and technology education. Accordingly the process can be developed and refined to proficiency and is both a creative and disciplined intellectual activity that can be applied to many different situations but which has particular resonance in those areas of design and technology education that have a focus on decision-making and problem resolution focussed on all pervading human contexts.

Whilst having aligned critical thinking with a philosophical approach, it is important to also identify the psychological dimension, which is a key aspect of critical thinking. In particular in relation to this chapter are aspects of social psychology, which may relate to prejudice, bias, delusion and self-deception when faced with decision-making in design contexts. For example, asking a student to design an item for another person in an unfamiliar context can reveal a whole series of misconceptions, unintended bias, particular emotions and strong beliefs about the new context. As such the issue is not just about getting students to think, as it is almost impossible not to think, but to challenge students to examine their thinking in familiar and unfamiliar contexts. Equally the challenge is not just to get the student to think about perceived solutions. The challenge is to get students to think about their understanding of the context they are working within and the ways they think they are thinking. As a starting point, students should also be asked to consider different ways of thinking and to reflect upon what non-creative, irrational and uncritical responses to unfamiliar contexts might look like. For example, asking a student to design something for a disabled person using the starting points above deliberately positions their thinking at the direct opposite end of the spectrum from where you want them to be at the end of their thinking processes.

Design and technology critical thinking therefore oscillates between philosophical and psychological aspects of thinking, and it is useful to consider how Paul and Binker (1990, p. 551) broad (edited summary) overview of these different types of thinking, bounded by criticality, can be characterised (Table 1).

At this point it is important to reinforce that critical thinking and decision-making are complex and difficult to untangle as, in addition to the philosophical and psychological dimensions above, critical thinking is also interconnected with our biological, social, political, theological, historical and cultural make-up and values. As a consequence, learners (and teachers) are susceptible to involuntary and unconscious cuing (Tversky and Kahneman 1983) recognised as an 'apophenic' state whereby we have an inclination to make spontaneous perception of connections and make meaningfulness of unrelated phenomena (Carroll 2011) when making what appears to be straightforward decisions. As such 'critiquing' becomes 'critical' not only when making decisions on how to proceed but also in attempting to fully understand both the starting context of the perceived problem and the end resolution point of a perceived solution. Therefore in returning to the previous example, whilst facilitating a student's engagement in challenging perceptions and unfamiliar contexts, we have to be 'conscious' of not drawing unsustainable conclusions and connections whilst equally critiquing any proposed resolutions to perceived problems.

Such problem resolution through critiquing lends itself to dialectic processes that promote the development of autonomous rational individual selves (Vygotsky 1987). Such rationality relies however on the challenging of the dominance, internalisation and prevalence of persuasive cultural tools. For example, our everyday communication, considered as collection of cultural tools, can distort our understanding and interpretation of design and technological problem framing. As such the language and signals we use are bound by emotion, culture and history and can distract us from truly understanding the contexts we are operating in. As a consequence we need to 'remove the voices (the partitioning of voices), remove the intonations (emotional and individualizing ones), carve out abstract concepts and judgments from living words and responses, cram everything into one abstract consciousness' (Bakhtin 1986, p. 147). Through removing some of the metaphorical noise, 'that's how you get dialectics' (Ibid.).

Hegel's (1931) triadic dialectic approach based around the concepts of thesis, antithesis and synthesis therefore provides a useful framework to construct critical thinking in design and technology education in order to challenge those assumptions

Table 1 Edited version of Paul's (1990) distinct contribution of psychology and philosophy to critical thinking

	Cognitive psychologists	Philosophers
Approach to thinking	Approach thinking descriptively	Approach thinking normatively
Modes of thinking studied	Focus on expert versus novice thinking, intradisciplinary thinking and monological thinking	Focus on rational reflective thinking on interdisciplinary thinking and on multi-logical thinking
Value emphasis	Emphasise the value of expertise	Emphasise the values of rationality, autonomy, self-criticism, open-mindedness, truth and empathy
Role of values and thinking	Separate the cognitive from the domain of a value choices of the thinker and the overall worldview of the thinker (at least when discussing basic mental skills)	Emphasises the role in thinking of values and the overall conceptual framework of the thinker, hence the significance of identity and assessing points of view and frames of reference
Place of dialogue	Play down the significance of the dialogic and dialectic thinking	Play up the significance of dialogic and dialectic thinking view debate and argumentation as central to rational thinking
View of the affect	Underemphasise the effective obstacles to rational thinking, fear, desire, prejudice, biased, vested interest, conformity, self-deception, egocentrism and ethnocentrism	Emphasise the effective obstacle to rational thinking (this emphasis is correlated with the emphasis on the philosophical idea of becoming a rational person)
Classroom climate	Play down the need to develop classrooms as communities of enquiry where dialogic and dialectic exchange is a matter of course	Play up the need to develop classrooms as communities of enquiry where students learn the art of analysing, synthesising, advocating, reconstructing and challenging each other's ideas
View of the teaching process	Give more weight to the significance of teaching as embodying step-by-step processes	Play up the significance of biological approaches that involve much criss-crossing and unpredictable backtracking

listed. A dialectic process is therefore inherently creative and design orientated as the process involves resolution and refinement through dialogic enquiry. Essentially this is exemplified through the modelling of ideas, as part of a design activity, through which we are able to critique both the externalisation of our thinking and associated language. Through exposing our thinking (e.g. through modelling), we can therefore engage in dialectic and dialogic enquiry and reflection that facilitates critical thinking. Sternberg sharpens the dialectic association and argues that creativity forms the 'antithesis' element of the dialectic process through the questioning and often opposing societal agendas, as well as proposing new ones (2001 p. 360). Central to this is recognising a broad definition of intelligence, which is acknowledged as the ability to adapt to the environment (Sternberg 2000). These principles are central to a design and technology education that is focussed upon improving the notional quality of life, which requires a *synthesis of the dialectic through the* balance of intelligence and creativity *to achieve both stability and change within a societal context (ibid).*

Therefore critical thinking in design and technology education, particularly within the context of decision-making and problem resolution focussed on human contexts, should be through a reciprocal process of dialectic reasoning leading to critique. Such epistemic cognition requires individuals to:

- Reflect on the limits of their own and others knowing and understanding
- Critique the certainty of their own and others knowing and understanding
- Question the criteria used to confirm their own and others knowing and understanding

Critical thinkers are therefore required to be intellectually and dialectically curious. However whilst there might be a tendency to consider that critical thinking is perceived as an implicit act within design and technology education, Paul (1992) argues that typical school instruction does not encourage the development of critical thinking with lack of distinction between critical thinking and content coverage. As previously noted the relationship of design and technology education to critical thinking should be considered as an integral one as the process of critiquing is essential in order to challenge everyday implicit assumptions, cognitive illusions and unsustainable fallacies. In previous work (Spendlove 2010), I have therefore challenged such assumptions that designing is a conscious, intuitive and rational act positing that as design thinkers we are prone to the cognitive and cultural distortions listed above. The subsequent translation of this into practice in the classroom has been through the application of a DT IDEAS pedagogical strategy (Spendlove 2013, 2014, 2015) that will be discussed in further detail at the end of this chapter.

3 Critical Thinking About design and technology Education and Pedagogy

In this section I want to extend the context for critical thinking located previously (Fig. 1) within the interrelationship of three appropriate domains of critique, namely, critical design theory (Dunne and Raby 2001; Bardzell and Bardzell 2013), critical theory of technology (Feenberg 2008) and critical pedagogies (Apple 1990; Giroux 1994), with each situated within a broader social and educational context. In examining these areas of 'criticality', I am attempting to identify the common intersection where critical thinking should thrive, specifically the design and technology education learning environment.

Central to the intersection of the three domains is the recognition that design and technology education and the pedagogical practices employed are acts that are politically, historically and culturally value laden. Therefore engaging in a process of 'critical thinking' reveals the location of such values and ideologies. Consequently, critical thinking rejects the neutrality of design and technology by drawing upon the associated critical theories whilst also engaging in a metacognitive process of reflection upon such thinking. Specifically this is illustrated by Feenberg who highlights that 'the values of a specific social system and the interests of its ruling classes are installed in the very design of rational procedures and machines even before they are assigned specific goals' (2011, p. 15). Likewise *critical design* is aimed at 'leveraging designs to make consumers more critical about their everyday lives, and in particular how their lives are mediated by assumptions, values, ideologies, and behavioural norms inscribed in designs' (Bardzell & Bardzell, p. 1).

In a broader educational context, Apple identifies how 'schools are an important part of a complex structure through which social groups are given legitimacy and through which social and cultural ideologies are re-created, maintained, and continuously built' (1986, p. 9). At the heart of critical pedagogy is the aim to critique oppressive and socially unjust institutions and practice. Likewise the critical theories employed related to design and technology seeks to critique our relationship with products and services in order to challenge preconceptions of power and influence. Critical thinking extends this by both identifying unreliable assertions and influence whilst also questioning the basis and reliability of such decision-making.

Whilst critical pedagogy can be considered as distinct from critical theory, the unique nature of design and technology education means that collectively each critical domain can be enacted within a liberating learning experience. From this we can begin to identify that the enactment of design and technology in an education context is far from neutral and represents a place of social, political, theological and cultural ideologies played out and represented in the choice of curriculum, the teaching methods and as decisions encapsulated in students' judgement making related to artefacts, systems and environments. Such realities are often not revealed or discussed, and as a consequence teachers are often 'delivering' a curriculum, and students are engaged in activities, unaware of the complex interplay of the underlying political, social and cultural influences on their decision-making.

Critiquing through critical thinking therefore offers teachers (and students) a lens to begin to grapple with ways of understanding 'how the kinds of cultural resources and symbols schools select and organize are dialectically related to the kinds of normative and conceptual consciousness 'required' by a stratified society' (Apple 1990, p. 2). Whilst such symbols and resources may be evident in the pedagogy and dialogue within the school environment, they are also inherent in the decisions we make related to the products and systems we create as part of design and technology education focussed upon notional 'progress' and 'improvement' for society. Such artefacts and systems are therefore ultimately value laden and represent the often hidden interests and beliefs that become manifest in the outcomes of teacher and student activities. As Bourdieu (1977) suggests, such institutions (educational, industrial, technological) as a result reproduce social structures and as such preserve the dominance of those that they serve. Such critiquing does not however merely represent a dystopian view, as adopting such dystopic positions would simply repudiate technology without question (Feenberg 1999). Therefore developing a critical perspective that neither falls into naive technological/educational optimism or blind faith or rigid technological/educational determinism and technophobia (ibid) offers a potential way forward.

The reality is that design and technology education offers a powerful context to question assumptions about civil liberties, political and economic power, society, poverty, media, consumption and wealth as each is implicit and embodied within the pedagogy of the teacher and the decision-making of the students. Teachers of design and technology should therefore not be circumspect in exploring these topics as at the centre of the subject are moral and ethical dilemmas and considerations. As Giroux states we need to rethink the 'conditions that make academic labour fruitful, engaging, and relevant' (McLean 2015, p. 306). In this context I want to pose three questions that teachers should ask in order to engage themselves in a critique of design and technology education. These include:

1. What is my understanding of design and technology and the wider social, economic, cultural and political influences on how my understanding is informed?
2. What is my role as a design and technology educator and to what extent do I offer students the opportunity to question the values I engage them with?
3. What aspects of my pedagogical strategies offer liberating opportunities for my students?

The reason for asking the above questions is to provoke reflection on how we can formalise the location of critical thinking within a design and technology experience. Without doing so we risk encouraging children not to think, not to question and to create superfluous, wasteful, unsustainable items that have little educational value and that are difficult to justify.

4 Critical Thinking About Thinking

Earlier in this chapter, I referred to the definition that I was adopting for critical thinking which was based upon judgements made through 'discernment in thinking'. Whilst such a definition may appear straightforward, built upon the rationale of robust decision-making, the realities are far from this. Earlier in this chapter, I also posed questions around the epistemological basis of our beliefs and views we hold relating to knowledge and understanding. As indicated, part of critiquing therefore involves discernment *in* thinking; however a second element involves discernment *of* thinking as in how we engage in processes of metacognition to think not only about *what* we think but also about *how* we think. Such metacognitive processes involve considering:

- How heuristic flaws and how our emotions can guide and mislead our decision-making processes.

- How critical 'design' thinking resists heuristic flaws, cognitive limitations or at least acknowledges awareness of them.
- The ethics of exploiting the heuristic flaws and cognitive limitations of others within design and technology education activities.

The reality is that we, as humans, are prone to over-reliance upon 'intuition', 'gut feeling' and simple 'rules of thumb'. Such fallibilistic epistemology (Siegel 1988) is based upon tentative beliefs held in the context of the currently available evidence and is often based upon 'quick thinking' and the result and basis of heuristic shortcuts which are often steered by our 'feelings'. However what we feel about something informs what we think (rather than our thinking informing our feelings) (Damasio 2008), and as such our intuition is used when we operate quasi-automatically and with reasonable proficiency (Pigliucci 2012). Such thinking is based on an instinctive sense that something is right, a heuristic shortcut, but which is prone to significant errors when engaged in decision-making particularly when operating under a cognitive load. Often such thinking can be referred to as 'tacit' representing uncodified, complex and implicit knowledge. Critical thinking allows reflection on the tacit and fallibilistic in order to get closer to a perceived ideal truth – based upon: '(1) avoiding atomistic view of logical errors in individual reasoning, (2) a concern about self-deception with respect to reasoning, and (3) the disposition of a person in a given context to have a reasonable doxastic attitude' (Ikuenobe 2001, p. 330).

More specifically in the process of design decision-making, avoidance of engagement with critical thinking and metacognition can be through the adoption of purely optimistic strategies. Optimism bias is recognised as a key survival strategy as we mentally project forward and identify our future needs. However optimistic bias is prone to errors which Sharot (2011) cites as the 'superiority illusion' in that we tend to think we are better than we are through failure to engage in critical thinking processes (Sharot gives an example of a survey of driving where 93 per cent of respondents indicated they were above average in driving ability, which would be statistically impossible). Whilst identifying that many limitations exist in relation to our ability to make 'design' decisions in the interest of ourselves, there is also the alternative side of this discussion, which relates to the exploitation by designers of the limitations of cognitive processing in others. Kahneman (2011) identifies a 'focussing illusion' where we misjudge the potential impact of certain circumstances. As such designers and marketing specialists exploit consumer demand by offering a better future using a combination of focussing illusion and visual illusions whilst thriving on an optimism bias, manipulating consumer emotions and thriving on consumers' heuristic shortcuts and cognitive limitations. Such exploitation raises the question of the ethics of the designer which would appear to be a key discussion that should take place in design and technology education programmes. Resolutions of such issues are far from straightforward; however the engagement in such cognitive dissonance and metacognition would seem essential features of agency, critical thinking and critiquing within design and technology activities.

5 Pursuit of Self-Understanding and Personal Agency: Discernment In and Of Thinking

Previously I have emphasised that the design and technology teaching environment should not merely be a passive location where knowledge and skills are transferred. Also that critical thinking involves discernment *in* thinking and discernment *of* thinking, thus by engaging through a process of critiquing, we can begin to see the complexity of the environment in which we teach and facilitate learning. In doing so we can also reveal the opportunities that exist for examining the various dimensions of power and reproduction that operate in a design and technology learning context that previously may have been viewed as benign. As Bourdieu (1984) points out, however it is important to recognise that such power is not only exerted culturally, economically and politically but also through intellectual and pedagogical discourse.

Establishing approaches aimed at reconceiving the 'design thinking' elements of design and technology activities through a critical pedagogical, designerly and technological dialectic therefore offers a way to view contradictions and assumptions often inherent in education. A critiquing approach to design and technology education may therefore adopt a critical dialectic reasoning focussed upon developing a 'meta-awareness'. Such awareness develops through 'an active process of first decoding reality, only to recode through the envisioning of alternative structures' (Au 2009, p. 221). This can be achieved in the design and technology learning environment through:

- Clearer conceptualisation of identification and engagement with problem owners
- Reconceiving contexts and solutions in authentic ways
- Questioning assumptions of neutral pedagogical practice
- Exploration of value-laden outcomes and the ethics of design and technological decision-making
- Application of specific critical design thinking skills

Whilst a central feature of the above list relates to promoting criticality through self-reflection (Schön 1987) and metacognition, a further dimension relates to the recognition and pursuit of self-understanding, 'agency' (Bandura 2001), and acknowledgement of Bourdieu's concept of 'habitus' related to the perceived 'autonomous' decision-making of agents, as in the teachers and students. Habitus (Bourdieu 1984) is recognised as the adoption of socialised norms and psychological tendencies that guide our behaviour and everyday thinking. As such habitus can be considered as how culturally society becomes 'deposited in persons in the form of lasting dispositions, or trained capacities and structured propensities to think, feel and act in determinant ways, which then guide them' (Navarro 2006, p. 16).

Agents (e.g. teachers and students) therefore have a habitus manifested psychologically and emotionally in their intuitive 'feel for the game'. This is not so much a 'state of mind as a state of the body, a state of being. It is because the body has become a repository of ingrained dispositions that certain actions, certain ways of behaving and responding, seem altogether natural' (Bourdieu and Thompson 1991, p. 13). Such intuition can be misleading and ultimately constraining particularly when operating in a creative and designerly context. As indicated by Stevens (1995), a paradox occurs as habitus 'does not determine, but it does guide. Individuals are both completely free and completely constrained...' (p. 112).

Within the context of this discussion, the habitus of the agents operate in competing, contrasting and interconnected ways. Therefore the various agents are represented as multiple identities within interconnected 'fields'. Such fields can be considered as networks 'or a configuration, of objective relations between positions. These positions are objectively defined, in their existence and in the determinations they impose upon their occupants, agents or institutions' (Bourdieu and Wacquant 1992, p. 72). As a consequence and as previously highlighted, the very act of students designing and notionally problem-solving is constrained by a whole range of cultural, historical and psychological factors that are both institutional and biological but which are often manifested as tacit or intuitive thinking. Such tacit preconceptions and unconscious cueing relate to 'self-evident' and often perceived common sense (Watts 2011) which offers an unreliable guide to problem resolution, yet we rely on this mode of thinking virtually all the time to the exclusion of other methods of thinking. As such, exposing students (and teachers) to critiquing and dialectic enquiry reveals how their decision-making and 'volition' (Ankiewicz 2013) are ultimately constrained.

A final concept to be considered in the context of critiquing thinking is Bourdieu's (1977) understanding of obvious and self-evident beliefs known as 'doxa', which is considered as the combination of both orthodox and heterodox norms and beliefs. Doxa therefore represents the assumptions and 'adherence to relations of order which because they structure inseparably both the real world and the thought world and are accepted as self-evident' (Bourdieu 1977, p. 471). An unfortunate consequence of such self-evident acceptance is an overriding antidote of avoiding engagement with complexity and reflection through an over-reliance upon 'intuition', 'gut feeling' and simple 'rules of thumb' often resulting in heuristic flaws.

Such instinctive feelings have been shown to inform us what we think and not the other way around (Damasio 2008); therefore intuition or deliberate practice is when we operate semi-autonomously. However an instinctive sense (considered a heuristic shortcut) that something is right can also be misleading as our instincts are prone to errors particularly when engaged in complex decision-making when operating in stressful circumstances. The perceived overcoming of such limitations can be considered to be achieved through the adoption of purely optimistic strategies that manifest as an 'optimism bias', used as a key survival strategy in that we mentally project forward and identify our future position with little interrogation of such optimism.

6 An Example of Critical Thinking and Critiquing: DT IDEAS

To overcome the cultural and psychological limitations and constraints described in this chapter requires a sense of agency achieved through 'intentionality and forethought, self-regulation by self-reactive influence, and self- reflectiveness about one's capabilities' (Bandura 2001, p. 22). Whilst such a statement can appear quite daunting, the reality and enactment of such a pursuit are most certainly achievable within the design and technology learning environment. As previously indicated such qualities, I would argue, are essential when engaging in genuine critical design thinking processes as part of any educational experience. In attempting to develop a design thinking pedagogy based around these concepts, a series of lessons were designed to challenge student perceptions of 'realities' to be borne out through reflection upon the differences between observed and perceived phenomena.

The lesson (just one of a series of lessons) being reported below was structured around the five broad themes of the acronym DT IDEAS (Table 2) with each letter of the acronym focussing upon an aspect of critical design thinking, for which an interactive teaching episode of 'teaching and learning' was planned. Therefore the lesson contained five sections (each related to the letter of IDEAS) of around 12 min each.

One example which can be used to illustrate the focus on critical thinking was through students reflecting upon their decision-making which related to the 'D' of IDEAS when designing in perceived familiar contexts. In this particular example, students were asked to design for a 'newly married couple'. Once they had designed a response to the design context, further information about the context was revealed to them to illustrate how their decision-making was constrained by preconceived cultural expectations and social norms. For example, in this scenario the newly married couple were shown to be an elderly couple challenging the largely preconceived idea of a young heterosexual couple that was predominantly being held in students' minds. Through this, students were able to examine their preconceived views and decision-making process, whilst teachers were able to illustrate to the student how such heuristic shortcuts can distort their thinking in everyday life. By engaging in a metacognitive process, students were being challenged to reconsider their assumptions and epistemological fallibility. As part of the lesson students were also exposed to psychological principles of 'anchoring'

Table 2 Acronym used and cognitive focus

D = Design
T = Thinking
I = Illusions
D = Decisions
E = Emotions
A = Anchoring
S = Self-serving bias

and 'self-serving bias' whilst also considering how their emotions influence their thinking and how cognitive and visual illusions distort the way we see and think about the world.

The rationale, research study, methods and findings for this case study are reported more fully elsewhere (Spendlove 2013, 2014, 2015); however a key focus of the lesson was about students being able to apply and recall the IDEAS acronym whilst reflecting on its usefulness in challenging their thinking. Results from the study, also reported elsewhere, indicated that students valued being exposed to the limitations in their thinking with a useful quote from one student (capturing what we had hoped to achieve) being that the process had 'opened my eyes to something I was blind to'. From the short input students were very quickly both able to recall and apply the acronym as well as being able to reflect on their cognitive limitations when engaged in critical thinking when placed in design decision-making contexts. Such reflection illustrates the inherent value derived from learning opportunities that illustrate how such cognitive limitations can be exposed in an accessible manner in order to potentially improve future decision-making when engaged in 'everyday' designing. Likewise the teachers engaged in the co-planning of the activity were challenged with considering their own preconceptions both relating to the content and context of the lesson and the associated pedagogies related to increasing student autonomy. Whilst the analysis of the lesson was overwhelmingly positive, both from the students' and teacher's perspectives, perhaps the most encouraging aspect was that several months after the DT IDEAS lesson, despite the short input, students were still referring to the acronym, and the class teacher was able to refer back to the discussed cognitive limitations as students worked through the decision-making aspects of their coursework.

The example above embodies the principles that run throughout this chapter and represented a key focus upon developing 'agency' within a critical design thinking framework, as in the intentional ability to exercise some control over one's thinking, environment and subsequent existence and action. The explicit nature of such critical thinking is unequivocally linked to understanding the extent to which we are responsible for our actions, and we therefore use our 'agency' to make 'metacognitive judgements about whether or not we were in control' (Miele et al. 2011, p. 3620). Agency is thus achieved through 'intentionality and forethought, self-regulation by self-reactive influence, and self- reflectiveness about one's capabilities' (Bandura 2001, p. 1), and such qualities I would again argue are essential when engaging in genuine critical design thinking processes.

7 Some Conclusions

Teachers who are highly accountable, whose reputation and performance are often measured through the perceived success of their students' assessed performance, will often, despite their best intentions, feel constrained and as a result provide their students with a benign and impoverished design and technology experience.

Such constraining of creative opportunities in learning experiences can lead to oppressive practices where students are conditioned into a response necessary for meeting a notionally 'correct' view of predetermined knowledge consumption and reproduction. This modus operandi has increasingly dominated much of teachers' pedagogical practice, when one of our core goals as educators should be to maximise the space for students to be creative and successful learners. Such a space offers genuine opportunities to energise and mobilise students to be informed and genuinely critically engaged to challenge the 'dead zones of the imagination' (Giroux, p. 307) that have come to exist in schools.

The design and technology educator therefore exists in a paradoxical world often distracted by debates on where design and technology exists on a vocational and academia spectrum and consequently is often diverted from meaningful substantive critical reflection about the significant contribution of design and technology to a broader education provision. Regardless, the role as an educator should be to engage students in critical thinking and dialectic reasoning challenging the often privileged knowledge that proliferates and that is manifested through reproduction of existing knowledge and practices. Bourdieu and Thompson (1991) refers to this as a 'heretic break', an escape hatch from reproduction of orthodox practices. Therefore new opportunities exist through reconceiving the 'critical thinking' elements of design and technology activities by creating opportunities for students to engage in:

- Participatory co-creation activities
- Genuine engagement with problem owners
- Working in authentic contexts and resolutions
- Applying specific critical 'design thinking' skills in context
- Engaging in processes of criticality and metacognition
- Considering the ethics of exploitation of cognitive limitations of others

Critical thinking, drawing upon critical theories and critical 'transformative pedagogies', ultimately facilitates the critiquing of epistemological beliefs and the engagement in wider socially responsible contexts. Equally, through questioning 'how ubiquitously the unspoken and sometimes unconscious beliefs about the nature of knowledge and learning' (Schommer-Aikins and Hutter 2002, p. 17) shapes our everyday thinking, we are able to critique the assumptions that pervade our everyday existence. Some may consider such discourses not to be within the realms of design and technology education; however my belief is firmly that such critiquing is a central and essential feature of any high-quality design and technology education experience and is integral to the future development of the subject.

References

Ankiewicz, P. (2013). A theoretical reflection on the implications of the philosophy of technology for classroom pedagogy. *Suid-Afrikaanse Tydskrif vir Natuurwetenskap en Tegnologie, 32*(1), 9-bladsye.

Apple, M. W. (1986). *Teachers and texts: A political economy of class and gender relations in education*. New York: Routledge & Kegan Paul.
Apple, M. W. (1990). *Ideology and curriculum* (2nd ed.). New York: Routledge.
Au, W. (2009). Fighting with the text: Critical issues in the development of Freirian pedagogy. In M. W. Apple, W. Au, & L. A. Gandin (Eds.), *The Routledge handbook of critical education* (pp. 83–95). New York: Routledge.
Bandura, A. (2001). Social cognitive theory: An agentic perspective. *Annual Review of Psychology, 52*(1), 1–26.
Bakhtin, M. (1986). *Speech genres and other late essays*. Austin: University of Texas.
Bardzell, J., & Bardzell, S. (2013, April). What is critical about critical design? In *Proceedings of the SIGCHI conference on human factors in computing systems* (pp. 3297–3306). ACM.
Bourdieu, P. (1977). *Outline of a theory of practice (Vol. 16)*. Cambridge University Press.
Bourdieu, P. (1984). *Distinction: A social critique of the judgement of taste*. Cambridge, MA: Harvard University Press.
Bourdieu, P., & Thompson, J. B. (1991). *Language and symbolic power*. Cambridge, MA: Harvard University Press.
Bourdieu, P., & Wacquant, L. J. (1992). *An invitation to reflexive sociology*. Chicago: University of Chicago Press.
Carroll, R. (2011). *The skeptic's dictionary: A collection of strange beliefs, amusing deceptions, and dangerous delusions*. Hoboken: Wiley.
Damasio, A. (2008). *Descartes' error: Emotion, reason and the human brain*. New York: Random House.
Dunne, A., & Raby, F. (2001). *Design noir. The secret life of electronic objects*. London: Birkhauser.
Ennis, R. (1989). Critical thinking and subject specificity: Clarification and needed research. *Educational Researcher, 18*(3), 4–10.
Feenberg, A. (1999). *Questioning Technology*. Routlege.
Feenberg, A. (2008). From critical theory of technology to the rational critique of rationality. *Social Epistemology, 22*(1), 5–28.
Feenberg, A. (2011). Lecture presented to the course on digital citizenship, IT University of Copenhagen, 2011 agency and citizenship in a technological society http://www.sfu.ca/~andrewf/copen5-1.pdf
Giroux, H A. (1994). In Kerry S. Walters (Ed.), *Toward a pedagogy of critical thinking. Re-thinking reason: New perspectives in critical thinking* (pp. 200–201). Albany: SUNY Press.
Hegel, G. W. F. (1931). The phenomenology of mind (2nd ed., J. B. Baillie, Trans.). London: Allen & Unwin. (Original work published 1807).
Ikuenobe, P. (2001). Questioning as an epistemic process of critical thinking. *Educational Philosophy and Theory, 33*(3–4), 325–341.
Kahneman, D., Thinking, F., & Slow, F. (2011). *Straus and Giroux*. New York.
Kumar, V. S. (1996, April). *Computer-supported collaborative learning: issues for research*. In Eighth annual graduate symposium on Computer Science, University of Saskatchewan.
McLean, H. (2015). Neoliberalism and the attack on education: An interview with Henry A. Giroux. Alternate Routes. *A Journal of Critical Social Research, 26*, 305–311.
Miele, D. B., Wager, T. D., Mitchell, J. P., & Metcalfe, J. (2011). Dissociating neural correlates of action monitoring and metacognition of agency. *Journal of Cognitive Neuroscience, 23*(11), 3620–3636.
Navarro, Z. (2006). In search of a cultural interpretation of power: The contribution of Pierre Bourdieu. *IDS Bulletin, 37*(6), 11–22.
Paul, R. W. (1992). Critical thinking: What, why, and how? *New Directions for Community Colleges, 1992*(77), 3–24.
Paul, R. W., & Binker, A. J. A. (1990). *Critical thinking: What every person needs to survive in a rapidly changing world*. Rohnert Park: Center for Critical Thinking and Moral Critique, Sonoma State University.

Pigliucci, M. (2012). *Answers for Aristotle: How science and philosophy can lead us to a more meaningful life*. New York: Basic Books.
Schommer-Aikins, M., & Hutter, R. (2002). Epistemological beliefs and thinking about everyday controversial issues. *The Journal of Psychology, 136*(1), 5–20.
Schön, D. A. (1987). *Educating the reflective practitioner: Toward a new design for teaching and learning in the professions*. San Francisco: Jossey-Bass.
Sharot, T. (2011). *The optimism bias: A tour of the irrationally positive brain*. New York: Random House LLC.
Siegel, H. (1988). *Educating reason*. New York: Routledge.
Spendlove, D. (2010, June 17–19). *The illusion of knowing: Towards a curriculum of unknowing*. Paper presented at the Technological Learning & Thinking Conference. University of British Columbia, Vancouver.
Spendlove, D. (2013, December). *Rethinking design thinking in technology education*. Paper Presented at PATT 27 Technology Education for the Future: A play on Sustainability: The University of Christchurch, New Zealand.
Spendlove, D. (2014, October). *Using an IDEAS acronym to rethink aspects of 'design thinking' as part of 'everyday' designing*. Paper presented at Emerging Practices Conference, Tongji University, Shanghai.
Spendlove, D. (2015). *Challenging design thinking in the classroom*. Paper presented at the 29th PATT Conference: Plurality and complementarity of approaches in Design & Technology Education. Palais du Pharo Marseille – France April 8th, 2015. Presses Universities De Provence p. 373–379.
Sternberg, R. J. (Ed.). (2000). *Handbook of intelligence*. New York: Cambridge University Press.
Sternberg, R. J. (2001). What is the common thread of creativity? Its dialectical relation to intelligence and wisdom. *American Psychologist, 56*(4), 360.
Stevens, G. (1995). Struggle in the studio: A Bourdivin look at architectural pedagogy. *Journal of Architectural Education, 49*(2), 105–122.
TNTP (2015). *The mirage: Confronting the hard truth about our quest for teacher development*. Retrieved from http://tntp.org/assets/documents/TNTP-Mirage_2015.pdf
Tversky, A., & Kahneman, D. (1983). Extensional versus intuitive reasoning: The conjunction fallacy in probability judgment. *Psychological Review, 90*(4), 293.
Vygotsky, L.S. (1987). In N. Minick (Ed. & Trans.), *The collected works of L.S. Vygotsky. Volume 1. Problems of general psychology. Including the Volume Thinking and speech*. New York: Plenum.
Watts, D. J. (2011). *Everything is obvious: How common sense fails us*. New York: Atlantic Books.

Alternative Knowledge Systems

Mishack T. Gumbo

Abstract In line with the theme of this book, this chapter critiques Design and Technology Education (D&TE) in as far as accommodating indigenous knowledge systems (IKS). D&TE should reflect different knowledge forms existent in the real world, which are informed by different cultures and contexts. Indigenous knowledge forms, in particular, have proven to sustain societies around the world, which have from one generation to the next depended on such knowledge forms. Such knowledge forms are dominantly practical in nature, thus suitable to be regarded as technology. Backed up by the relevant literature, I explore alternative knowledge systems, discuss IKS and their characteristics, examine technology as an aspect of culture, survey few indigenous knowledge forms for D&TE and map out IKS characteristics to the teaching of D&TE. Critiquing the current linear D&TE helps to create awareness about the need to transform it so that indigenous students can be accommodated and learn about what they know. This transformation will also help non-indigenous students to become aware of and be introduced to the alternative forms of knowledge in their learning. D&TE teachers and other stakeholders will as well be helped to look to other forms of knowledge in their practice.

Keywords Indigenous knowledge systems • Design and Technology Education • Alternative knowledge systems • Culture • Critique

1 Introduction

Indigenous knowledge systems (IKS) can act as a form of critique on the acceptance and development of appropriate technologies. What are the implications of this consideration? A critique of Western knowledge systems (WKS) and technological practices is offered in this chapter as a response to this question. This question is crucial because:

M.T. Gumbo (✉)
University of South Africa, Pretoria, South Africa
e-mail: gumbomt@unisa.ac.za

PJ. Williams, K. Stables (eds.), *Critique in Design and Technology Education*, Contemporary Issues in Technology Education, DOI 10.1007/978-981-10-3106-9_6

> Developing societies have often relied on western or Eurocentric knowledge as a consequence of colonization process, intellectual imperialism, as well as forces of modernization and its dependencies, as well as globalization rhetoric. Currently the popular version of universal western knowledge is wrongly promoted as global knowledge (Selvadurai et al. 2013, p. 97).

The synergy between human thought and tool use has been a successful adaptation, allowing for the spread, sustenance and progress of humans across the globe. The influence of culture on technology and vice versa is a strand in the history of humankind, and unarguably social. However, the technological knowledge that is being taught to students and how it is taught have generally not been critical of the domination of Western approaches despite efforts by scholars who encourage other forms of knowledge. For example, Manitoba Education and Youth (2003) cites Ralston Saul who claims that the Canadian nation is built upon Aborigines, Francophones and Anglophones, yet the Canadian society has ignored and continues to ignore the contributions that Aborigines make towards the development of Canada. The thesis of this chapter is thus to add to the body of knowledge that critiques the dominance of conventional approaches to technological knowledge. The chapter has five objectives: to explore alternative knowledge systems, to discuss IKS and their characteristics, to examine technology as an aspect of culture, to survey a small number of indigenous knowledge forms for D&TE, and to map out the IKS characteristics to the teaching of D&TE. My critical views do not suggest devaluing of some aspects of WKS nor glorifying all that is IKS. In the chapter I am using IKS and alternative knowledges interchangeably.

2 Alternative Knowledge Systems

In the last two centuries, education and thinking have been shaped by a world view that is defined by the analytical knowledge philosophy or method. Analytical philosophy (refer to chapter "Philosophy as Critique") proceeds via analysis to understand the composition of its subject matter out of simple components (Longworth 2015, p. 1). It is a prominent philosophy in most Anglo-American university philosophy departments (Longworth 2015, p. 1). This Platonian and later Descartes' and his followers' world view re-emerged in nineteenth to twentieth century Europe after it was coffined by the Kantian movement which was active especially in Germany (Longworth 2015). Franz Brentano and C.S. Pierce forayed it in the nineteenth century, but it was brought to prominence by G.E. Moore and Bernard Russell in the twentieth century as a pinnacle of Anglo-American philosophy (Longworth 2015). It played a significant role in technological advancement in some nation states. However, philosophical enquiries in D&TE have highlighted diverse ways of knowing that can address ill-defined problems of society and challenges of the twenty-first century (Gumbo 2015; Fleer 2015; Maweu 2011; African Technology Policy Studies Network (ATPS) 2010). Content-wise and pedagogically, diverse ways of knowing suggest an open platform to D&TE to critique the analytical hegemony in consideration for alternative ways in line with alternative frameworks explored in chapter "Critique of Technology".

In this chapter, IKS provide alternative ways of knowing which have been underexplored. IKS promise to offer alternative ways to address societal problems alongside WKS. There is a growing consensus among proponents of IKS, that some solutions to the problems facing Africa, for instance, lie in the need to understand the dynamics of indigenous knowledge (Maweu 2011). For example, researchers have realised that indigenous agricultural practices are cost-effective and pose less production risks and environmental degradation (Bamigboye and Kuponiyi 2010, p. 39). Looking only to exogenous solutions worsens the situation in indigenous contexts. The decline of production of a version of rice called *Oryza sativa* in Nigeria was as a result of externally adopted production methods (Bamigboye and Kuponiyi 2010). This situation is one example which strengthens the case to consider IKS in D&TE.

Opponents of IKS cast doubt about their viability (Maweu 2011), and that motivates the critique agenda. The west elevates WKS above IKS, masquerating knowledge as power. Weiler (2011: 2) writes: 'different forms and domains of knowledge are endowed with unequal status'. Weiler outlines three ways through which unequal power is dispensed: (1) natural sciences have traditionally lead and relegated other forms of knowledge to 'lower ranks of prestige' (p. 2; also see Prasad 2006, p. 219); (2) institutional arrangements have been assigned knowledge production function, e.g. prestigious American research universities such as Max Planck Institutes and Grandes Écoles, which organise the order of knowledge according to prestige, resources and influence; (3) subtle treatment and positioning of professor and student, institute directors and staff, senior and junior faculty and sometimes administrators and faculty in knowledge institutions. These hierarchies attract questioning (Weiler 2009, 2006) and bring to light the subject of critique. It is in this light that D&TE is brought to question.

Michael Foucault acuminated the issue of the linkage between knowledge and power (Pitsoe and Letseka 2013; Weiler 2006; Popkewitz and Brennen 1998). Weiler (2009, 2006) adds to Foucault, that contemporary discourse on knowledge reveals three deficits: absence of critical view of knowledge; obliviousness to knowledge politics; inadequacy to envisage higher education's structural changes to free them from the current knowledge culture. Thus, there is a need to critique deficiencies in D&TE.

There is intimate and consequential linkage between knowledge and power (Weiler 2011, p. 1), expressed through technoscientific landscape in which the status of periphery, i.e. non-Western society, is defined as recipient who depends on the centre, i.e. Western societies (Prasad 2006). Until IKS are elevated to the level of WKS in D&TE discourses, indigenous students will be switched off and non-indigenous students will not learn about alternative knowledges.

Knowledge power position is deep seated in societal and cultural discourses. There is deliberate tendency for those that are in power to include or exclude other discourses, meanings, claims, rights and positions (Pitsoe and Letseka 2013, p. 24). Discourse is thus a social construct which can be manipulated by hegemonics (Pitsoe and Letseka 2013). Pitsoe and Letseka (2013) borrow from Foucault's claim, that discourses have a control function because they determine what can be said and thought, and who can speak, when and with what authority. Teachers of D&TE should accord students equal opportunities to exercise their power through learning activities. Discourses are inherent in societal institutions as well, and they can be expressed in written or oral form (Pitsoe and Letseka 2013). Pitsoe and Letseka (2013, p. 24) succinctly sum up their discussion of Foucault's *Discourse and Power: Implications for Instructionist Classroom Management* as follows:

> Control of knowledge is a form of oppression–only certain groups have access to certain knowledge. Those in positions of power are responsible for the assumptions that underlie the selection and organisation of knowledge in society.

It is clear from Pitsoe's and Letseka's claim that those who subscribe to hegemony relegate other knowledges to the backstage. Maweu (2011) blames this relegation on globalisation whose emphasis has been on science and technology to the detriment of indigenous knowledge. Maweu maps out differences between Western knowledge forms and indigenous knowledge forms, which illuminate the critique being advanced in this chapter. Maweu (2011, p. 38) cites Kaplan and Kaplan to explain the sense in which indigenous knowledge is lowered – it is often regarded as primitive, unscientific and cultural, lacking objectivity and credibility, closed and unsystematic. It is lowered for its context-baseness, rootness in a certain social group in a particular setting and certain time, holism and unanalytical, oral recording and transmission. Western knowledge, on the other hand, often stems from an epistemic framework (compare with alternative frameworks in chapter "Critique of Technology"); strives for universal validity; is perceived as contemporary, objective, universally true and thus credible, open, systematic, reductionist and analytical; and advances by building rigorously on prior achievements. Thus, according to Prasad (2006, p. 220), the impact of Western technoculture extends well beyond analyses of development and diffusion of technology; it disregards the fact that the heartbeat of globalisation and civilisation owes its origin to indigenous knowledge (Maweu 2011). For example, there is growing interest in indigenous ecological knowledge and technology due to its promising contribution towards the conservation of biodiversity (Maweu 2011). I suggest, then, that learning about IKS should be promoted through D&TE because indigenous knowledge can expand its understanding and enable students to contribute critical views towards globalisation.

> Globalization should therefore be effectively used to assimilate and enrich indigenous knowledge into the "global village" knowledge systems in order to debunk the belief that the western oriented knowledge system is the only viable approach to the global environmental crisis. (Maweu 2011, p. 43)

The discomfort that I have with Maweu's above quotation is the involvement of 'assimilate', which I have problematised elsewhere (Gumbo 2003, 2001). The current approach to globalisation and education is predominantly Western. To assimilate indigenous knowledge would mean being uncritical about the status quo and the deadening of IKS. I prefer the term integration which I motivated in my other works (Gumbo 2015, 2012, 2003).

3 IKS and Their Characteristics

Technology is as old as human existence. Communities around the world continue to develop their own survival modes and tools based on knowledge systems. The reality is thus that various groups of people in different parts of the world perceive and relate to environment in their own peculiar ways (Maweu 2011, p. 35). These knowledge systems have now come to be termed IKS and have aroused interest of scholars who research in the field (e.g. Gumbo 2014; Ngara 2007; Nsameng 2006; Masango 2006; Emeagwali 2003). IKS are distinct from the largely analytical ways of generating, recording and transmitting knowledge. For one, IKS are generated in a space-time context of a community and are subject to adaptive and dynamic evolution. They also form the community's identity that encompasses skills, experiences, values and insights. The space-time context of IKS does not preclude the fact that other knowledge forms, particularly in engineering, are not generated in a space-time context (Meijers and de Vries 2009).

Indigenous knowledge is defined by UNESCO (2007, p. 6) as the cumulative and dynamic body of knowledge, values, practices and representations related to the natural world and possessed by peoples with close ties to their natural milieu. I would add to the natural world the human-built world. These sets of knowledge are widely recognised as essential building blocks for sustainable development and conservation of biological and cultural diversity and are fundamental to sustaining rural livelihoods, identity and well-being (UNESCO 2007, p. 6). Owusu-Ansah and Mji (2013, p. 1) define indigenous knowledge as experiential knowledge which is based on a world view and culture that is basically relational. According to Owuor (2007), indigenous knowledge can be defined as multifaceted bodies of knowledge, practices and representations which people with long histories of close interaction with the local natural environment have developed and maintained. This multifacetedness confronts the analytical philosophy. In D&TE a critical stance is necessary to ensure consideration of these bodies of knowledge (compare with chapter "Critiquing as Design and Technology Curriculum Journey: History, Theory, Politics and Potential" on Critique in D&TE).

Indigenous, as an adjective, denotes the idea that this kind of knowledge belongs to peoples from specific places with common cultural and social ties (Maweu 2011). A particular community possesses knowledge whose content may be as broad as human experience (Bhola 2002, p. 11). According to the World Intellectual Property Organization (WIPO) (2001, p. 25), indigenous knowledge is defined through tradition-based literacy, artistic or scientific works, inventions, performances, scientific discoveries, designs, marks, names and symbols and all other tradition-based innovations and creations resulting from intellectual activity in the industrial, scientific, literary and artistic fields. The incorporation of the term innovations in the list denotes dynamism that is involved in this type of knowledge which is brought about intergenerationally in step with changes in time and space. WIPO (2001) states in this regard that indigenous knowledge changes over time as generations adopt new forms of this knowledge depending on environmental conditions. It is thus adaptable, based on skills, abilities and problem-solving techniques (WIPO 2001). Categories of this knowledge are elaborated by Gumbo (2015, pp. 61–62), e.g. food technology, metallurgy and astronomy. This categorisation is only for purposes of distinguishability; otherwise, because of the principle of holism that is enshrined in IKS, these fields are not approached as compartmentalised, but interdisciplinarily and transdisciplinarily.

From an African perspective, for instance, it is almost impossible to have a good grasp of the concept of IKS without relating them to the true African world view. The African world view denotes wholeness, community and harmony which are deeply rooted in cultural values (Owusu-Ansah and Mji 2013). By drawing from the principle of Ubuntu, this world view finds its true meaning in the popular yet important saying, *a person is because the community is* – a person finds meaning for his or her existence in the context of the collective; isolated or aloof from the community, this meaning stands the risk of being lost. This view permeates the profundities of knowledge production and management – the acquisition of knowledge is collective and community oriented (Owusu-Ansah and Mji 2013, p. 2). Hence, the pursuance of knowledge from an indigenous perspective is a value-laden enterprise, which should be defended by critiquing programmes which were designed from analytical stances.

3.1 Characteristics of IKS

Surveyed literature (such as Gumbo 2014; Ngara 2007; Nsameng 2006; Masango 2006; Emeagwali 2003) reveals the following characteristics of IKS elaborated on:

Collectiveness: A person can only find true meaning by viewing himself through the community, the membership of which is by cultural ties and values. Meijers and de Vries (2009) allude to this fact, although they do this in the engineering professional context and from a research point of view.

Holism: True understanding of nature is achieved through viewing it as an integrated whole, i.e. ecosystemic view. Thus, knowledge is not a linear, logical compartmentalisation of things, but, rather, it is integrated and interdisciplinary.

Cocreative orientation: Knowledge is cocreated and is community owned rather than individualised, with elders being the libraries of such knowledge, i.e. they possess the richness of indigenous knowledge.

Cooperative approach to problem-solving: Problems are being attended to in a collective, hence lekgotla (tribal meeting where issues are addressed) which is participatory in nature, while seniority by eldership is observed.

Experiential knowledge: The young are mostly taught through observation by keeping them close to elders engaged in activities of the day. Elders harbour tacit knowledge (embodied in persons, Meijers and de Vries 2009) which they attempt to impart through demonstration. Thus, education mostly happens in a real context through experience, demonstration and observation.

Orality: Knowledge is mostly shared or transmitted through oral communication.

Ubuntu: Knowledge about core cultural values is highly valued in which the young are taught respect, responsibility, unity and so forth. In most indigenous communities, this happens institutionally, e.g. in initiation schools.

Spirituality: Part of indigenous knowledge is held as sacred as it is divinely revealed by the Creator. Knowledge about nature can thus not be divorced from the Creator, and this facilitates perpetration of a moral responsibility over nature which is taught even to the young through expressions, idioms or riddles.

Values: Technological knowledge is not value-free as it cannot be divorced from the cultural and value system of indigenous community concerned. It is also applicable as part of technological knowledge in terms of normativity, e.g. 'I know that this is a good hammer' (Meijers and de Vries 2009).

Complexity: Indigenous knowledge's rich complexity is found in ceremonies and rituals, i.e. dance, music, storytelling, folktales, epic, poetry, recitation, demonstration, (word) games, sport, praise, riddles, reasoning, puzzles, tongue-twisters.

These characteristics have crucial implications for pedagogy in D&TE. I will come back to them towards the end of the chapter.

4 Technology as an Aspect of Culture

There is a close link between technologies and environment within which they are generated and used. The section addresses the multilayered, complex and dynamic associations of technology and culture, which have a bearing on D&TE.

4.1 Culture Informs Design

Culture is the driving force behind technology because it 'fuels and inspires technological accomplishments in human society' (Ogunbure 2011, p. 87). According to Moalosi et al. (2005, p. 1), 'the concept of culture and design compliment each other and one is inconceivable without the other'. Culture encompasses all aspects of life which give definition to human membership in society. Since culture is values and norms people have which make them live in a particular way, it is the sum total of all things that refer to religion, the origins of people, symbols, languages, songs, stories, celebrations, clothing and dressing, food productions, kinship, the interpersonal relationships, political and economic systems and all social relationship and all expressions of life past and present (Gumbo 2012, 2001; Ogunbure 2011; 2003; Custer 1995). Culture is the core of technological development (Ogunbure 2011). Hence, placing design in D&TE out of the bounds of culture is unfathomable.

According to Ogunbure (2011), culture can be viewed as the pattern of behaviour that enables people to live in social groups and to learn, create and share. It distinguishes one human group from another. It follows, then, that in indigenous communities where group cohesion is valued, members of these communities adopt a group approach to addressing their problems and needs; hence they collaborate and design group-based solutions to their problems. For instance, the Tswana men (also common in other African cultures) have a way of slaughtering a cow for a wedding function in a group which uses specialised skills, resources and processes. They tie its horns to the tree trunk with leather ropes, stab it with a knife on the grove behind the head. As soon as it falls they cut its neck and carefully de-skin it such that the leather is clean of meat on the inside. Then they dismember its parts based on a specialised knowledge about what should be done with each part, e.g. the head will be presented to the uncle of the bride. Young men should always be around when this happens to be taught knowledge and skills. Group work is currently a dominant pedagogical strategy in D&TE (Mehrotra et al. 2009) and can thus be extended to encourage students to express their design ideas informed by their cultural backgrounds.

4.2 Technological Artefacts as Expressions of Culture

Artefacts are diverse cultural expressions (Custer 1995, p. 223) and become creations that represent, inspire and define a given culture. Thus, one should begin with imagination and culture and then consider and appreciate the wonderful diversity that has been created (Custer 1995, p. 224). Instead of form being a distinguishing criterion (e.g. machine, tool, artwork, score of music), emphasis should be placed on the ways in which the values, priorities and needs of various cultures take form through the creative energy of their people (Custer 1995, p. 224). Limiting the understanding of these indigenous technological artefacts only to form is unfortunate because it renders them to museumisation, which results in a shallow conception of artefacts. Instead, any technological advancement should inspire skilful exploration of a people's culture that harbours knowledge systems powering such advancement.

Ogunbure (2011, p. 87) argues that 'every technology within a social praxis is a product of culture' and hinges his argument on the fact that culture is a phenomenon which undergirds all the material and non-material expressions of a people. Obikeze (2011) refers to these expressions as tangible or intangible devices, formulations and techniques which fulfil some need or provide some service for humankind in a given environment. Three categories of these technologies include material (physical) technology, e.g. bows and arrows, ploughs, looms, laboratories, machines and electronic devices; social technology, e.g. methodologies, techniques, organisational and management skills, bookkeeping and accounting procedures and negotiating and counselling techniques; communication technology, e.g. language, signs and symbols, drumming and the internet.

These culture products are in turn organised according to goods and services and can be further subdivided into material goods, e.g. soap, food items such as maize, houses and ornaments; social goods, e.g. values, norms, customs, motherhood, priesthood and friendship; and intellectual goods, e.g. ideas, abstract concepts, names, terminologies, cognitive knowledge and idioms. In this way, Obikeze (2011) declares that technology is any human-made or culture-generated devices, formulations or organisations utilisable for the purpose of producing or creating needed goods and services. In a more elaborative way, technology refers to the knowledge, technical skills and resources available in the community and environment that they occupy, which people use to meet their needs or wants to adapt the environment for desirable human living. Technologically, then, culture is viewed as a totality of the way of life evolved by a people in their attempt to meet the challenges of living in their environment (Ogunbure: 2011, p. 88).

Ogunbure (2011, pp. 89–96) goes further to define technology from a cultural viewpoint. Intellectually, technology finds its expression in harnessing nature in its entirety for humankind's development and sustenance. Humans are constantly thinking about the best way to do things in order to suit their needs while drawing inspiration from the cultural mindset within a social praxis. Technology is also the intermediary between humans and the vast resources available in environment being exploited. Educationally, the '-ology' in the term *technology* presents technology as the systematic study and development of techniques for making and doing things, i.e. machines (products and by-products), methods and processes that involve knowledge, skills and resources in the environment inhabited by humankind. Humans systematically apply their technological knowledge to produce goods and provide services in order to achieve perceived socio-economic systems. Technology is thus often associated with the hardware of production knowledge about machines and processes, even though that does not wholly define what technology is – technology can also be expressed non-materially. Having said this, it is imperative to attempt to understand technology from a cultural and indigenous perspective by asking crtitical questions – what are the prevalent technologies? What is the nature of the problems or needs in that context? How do people organise themselves to address these problems or needs? What designs and processes do they opt for to solve the problems or meet needs? These questions raise a need to incorporate indigenous technologies from these technological categories in the D&TE content. This anthropocentric view of technology can however not be accepted without critique. An ecological perspective should be considered to balance the anthropological one. The example given above about the potential contribution of indigenous technology towards the conservation of biodiversity suggests that the integration of indigenous forms of technology in D&TE can help to develop the attitude of environmental and sustainable development.

When one considers a technology known to and used by a particular nation, one can declare that, technology consists of a series of techniques, and thus the technology available to a particular nation is a sum total of all the techniques that such nation knows about, and could acquire, while the technology in use is a subset of techniques it has acquired and mastered. Ogunbure refers to Nigeria and states that a nation such as this does not possess the technology of manufacturing cars, hi-tech equipment and other sophisticated telecommunication equipment because it does not possess the knowledge of the techniques, procedures and the enabling environment required for advancing such capabilities. This, however, does not mean that Nigeria is devoid of the knowledge of automobile manufacturing. This may also not be true when considering the manufacturing of batik, adire, aso-oke and other indigenous clothing which are particular products of the people's cultural experience. One needs to stretch one's thinking to notice even the biotechnological forms existent in the country – alcohol beverages, gin from palm wine, soap from palm oil, cheese made through bacteria cells that feed on goat milk and meat preserved by extreme heat in different forms to produce varieties like suya and killichi among the Hausa people. Among the south-western regions of Nigeria, smoked fish, meat, roasted beef and seafoods, fried melon seeds and fried cassava flakes involve local agricultural technology to sustain the livelihoods of the Yorubas. The dominance of textile technology, e.g. adire among Yorubas in Nigeria and Kente in Ghana, cannot rule out a need to recognise the local technology in these forms. These localised forms of technology are important to sustain the livelihoods in indigenous contexts.

5 Indigenous Knowledge Forms to Consider in D&TE

Technological evolutions in textiles, health care, agriculture and environment are used to illustrate examples of existing forms of knowledge that could be integrated in D&TE.

5.1 Textile Technology

Archaeological studies suggest that a first textile was felt, i.e. it was a non-woven cloth which was produced by condensing and pressing woollen fibres and that it was at first prevalent in Egypt, India, Turkey and China (Biselle 2009). Notably, China has been the key role player in textile technology since ancient times. Evidence of this can be found in textile fragments and scraps of silk found from between 5000 and 2700 BC (Biselle 2009). Trade in textiles stretched as far as Rome and Iran and thus influenced textile technology in medieval Europe (Biselle 2009). It was from this that England, Italy, France, Spain, Germany and Scandinavia developed sophisticated clothing markets (Biselle 2009). However, gatekeeping of this trade was ensured by the west with a determination to dominate the market. This would not be a desired approach at least in D&TE teaching where students should be taught to collaborate and share. One would have expected the west to welcome the African trade and integrate indigenous textile forms into its own for a collaborative way forward.

Another notable and remarkable case in the textile industry is that of Nigeria's unique cultural expression, technology and entrepreneurship, as related by Saheed (2013). Saheed employed survey and questionnaire methods to trace the evolution of traditional adire production and its uses among the people of Egbaland in Ogun

State and Nigeria as a whole, which supplies different vocations such as designers, pattern makers, dyers, tie and dye experts, cloth and brocade sellers, merchandisers, distributors, oloolu (local ironers) and other related menial jobs. Saheed's findings revealed that adire making has undergone innovation which has helped in creating incremental wealth and generating employment for the people. Given its multi-ethnic and diverse culture, the art of cloth making, known as adire, reflects the culture from which it is made. The essence of culture remains as it is handed down from generation to generation, making it stronger and stronger. Adire textile is therefore an integral part of the culture and cultural heritage of the people of Egba kingdom in south-western Nigeria. Women use a variety of resist dye techniques to produce various designs of adire textile craft for the local and national market. Adire textiles represent a case about a need to look to and consider alternative forms of technology.

Saheed relates the technological process that is involved in the making of adire textiles which entails sourcing, tying, dying, drying and planking and packing the fabric. The fabric is sewn into the gorgeous designs by the tailors. Pieces of cloth are dyed into different designs at the discretion of the owners. Designers make a basic sketch on the fabric after they have decided on the design. The Guinea brocade is the most preferred material for the batik production. The brocade is dipped in cold water to remove the factory's thickness of the cloth. A slate of candle is put in a large pot to prepare wax by putting it on the fire to melt. A wooden or foam stamp is dipped into the melted candle wax and stamped on the guinea in horizontal or vertical form which brings out a desired pattern or design and left to dry. After drying, the dye is prepared by first pouring hot water into a pot and adding caustic soda in the correct proportions. The mixture is stirred until it bubbles, after which the dye of choice is added and stirred. The wax cloth is dipped or immersed into the mixture, making sure that all parts of the cloth are dyed. The excess dye is rinsed off and the cloth is dried. The dyed cloth is soaked for 30 min and rinsed and starched. Then cloth is then planked, i.e. local ironers fold the cloth with a log of wood over a wooden slab. The cloth is finally packed, which entails sealing the fabric in transparent polyethylene coverings for onward dispatch to the shops where customers come to buy it.

An effort has been made by indigenous nations to market their textile products abroad because they attract international market through trade and technological processes which were displayed in the textile production. In Europe, English workers were experienced in flax and silk (Gekas 2007). However, the same methods were not always applicable to the preparation of cotton for dyeing even if the right ingredients such as indigo and madder were imported (Gekas 2007). This could well represent tacit knowledge which is not easily transferrable. For a solution to this challenge Europe depended on Indian workers' weaving and finishing, i.e. printing and dyeing (Gekas 2007). This situation necessitated the importation of technical knowledge from the East (Gekas 2007). However, the Europe-based structures were not going to like the penetration of its local market by indigenous materials as cited above. The Ottoman Empire, which specialised in cotton fabric, decided to block India's influence and penetration of Europe ultimately, through its protectivism; it limited the importation of cotton materials for the European market (Gekas 2007). This provides an example about blocking thriving of IKS, which deserves critiquing in educational engagements and D&TE in particular.

5.2 Health Technology

With regard to health, from a World Health Organization perspective, technology is understood to be an association between methods, techniques and equipment together with the people using them (Cohen 1989, p. 105). Specifically, appropriate health technology is technology that can be adapted to people's needs, acceptable to them and maintained by them to ensure self-reliance and affordability (Cohen 1989, p. 105).

Primary health-care problems in developing countries or rural areas are of a biomaterial nature, thus seeking biotechnological solutions compared to problems in developed countries or urban environments (Cohen 1989), which are predominantly of a chemical nature, thus requiring chemical technological solutions. For instance, this could apply in the contamination of water in these different contexts. In providing the solutions in developing countries, a team-based approach is the best practice model because no single entity can solve all health-care problems by itself (Bauer 2012).

The Canadian Aboriginal medicine wheel makes an interesting example. The circle is divided into four quadrants according to the four cardinal points each representing one of the four directions (Mehrotra et al. 2009, p. 10–11):

- North: spiritual elders
- East: emotional children
- South: physical youth
- West: mental adults

Four and *balance* are very important aspects of meaning in the wheel. The four aspects of an individual which are spiritual, emotional, physical and mental are represented, and they should maintain balance to achieve a healthy life. If one of them suffers, the other three will be affected as well. For example, if a person has a condition such as sugar diabetes (physical), his moods may swing time and again (emotional), his thinking may be affected as well (mental) and he may feel down spiritually.

Networked information and communications technologies are transforming business models and production processes in the medical fraternity. Telemedicine is now being introduced to overcome traditional barriers of time and place (Bauer 2012). Informatics and analytics are allowing providers and payers to reduce costs of treating the most expensive patients (Bauer 2012). These future directions are basically of a Western origin. To what extent will they embrace indigenous practices is the question that we should concern ourselves with. The investigation approach that characterises D&TE could be used to critically assess the viability of IKS within these developments and the preparedness of D&TE designers and implementers to accommodate them.

5.3 Environmental Technology

Lastly in this section I look at technological solutions as applied to challenges of environmental management. According to Kelsey (2003, p. 2), the privileged status of expert information marginalises public knowledge, thus limiting the public's participation. Authoritarian ideology underpins the science-first model (Kelsey 2003, p. 3). To confront this orientation, Kelsey (2003) uses the Eastern Ontario Model Forest and Ashkui Project in Labrador, which prioritise building onto the elders' indigenous knowledge and constructively engage inclusivity in decision-making about environment activities. Imposition of external solutions to the environmental problems which leave out the rich indigenous knowledge is undesired. Kelsey (2003, p. 7) employs the term hit-and-run strategy to explain a tendency for scientists to enter a community, extract information and disappear. According to Kelsey (2003, p. 7), it is a big mistake to come in with a fixed research design and to try to add traditional knowledge to confirm the results, which undermines ownership by the community.

6 IKS-Informed Educational Practices in D&TE

6.1 Need to Integrate IKS in D&TE

The term 'Western', often used in the place of 'modern', refers to the ideas and practices whose origins can be traced to European traditions of knowledge, teaching and learning. Modern school education is concerned with learning and acquisition of knowledge that comprises abstract, decontextualised formal concepts (Fleer 2015) independent of the situations in which it is learned and used. IKS, on the other hand, integrate the generation, use and transmission of knowledge. The possibilities of defining technology in greater alignment with IKS are discussed to provide basis for critiquing D&TE. The implications of social constructivism and similar theories in D&TE are also critiqued.

It is interesting to note that a corpus literature such as Maluleka et al. (2006) acknowledge that technology is defined contextually. This means that context plays a very crucial role in the meaning of a concept. Ironically, however, is the realisation that D&TE curriculum being offered to the nations around the world seems to be very much driven by the universalist and industrial ideology which invades and erodes instead of integrating IKS. This ideology is informed by how technology is defined in its context. Fleer's (2015) treatment of this issue buttresses my point by weaving her argument through the literature that supports her thesis, mapping out the disjuncture that is drawn between the cognitive or abstract treatment of the content and pedagogy of D&TE and the practical – curricula taught in indigenous environments are mostly packaged from Western ideological and cultural orientations and run at odds with indigenous environments. We should include IKS in order to make the curriculum relevant to indigenous contexts.

Due consideration should be given to alternative theoretical perspectives such as sociocultural constructivism, communities of practice, southern theory and blended knowledges (Yishak and Gumbo 2015; Fleer 2015; Wahyudi 2014; Yishak and Gumbo 2012; Wenger 1998). When we consider the southern theory (ST) in the pool of these alternatives, for instance, we notice that it promotes multicentred social science perspectives, social science function of critiques, social sciences that produce many forms of knowledge and social science that is relevant to democracy (Wahyudi 2014, pp. 230–231). ST critiques the periphery-centre struggles when it comes to the treatment of knowledge forms. The northern theory (NT) promotes Western knowledge and places it at the centre (superior) and pushes indigenous knowledge to the periphery (inferior). This is unacceptable by ST. The alternative approaches mentioned above can jell comfortably with the indigenous principles of communalism, cultural values, Ubuntu, team approach to problem-solving, holism and so forth. Bearing in mind these alternatives, reference can be made to the synopsis of the D&TE curriculum that integrates indigenous technology, which I outlined in terms of the goals, content, learning support materials, pedagogy and assessment (Gumbo 2015, pp. 70–71).

The definition expanded from Obikeze (2011) and Ogunbure (2011) above helps us to understand the meaning of technology from an indigenous perspective. The scope of the definition also provides the content aspect that can be considered for integration in D&TE. I caution, though, that Obikeze's, Ogunbure's (in this chapter) and my (Gumbo 2015) categorisation of indigenous technologies according to their specialisation fields do not suggest teaching them in silos. The principle of holism should undergird the pedagogical approach. Currently, D&TE in Indian and South African contexts aims to develop students in terms of the knowledge and a range of skills objectives (Mehrotra et al. 2009; Department of Basic Education 2011) especially when they engage in a design project which is managed by the teacher, i.e. investigate, design, make, evaluate and communicate. These objectives target the conceptual knowledge, procedural knowledge and investigation, collaboration, group work, teamwork, cutting, communication and many other skills. While I am calling for the overhaul of D&TE such that it ensures integration of IKS, in the meantime a compromised situation can be to bring in aspects of IKS into the current D&TE. It is for this reason that, as promised, I map out the characteristics of IKS to the teaching of D&TE. These characteristics are applicable especially in diverse cultures. I believe that there is hardly a pure monocultural context anymore due to people moving around and thus interacting with others especially for economic reasons. There are therefore diasporan children who attend school in Western contexts who are endowed with indigenous knowledge from their own contexts. In previously colonised contexts such as South Africa, many schools still disregard IKS. That denies non-indigenous students the opportunities to learn alternative knowledge forms from their own, and this disadvantages them in terms of understanding indigenous contexts in which they will work ultimately. Besides, non-indigenous students need to be prepared to function in diverse cultural contexts. It is in this light that I think that adopting these characteristics in the teaching of D&TE is applicable in these explained contexts.

6.2 A D&TE Pedagogy That Is Informed by the IKS Characteristics

The critique position kept this far in this chapter avails an opportunity to rethink the current influence of the analytical knowledge philosophy that WKS are associated with. On the other hand, this critique introduces IKS which should be provided space alongside and integrated with WKS in D&TE. This section provides a solution as to how the characteristics of IKS discussed above can provide a springboard for a transformed D&TE curriculum and pedagogy.

Collectiveness: D&TE students can be organised in teams or groups to collaborate and cooperate in tackling the design projects. They can be taught teamwork and how they can be resourceful to one another and share responsibilities of completing the task. A study by Mehrotra et al. (2009) revealed that it should not be taken for granted that students will work together especially when they are from different sociocultural contexts. But the teacher can use collectiveness to reinforce Vygotsky's theory of social constructivism (Mehrotra et al. 2009).

Holism: Different knowledge and cultural backgrounds of learners can be a powerful teaching tool for D&TE teachers to promote in students. Holism can be achieved through these students learning from each other the diverse knowledge and skills as they execute their tasks.

Cocreative orientation: Students stand a good chance to cocreate in a negotiated manner their design ideas as they are contributed from different knowledge and technological backgrounds.

Cooperative approach to problem-solving: D&TE teachers can learn about how the lekgotla principle is being practised and adopt it to manage the students' design projects. Since design activities rely more on discussion and negotiation, this principle can come in handy in this regard. In lekgotla, a group (mostly of men in their seniority of leadership) sits in a circle to deliberate on the issues of the community. The most senior person (can be the king) starts the discussion and invites all to participate as he asks the next most senior person to contribute his opinion. This continues down to the least senior person. For purposes of teaching D&TE, the teacher may focus more on the value of the circular learning model implied in this principle, which ensures the participation of all students. The model is important for students to discuss about their design projects.

Experiential knowledge: Students have an opportunity to consult members of the communities about their design ideas, especially elders who can share rich knowledge. Students can also bring to the design project the knowledge that they have amassed from their indigenous contexts. Mehrotra et al.'s (2009) study revealed a student who was helped by her mother in a puppet project for her group. In another group on the windmill project, a student moved around to observe how the blades were shaped and benefited from that for his group (Mehrotra et al. 2009).

Orality: A D&TE teacher should take cognisance of the learning styles of indigenous students, with orality as being the prominent style. Orality speaks well to the communication aspect of the design process – investigate, design, make, evaluate and communicate. Presentation styles that the teacher might prescribe should include orality.

Ubuntu: Teaching D&TE may be a messy task if students will not be taught respect for the teacher and their peers. With emphasis on unity and teamwork, the D&TE class or lecture should be an enjoyable atmosphere.

Spirituality: In the D&TE laboratory/workshop/studio in particular, students are taught to exercise responsibility. This is where IKS can help emphasise this responsibility. Thinking in a responsible manner in how they approach nature and the use of resources in the context of their learning activities can be enhanced by sustainable development thinking.

Values: D&TE teachers should accord students the liberty to showcase their culturally informed thinking and designs so that they design solutions that will encourage them to give priority to problems in their authentic contexts.

Complexity: In line with the approach to values above, the D&TE teacher should not tie students down by specialisation but expose them to the complexity of technological knowledge.

7 Conclusion

WKS thrive and have been shaped by the analytical knowledge philosophy which presents a skewed propagation of technology, thus sidelining indigenous forms of technology. In this chapter I have critiqued this orientation in favour of IKS and indigenous technology in particular. The analytical knowledge philosophy which has contributed to the linear approach in D&TE motivates the critique exercise in this chapter and the book as a whole. IKS promises to offer desired solutions to societal problems of separate educational development. In the introduction I made a statement about my non-intention to devalue some aspects of WKS. Credit goes to where WKS have been able to solve some of these societal problems. What I was rather concerned about in this chapter is the non-integration of IKS. Hence, my argument that IKS should be integrated in discourses about issues surrounding the future of D&TE. Most importantly, we should envisage a culturally inclusive D&TE curriculum, which offers students equal opportunities and multiple perspectives which can facilitate and broaden their understanding of technology. The alternative perspective which IKS offers can thus enrich the current attempts to seek solutions to the societal problems while at the same time ensuring dignity of all knowledge forms and accommodation of indigenous cultures.

References

ATPS. (2010). *The african manifesto for science, technology and innovation*. Nairobi: ATPS.

Bamigboye, E. O., & Kuponiyi, F. A. (2010). The characteristics of indigenous knowledge systems (IKS) influencing their use in rice production by farmers in Ekiti State, Nigeria. *Ozean Journal of Social Sciences, 3*(1), 39–50.

Biselle, M. (2009). *China's role in the global textile industry*. Retrieved March 1, 2015, from www.ceibs.edu/bmt/images/20110221/30207.pdf

Bauer, J. C. (2012). *Science and healthcare technology evolution: Past, present, future*. Annual meeting. Retrieved March 1, 2015, from www.nefloridacounts.org/javascript/htmleditor/uploads/Bauer_HPCNF_JAX101812_Handout.pdf

Bhola, H. S. (2002). Reclaiming old heritage for proclaiming future history: The knowledge for development debate in African context. *Africa Today, 49*(3), 3–21.

Cohen, J. (1989). Appropriate technology in primary health care: Evolution and meaning of WHO's concept. *International Journal of Technology Assessment in Health Care, 5*, 103–109.

Custer, R. L. (1995). Examining the dimensions of technology. *International Journal of Technology and Design Education, 5*(3), 219–244.

Emeagwali, G. (2003). African indigenous knowledge systems (AIK): Implications for the curriculum. In T. Falola (Ed.), *Ghana in Africa and the world: Essays in honour of AduBoahen* (pp. 121–137). Trenton: Africa World Press.

Fleer, M. (2015). Theorising technology education from a cultural-historical perspective: Foundations and future imaginings. In P. J. Williams, A. Jones, & C. Buntting (Eds.), *Contemporary issues in technology education: The future of technology education* (pp. 57–75). Singapore: Springer.

Gekas, A. (2007). *A global history of Ottoman global textiles, 1600–1850*. Badia Fiesolana: European University Institute.

Gumbo, M. T. (2001). Multicultural education and its politics. *South African Journal of Education, 21*(4), 233–241.

Gumbo, M. T. (2003) *Indigenous technologies: Implications for a technology education curriculum*. Unpublished doctoral thesis, Vista University, Pretoria.

Gumbo, M. T. (2012). Claiming indigeneity through school curriculum with specific reference to technology education. *Africa Education Review, 9*(3), 27–46.

Gumbo, M. T. (2014). Elders decry the loss of Ubuntu. *Mediterranean Journal of Social Sciences, 5*(10), 67–77.

Gumbo, M. T. (2015). Indigenous technology in technology education curricula and teaching. In P. J. Williams, A. Jones, & C. Buntting (Eds.), *Contemporary issues in technology education: The future of technology education* (pp. 57–75). Singapore: Springer.

Kelsey, E. (2003). Integrating multiple knowledge systems into environmental decision-making: Two cases of participatory biodiversity initiatives in Canada and their implications for conceptions of education and public involvement. *Environmental Values, 20*, 1–16.

Longworth, G. (2015). Analytic philosophy. Retrieved August 27, 2015, from www2.warwick.ac.uk/fac/soc/philosophy/people/faculty/longworth/keyideasanalyticphilosophy.pdf

Maluleka, K., Wilkinson, A., & Gumbo, M. T. (2006). The relevance of indigenous technology in curriculum 2005/RNCS with special reference to the technology learning area. *South African Journal of Education, 26*(4), 501–513.

Manitoba Education and Youth. (2003). *Integrating aboriginal perspectives into curriculum: A resource for curriculum developers, teachers and administrators*. Manitoba: Manitoba Education and Youth.

Masango, M. J. S. (2006). African spirituality that shapes the concept of ubuntu. *Verbum Et Ecclesia Jrg, 27*(3), 930–943.

Maweu, J. M. (2011). Indigenous ecological knowledge and modern western ecological knowledge: Contemporary, not contradictory. *Thought and Practice: A Journal of the Philosophical Association of Kenya (PAK), 3*(2), 35–47.

Mehrotra, S., Khunyakari, R., Chunawala, S., & Natarajan, C. (2009). *Evidences of learning through collaboration in design and technology tasks in Indian classrooms*. Proceedings of the 3rd international conference to review research on science, technology and mathematics education. Mumbai, India.

Meijers, A. W. M., & De Vries, M. (2009). Technological knowledge. In J. K. B. Olsen, S. A. Pedersen, & V. E. Hendricks (Eds.), *A companion to the philosophy of technology* (pp. 70–74). West Sussex: Blackwell.

Moalosi, R., Popovic, V., Hudson, A., & Kumar, K.L. (2005). *Integration of culture within Botswana product design*. Paper presented at international design congress Yunlin, Taiwan.

Ngara, C. (2007). African ways of knowing and pedagogy revisited. *Journal of Contemporary Issues in Education, 2*(2), 7–20.

Nsameng, A. B. (2006). Human ontogenesis: An indigenous African view on development and intelligence. *International Journal of Psychology, 41*(4), 293–297.

Obikeze, D.S. (2011). Indigenous Knowledge Systems and the Transformation of the Academy in Africa: The CULPIP model. Retrieved February 11, 2015, from http://citeseerx.ist.psu.edu/viewdoc/download?doi=10.1.1.164.9357&rep=rep1&type=pdf

Ogunbure, A. A. (2011). The possibilities of technological development in Africa: An evaluation of the role of culture. *The Journal of Pan African Studies, 4*(3), 86–100.

Owuor, J. (2007). Integrating African indigenous knowledge in Kenya's formal education system: The potential for sustainable development. *Journal of Contemporary Issues in Education, 2*(2), 21–37.

Owusu-Ansah, F. E., & Mji, G. (2013). African indigenous knowledge and research. *African Journal of Disability, 2*(1), 1–5.

Pitsoe, V., & Letseka, M. (2013). Foucault's discourse and power: Implications for instructionalist classroom management. *Open Journal of Philosophy, 3*(1), 23–28.

Popkewitz, T. S., & Brennan, M. (Eds.). (1998). *Foucault's challenge: Discourse, knowledge and power in education*. Columbia: Teachers College Press.

Prasad, A. (2006). Beyond modern versus science debate: Analysis of magnetic resonance imaging research. *Economic and Political Weekly*, 219–227.

Saheed, Z. S. (2013). Adire textile: A cultural heritage and entrepreneurial craft in Egbaland, Nigeria. *International Journal of Small Business and Entrepreneurship Research, 1*(1), 11–18.

Selvadurai, S., Choi, E. A., Maros, M., & Abdullah, K. (2013). Shifting discourses in social sciences: Nexus of knowledge and power. *Asian Social Science, 9*(7), 97–106.

UNESCO. (2007). *Science in Africa: UNESCO's contribution to Africa's plan for science and technology to 2010*. Paris: UNESCO.

Wahyudi, R. (2014). "Democratic" online courses as "glocalized communication" in English language teaching and applied linguistics: A proposal. *Journal of Global Literacies, Technologies, and Emerging Pedagogies, 2*(3), 248–260.

Weiler, H. N. (2006). Challenging the orthodox of knowledge: epistemological, structural and political implications for higher education. In G. Neave (Ed.), *Knowledge, power and dissent: Critical perspectives on higher education and research in knowledge society* (pp. 61). Paris: UNESCO Publishing.

Weiler, H. N. (2009). *Whose knowledge matters? Development and the politics of knowledge*. Retrieved March 1, 2015, from http://web.stanford.edu/~weiler/Texts09/Weiler_Molt_09.pdf

Weiler, H. N. (2011). Knowledge and power: The new politics of higher education. *Journal of Educational Planning and Administration, XXV*(3), 205–221.

Wenger, E. (1998). *Communities of practice: learning, meaning and identity*. New York: Cambridge University Press.

WIPO. (2001). *Intellectual property and indigenous knowledge. 2001 Report*. Geneva. Retrieved February 11, 2015, from www.millenniumassessment.org/documents/bridging/papers/woodley.ellen.pdf

Yishak, D. M., & Gumbo, M. T. (2012). Indigenizing the school curriculum: An alternative approach for ensuring relevance to the context. *Indilinga – African Journal if Indigenous Knowledge Systems, 11*(2), 179–193.

Yishak, D. M., & Gumbo, M. T. (2015). A standalone, blended, or restructured indigenisation approach to curriculum? A critical perspective. *International Journal of AfricanRenaissance Studies – Multi-, Inter- and Transdisciplinarity, 10*(1), 60–82.

Part II
Critique in Design and Technology Education

Critiquing as Design and Technology Curriculum Journey: History, Theory, Politics and Potential

Steve Keirl

Abstract *Critiquing*, as a key component of Design and Technology (D&T) education, made its global debut 15 years ago in the redesigned South Australian curriculum. It has since gained international recognition for its validity for the education of all children. This chapter sets out the story of *critiquing* as Design and Technology curriculum phenomenon, and, while the story reports a personal research journey, it was the work of a dedicated team that brought the curriculum as a whole to fruition. Key episodes of the story address: curriculum research method as autobiography; the politics of D&T curriculum; the theoretical underpinnings of the critiquing innovation; its local, national and international contexts; the curriculum challenges its introduction was intended to resolve; and some consequent theorisation since its inception. In this story, 'Design and Technology' is seen as much more than a school 'subject'. It is argued that critical-ethical design and technological literacy is necessary if sustainable, democratic futures are to be achieved. *Critiquing* is fundamental to such literacy.

Keywords Critiquing • Critical pedagogy • Curriculum • Technological literacy • Design literacy

1 Introduction

When *critiquing* became central to the South Australian Design and Technology (D&T) curriculum in 2001, it was a world first. No D&T curriculum had ever made such a move – one that expected critiquing, along with designing and making, to permeate all D&T pedagogy and learning.

> The Design and Technology Learning Area is articulated through three strands. These reflect the processes of thinking and doing that constitute a quality education *common to any technology* (eg agriculture, architecture, information and communication technology,

S. Keirl (✉)
Goldsmiths, University of London, London, UK
e-mail: s.keirl@gold.ac.uk

PJ. Williams, K. Stables (eds.), *Critique in Design and Technology Education*,
Contemporary Issues in Technology Education, DOI 10.1007/978-981-10-3106-9_7

electronics, engineering, food, genetics, media, robotics, textiles, viticulture). The three strands are:

- Critiquing
- Designing
- Making

> These three strands *are interdependent and none of them is predominant.* Read alongside each other *they do not constitute a sequential process...* A quality Design and Technology education weaves the three into a dynamic and holistic learning experience *for all students.* (DETE 2001a, b. My italics)

Like many curriculum innovations, critiquing got a varied reception – locally, nationally and internationally. Reactions ranged from cynicism, through hesitant acceptance, to those who could immediately see its educational validity. Today, as this book attests, the story is different. Critique, as both noun and verb, is now recognised for its validity as a vibrant component of best D&T practice. *Critiquing* as documented in this chapter was developed as a properly theorised curriculum inclusion, intentionally problematic and pedagogically challenging. It was never intended as a vanilla version of product appraisal, as a tick-box reflection or as a casual curriculum add-on. As a defensible educational practice, critiquing is far more than this. Critiquing emerged from substantial research and it continues to demand research.

I begin by outlining some personal engagements with, and orientations towards, critiquing and curriculum research. I then describe the theoretical underpinnings of critiquing – so necessary for its advocacy and defence for inclusion in politically contested curricula. Moving through international, national and local settings, I set out the educational and geopolitical context for the (then) innovation before presenting an overview of the South Australian curriculum. This backgrounding is necessary for a proper understanding of critiquing's theoretical rigour and its educational potential for all. I close with a selection of theorisations, reflections and developments that have resulted from the original innovation. The chapter can't accommodate full detail of all aspects of this curriculum journey but sources are given wherever possible.

2 Curriculum Work as Personal Journey and Autobiographical Method

As leading international curriculum theorist, Pinar has said: 'Our pedagogical work is simultaneously autobiographical and political' (Pinar 2004/2008:4). I partly describe my own motivations thus:

> As a species we are unable to define or describe ourselves without reference to technologies. Our very existence is dependant on, and inter-dependent with, technologies. The quality of our co-existence with other species and the planet cannot be determined without technological critique. Why is it then, when the phenomenon of technology constitutes such

> a pervasive and hegemonic part of life on the planet, that it is so ill-addressed in education? This is the question that drives my own curriculum enquiry. (Keirl 2007d:77)

The genre of *The Critical* has engaged me since Philosophy of Education seminars with an excellent lecturer (Bernard Down) in the 1970s in England. Early appreciations of both *philosophy as critique* and *critical thinking* as keys to educational theory, policy-making *and* pedagogy were natural antecedents to my early-1990s introduction to *critical theory* at the University of Tasmania and, subsequently, to its application to curriculum, pedagogy and literacy at the University of South Australia. Whether in advocating *critical practice* in D&T through *critiquing* as curriculum dimension (Keirl 1997a, b), or in arguing for *ethical technological literacy* (Keirl 2006), theorising 'the critical as educational tool' has remained central to my curriculum research journey.

But what of the political? There are two realms here – the politics of education and the politics of technology. For me, they meet around questions of democracy and what should constitute a proper and defensible Design and Technology education for every child, across the planet. Education, and D&T as a component of it, is a political act (Layton 1994; Petrina 2000a; Keirl 2006, 2007c), and I believe that D&T must continuously critique itself in at least two ways. One is in how it is politically constituted as an educational field by external agencies as much as by its own players. I have attempted more recently to articulate a case for a role for D&T as agent against neoliberalism (predatory capitalism) in defence of sustainability (Keirl 2015a). Linked with this is the need for D&T to maintain its political sensibilities around how it contributes (or otherwise) to global democratic citizenship (Keirl 1999/2001, 2006). For me, these political necessities preface and inform any other D&T curriculum considerations. On technology, as Feenberg says: 'The fate of democracy is . . . bound up with our understanding of technology' (Feenberg 1999:vii).

Within all of this is the self using the theoretical and the philosophical as tools to advance and establish new practices. Pinar notes that: '(The) "theoretical relationship" with oneself can be explored and recast through autobiographical reflection, through conversation with oneself' (Pinar 2004/2008:251). Ultimately, the personal, the theoretical and the political cease to be mere categories. They become one's (and Design and Technology's) holistic engagements with the existential, as personal lived experience (van Manen 1990; Morris 1966/1990) as well as with one's coexistences (Keirl 2010). However, the self and the critical also interplay, and, as ever in research, considerations of bias and prejudice arise. I declare my bias when I argue my case *against* D&T as, for example, skilling, vocational education, gendered roles, preparation for a destructive and divisive economic system and more. I declare my bias when I argue my case for D&T *for* sustainable futures, democratic life, dynamic curriculum, ethics, design education and a critical disposition (Keirl 2001b).

I draw on Pinar's curriculum scholarship (2004/2008, 2007) and his 'method of *currere*' (the Latin root of 'curriculum') as '. . . an autobiographical method asks us to slow down, to remember even re-enter the past, and to meditatively imagine the future' (Pinar 2004/2008:4). This, he offers, is a matter of the reconstruction of self and society alike, and he talks of the 'nightmare' we are living and how we '...believe in education (yet) we see how powerfully schooling crushes it...' (Pinar 2004/2008:127). Pinar is not alone as a curriculum theorist who sees persons as central to education, who takes a critical stance (self-critical as well as socially critical) towards curriculum thinking and action, and who valorises the educator as intellectual activist (see, e.g. Freire 1972, 2001; Postman and Weingartner 1969/1971; Apple 1979; Giroux 1983; Goodson and Walker 1991; Blackmore 2002; Smith and Lovat 1991; Pinar 2007; Kincheloe 2008/2010; Darder et al. 2009a; Smyth 2011). All such curriculum workers eschew simplistic and instrumental conceptions of curriculum, and they subscribe to Pinar's articulation of 'curriculum as complicated conversation':

> The method of *currere* reconceptualized curriculum from course objectives to complicated conversation with oneself (as a "private" intellectual), an ongoing project of self-understanding in which one becomes mobilized for engaged pedagogical action – as a private-and-public intellectual – with others in the social reconstruction of the public sphere. (Pinar 2004/2008:37)

Critiquing is this chapter's focus and, despite 20 years of D&T curriculum work around critiquing and critical technological literacy, there remains much to achieve. I value Canby's 90-year-old insight not only for what it says about criticism but for its resonance with Technology Education's multiple players:

> One reason why football is more satisfactory than criticism is that there is only one ball. In criticism, too often everyone brings his (sic) own ball, and when he pushes it over the goal line thinks he has won the game. (Canby 1924/1967:226)

As part of D&T's own complicated conversations, *critiquing* cannot be seen as an isolated piece of a curriculum jigsaw. It must be understood *as permeating all* designerly and technological behaviours and circumstances. Critiquing is necessarily of the complex and of the holistic. It was never some isolated concept plucked from the air, chosen to be fashionable, or as a quick fix for a particular curriculum problem. Critiquing in D&T has run a 15-year journey grounded in a proper rationale and theorisation that continue to be revised and refined.

While some curricular innovations are trends that occur concurrently across the globe, others (such as critiquing) have arisen in a specific jurisdiction at a specific time under specific conditions. For D&T, there are two broad temporal considerations regarding critiquing: how it might apply in the immediate study space, for example, when designing; and how it contributes to learning for life – whether or not associated with design activity. It is invaluable '. . . both to designing and to the interrogation of the values and merits of extant technologies, products and systems. The nurturing of a critiquing disposition serves specialist Design and Technology Education and generalist education for democratic life equally well' (Keirl 2009).

3 Critical-Theoretical Underpinnings

The introduction of *critiquing* into the D&T curriculum was informed by critical theory whose global educational influence is most strongly evidenced in *critical literacy* and *critical pedagogy*:

> Critical theorists begin with the premise that *men and women are essentially unfree and inhabit a world rife with contradictions and asymmetries of power and privilege.* The critical educator endorses theories that are, first and foremost, dialectical; that is, theories that recognize the problems of society as more than simply isolated events of individuals or deficiencies in the social structure. (McLaren 1989/2009:61)

Two of critical theory's greatest protagonists, Freire (1972, 2001) and Habermas (1971), have been particularly influential in education. By suggesting that we hold three kinds of 'cognitive interests', Habermas (1971) opened a door to understanding the world as lived, how power is (ill-)distributed and how we can act. He wrote of '. . . the task of a critical philosophy of science that escapes the snares of positivism. The approach of the empirical-analytic sciences incorporates *a technical cognitive interest*; that of the historical-hermeneutic sciences incorporates *a practical one*; and the approach of the critically oriented sciences incorporates *the emancipatory cognitive interest* . . .' (Habermas 1971:308 My italics). Unsurprisingly, one methodology of the Habermasian approach is *ideology critique* tackling the values, beliefs and practices of particular dominant groups (Morrison 2001).

It is one thing to operate in and on the world at a technical level of facts and utility, another to be able to interpret the world and experience, and yet another to see it as emancipatory – building on both the technical and the practical but ultimately being liberated from (technological) structures and regimes that dominate our ways of being. All designed technologies are enactments of human decision-making, and they amount to being assemblages of values. Rampant consumerism, unsustainability, environmental destruction, multiple anti-democratic practices and more are sites for technological critique of just whose interests are being served and what our personal positioning and responsibilities might be.

Morrison (2001) shows how the three knowledge-constitutive interests can influence curriculum design. The 'rationalist/behaviourist' view of curriculum values the bureaucracy-driven, heavily tested, curriculum as instrument. A 'curriculum as practice' is humanistic, interpretive and pragmatic, it privileges understanding over outcomes, and its hermeneutic knowledge interest optimises genuine experiential learning. 'Curriculum as praxis' takes an 'existential, empowering and ideology-critical' approach that is emancipatory in nature. In advancing the emancipatory aspect, curriculum is problematised by all involved – not least the students. (Morrison 2001:218). Such an approach is implicitly political and necessarily controversial as it commits to challenging the dominant ideology (Keirl 2015a). Applying the critical lens of Habermas's knowledge interests to D&T curricula highlights how seriously limited many are – often (even intentionally) fulfilling only the technical interest.

Critical theory has been richly applied to education in many ways (McLaren 1989/2009; Emmitt and Pollock 1991; Kemmis 1991/2005; Comber 1992, 1994;

Aronowitz and Giroux 1993; Comber et al. 1998; Morrison 2001; Kellner 2003; Smith and Lovat 2006). Kemmis (1991/2005) points out that '... critical theorists themselves are suspicious of any "grand narrative" of history...', and he notes how '... critical theorists... have been both critical and self-critical...' (Kemmis 1991/2005:314). Giroux (1983), a leader in bringing critical theory to education, argued the importance of critiquing the positivist and functionalist rationality dominant in schools:

> Rather than celebrating objectivity and consensus, teachers must place the notions of critique and conflict at the center of their pedagogical models... Critique must become a vital pedagogical tool – not only because it breaks through the mystifications and distortions that "silently" work behind the labels and routines of school practice, but also because it models a form of resistance and oppositional pedagogy. (Giroux 1983:62)

3.1 The Critical Literacy Movement

Freire's (1972) landmark pedagogical critique showed how education in the dominant Western model maintains a status quo of conformity and control through disproportionate power distribution. Famously, he contrasted 'banking education' with 'problem-posing education' (for D&T, 'skilling' compared with 'design and action on the world'). Critical literacy theorists have shown how different interests were served by different approaches to literacy. For example:

> The link between literacy and economic rationalism has a long, if not altogether distinguished, history... There are powerful economic and political precedents and parallels for current social policy in Australia. In those countries with conservative governments... educational and social policies have stressed a binary approach to literacy: 'cultural literacy' based on the Anglo-colonial literary canon for an elite...and 'functional literacy' for everyone else (i.e. 'survival' skills for the emergent underclass). (Luke 1992:3)

Witness, then, Technology Education formulations that ensure elites of designers, programmers, architects and engineers alongside a mass of skilled and semi-skilled operatives. McLaren saw this as: '... reproducing dominant class interest directed towards creating obedient, docile, and low-paid workers' (McLaren 1989/2009:62). A significant literacy milestone came from the New London Group (NLG 1996), an international collaboration of ten literacy scholars, five of whom, including Luke, were Australian. Their influential work on 'multiliteracies' prompted speculation for Design and Technology education:

> Interestingly, on their journey of deliberation of the 'state of literacy pedagogy', this group make use of the 'key concept' of 'Design' and discover that "... as designers of meaning, we are designers of social futures – workplace futures, public futures, and community futures" (NLG, 1996:65)... The New London Group, in moving on from 'mere literacy' advance the idea of 'multiliteracies'. Thus we might accept multiple and changing meanings of technology rather than searching for a grail articulated through a single definition. (Keirl 1999a:75)

Concurrently, Petrina (2000b) gave *critical technological literacy* an appropriate mainstream debut. D&T still has a long way to travel to embrace the propositions and issues presented in that article. However, early formulations of a Habermasian critical technological literacy (Keirl 1996, 1997a, 1999b) fed into the South Australian curriculum development and were subsequently developed as an *ethical technological literacy* advocating critical curricula and pedagogies in the defence of democratic existences (Keirl 2006).

The end of the twentieth century had witnessed the growth of a global movement in critical theory and critical literacy theory, and the Australian chapter was vibrant – not least a strong South Australian grouping of critical literacy theorists. Meanwhile, globally, negative reactions emerged against such liberatory theorising – a political backlash from the right with its politics of the (uncritical, unproblematic) soundbite of the 'back to basics' kind (For critiques of this drive not only to control and shape education but to silence opposition, see, e.g. Aronowitz and Giroux 1993; Coomber et al. 1998; Lankshear 1998; Apple 2001; Smith 2003; Pinar 2003, 2004/2008, 2007; Reid 2004/2005; Darder et al. 2009b; Smyth 2011; and, on Design and Technology, Petrina 2000a, b, 2003; Keirl 2006, 2015a, b).

3.2 *Critical Pedagogy in Context*

Darder et al. (2009b) remind us that critical pedagogy must be understood as part of '... a long tradition of progressive educational movements and on-going struggles of reinvention ...', and they caution against any '... temptation to inadvertently reify and reduce critical pedagogy to a teaching "method"' (Darder et al. 2009b:19). Critical pedagogy is no more or less political in its aims and practices than that of neoliberalism whose agenda seeks to homogenise the educational experience of the masses; is competition driven; demands high-stakes testing for 'accountability'; treats education as a business; creates artificial decentralisation; severely constrains teachers' professionalism; promulgates a misperception of 'failure' of students, schools and teachers alike; and promotes curriculum determination by noneducational groups, nationally and internationally, in support of internationalised labour markets (Smith 2003; Keirl 2015a).

D&T's curriculum and pedagogy are embroiled in this dehumanising strangulation of education, and D&T has a simple option: to do very little (or nothing at all) and to have its lot determined in ways that advance the neoliberal agenda or to exercise some self-determination over its educational role and purposes. 'Self-determination' need not mean selfish determination. Properly understood and justified, any 'subject' has duties towards both specialist and general education. Year on year we are deepening our understandings of how *rich* D&T contributes to the education of all students in terms of identity, citizenship, capability, critical awareness and more. This is D&T's contribution to students, to society and to global democracy, and it is fundamentally a matter of ethics (Keirl 2006, 2015b). D&T's special role concerns the designed and made world and the host of accompanying issues demand that it has its own particular critical pedagogy – one that explores, exposes and declares multiple technical-instrumental, interpretive-hermeneutic and critical-emancipatory technological interests.

The global influences of critical theory, critical literacy and critical pedagogy were one politico-educational phenomenon of the late twentieth century. However other circumstances and considerations warranted the critiquing innovation too.

4 Global Influences on Technology Curricula

In the 1990s, issues of sustainability, globalisation and emergent technologies all achieved growing, if uncritical, attention (Keirl 2002b, c). Despite public concern, positive change remained negligible. Running through all of these is an obvious ethical thread but there was a democratic concern too. Almost all the issues were technology related. While societies, communities and individuals continued to be 'shaped' by technological developments, public engagements with technological design decision-making were as remote as ever. Democracy, itself a technology, was (and remains) threatened, being perversely linked to 'economy', while privacy was being negated, surveillance increased, and criticism and protest demonised (Keirl 2006, 2015a, b).

Technological developments (artificial intelligence; robotics; genetics; nanotechnologies; communications technologies, so-called 'social' media; and more) were not only emergent but increasingly convergent. The humanising of technologies and the technologising of humans together inform discussion of the end of humanity, post-humanism and transhumanism (Keirl 2015a, b). The 1970s and 1980s had seen the sanctioned decline of craft education in schools, while, today, its existential benefits invite educational argument for inclusion. Subsequently, consciousness and criticism became *topics du jour* with regard to democracy and technology. For example: '. . . not so long ago it was fashionable for social critics to condemn technologies as such . . . Increasingly, however, social criticism has turned to the study and advocacy of possible reconfigurations and transformations of technology to accommodate it to actors excluded from the original design networks' (Feenberg 2010:77). In 1999 Postman argued his (Enlightenment) case for the need for *scepticism* in general and in education in particular: 'Modern educators do not usually use the word, preferring something like "critical thinking". But in any case, they do not do much about it. There are several reasons why. The first is that it is dangerous . . . ' (Postman 1999/2000:159–160). In 1995, Sclove argued for: '. . . a democratic theory of technology . . . using political philosophy to develop prescriptions for technological design and choice . . . (and) . . . challenging the foundations of modern economic thought . . . (with the aim of helping) achieve citizenship in a future world of democratic technology' (Sclove 1995:x–xi).

Meanwhile, Saul (1995) had shown how decision-making in society had moved away from the individual and democracy towards conformity and corporatism, where decisions are made through constant negotiations between specialist interest groups. He outlined *The Great Leap Backwards*: '. . . our leap into the unconscious state beloved of the subject who, existing as a function in any one of the tens of thousands of corporations – public and private – is relieved of personal, disinterested responsibility for his (sic) society. He thus gives in to the easy temptation of embracing what I can only call the passive certitude offered by every ideology' (Saul 1995:37). He signals the importance of confronting reality and how this '. . . usually is a negative process. It is ideology that insists upon relentless positivism. That's why it opposes criticism and encourages passivity' (Saul 1995:38). And he defends '. . . (his) . . . Socratic right – to criticize, to reject conformity, passivity and inevitability' (Saul 1995:39). There is little in Saul's 'unconsciousness' thesis that is not directly applicable to our designed technologies and to our technologically mediated worlds.

4.1 Critiquing Technologies and Design

For assessments of technologies, a *Habermasian critiquing* cannot stop at the traditional positivist question of 'does it work?' (the technical level) but must go further into the hermeneutic, interpretive explorations that effect meaning-making and, further, into the critical-emancipatory engagements that facilitate the existential and the liberatory. Critiquing is not mere analysis: it is inward- and outward-directed interrogation; it debunks, demystifies and exposes power relations too. 'Critical theory (can be) a highly reflexive enterprise – it is never satisfied with asking what something means or how it works, it also has to ask what is at stake in asking such questions in the first place' (Buchanan 2010:100).

The ever-growing literature on philosophy and sociology of technology, and of design, provides rich grounds for D&T educational research. Here, I nominate just two authors. Feenberg's (1991) *Critical Theory of Technology* was followed by over 20 years of deep critical theorising of technology (e.g. Feenberg 1999, 2010). Offering important critiques of Heidegger and Habermas, he notes two 'substantive' theories of technology and argues that technology's pervasiveness in our lives is such that:

> . . . one can draw diametrically opposed conclusions: either politics becomes another branch of technology, or technology is recognised as political. The first alternative leads straight to technocracy: public debate will be replaced by technical expertise; research rather than the uninformed opinion of the voters will identify the most efficient course of action . . . In opposition to this technocratic trend, there is a grand tradition of romantic protest against mechanisation going back a century or more. (Feenberg 1999:2)

Concurrently, Fry developed his critique of design (Fry 1992, 1995). He explores craft as ontology, also adopting a Heideggerian critique of the technological displacement of humans. He talked of recentring '. . . the human maker that advanced technology decentres and displaces. In doing this working life is retained as a "lifeworld" in which the care for "earth" is lived as a practice of making with *care* . . . The notion of *care* goes beyond the common usage of the idea. For

Heidegger it became a key existential condition of being...' (Fry 1992:263). He subsequently writes of our 'ecological crisis' and how we have designed it, driven it and lived it and noting how 'Design can now more clearly be seen to ride the line between creation and destruction' (Fry 1995:190). He also notes that: '. . . design's acknowledged and celebrated forms have been attached to explicit economic functions and cultural appearances that *lack any ability to engage in critical reflection*, especially of design's impact on the social and the environmental fabric of our world' (Fry 1995:190–191. My italics).

4.2 International and National Technology Education Developments that Informed the Critiquing Innovation

For the field of (D&T) education itself, the 1990s offered some serious affirmations. Research, although disparate, was growing. At least three internationally oriented academic journals were established as were five international research conferences. A significant international study of innovations in Technology Education (Layton 1994) reported: '. . . *(an opportune) moment in curriculum history...* In many education systems around the world, irrespective of whether the country is low income and developing or high income and industrialised, the case for *technology as a component of general education* is under examination and is impelling specific curriculum innovations' (Layton 1994:11. My italics).

I add the emphases to point both to Technology Education's zeitgeist and to its legitimation as a dimension of education for all students ('general education'). Layton also reported that: 'School technology . . . is subject to a range of competing influences and the politics of technological literacy – who creates and controls the meanings of the phrase, how the imposition of meaning is attempted – is a central concern of technology education today' (Layton 1994:13). Furthermore, he documented the complex range of stakeholder interests and associated tensions at play in Technology Education around the world noting six groupings spanning a philosophical-political-social spectrum of interests of such breadth that even the most encompassing curriculum would find challenging to meet.

South Australia embraced these international issues: technology education as general education, contestation over technological literacy and stakeholder rivalry – as a genuine curriculum challenge. Early curriculum theorisations (Keirl 1997a, b) explored: (1) a practical application of the Habermas *knowledge interests*, (2) the development of a *critical technological literacy* and (3) the potential of *critique* as a 'partner disposition' to design.

Australia's federal system of government precludes national constitutional power over education which is the prerogative of the six states and two territories. However, a series of federal collaborations by the respective Ministers of Education (AEC 1989; MCEETYA 1999, 2008) established common *frameworks*, rather than prescriptive detail, for curriculum across the nation. 2010 onwards has seen the gradual emergence of a national curriculum (ACARA 2015a).

By 1994 eight common 'Learning Areas' had been agreed – each with its own 'statement' and 'profile' to describe the coverage and anticipated learning outcomes. This offered, for the first time, national status and a common language for Technology Educators, but, at the time, there were over 80 technology subject associations across the country, a bundle in every state and territory, competing for resources and curriculum space while also maintaining scepticism towards local or national partnerships.

Permeating these circumstances ran populist notions of what constituted 'technology' – what I had called 'orthodoxies of technology'. Today, they are less remarkable but still warrant researcher respect. They are that technologies must be *new*, technologies are *things*, technologies are *neutral*, 'technology' *equates* '*computers*', technology is *applied science*, technologies are *inevitable* and technology is *incomprehensible* (Keirl 1999a:76–77).

4.3 The South Australia Educational Climate

How then could the complexity described here be managed through good curriculum design? Apart from the international climates for the new curriculum, the educational climate of South Australia warrants description. *Critiquing* did not simply 'appear' in the emergent D&T curriculum nor was it ever fashion or fad. Some influence has to be attributed to distinguished critical literacy and curriculum theorists within and beyond the University and Education Department (e.g. Boomer 1989/1999a, b; Comber 1992, 1994; Comber et al. 1998; Johnson and Reid 1999; Smyth et al. 2000; Smyth 2011).

In response to Australia's emergent economic instrumentalism, and advocating innovative constructivist educational practices (over transmissive ones), Boomer wrote: 'Constructivist teachers treat children as if they have brains. They demand students learn to plan and design and construct their own understandings, assisted, of course, by excellent demonstration and instruction at the point of need' (Boomer 1989/1999b:78–79). Reflective D&T practitioners knew that they were already ahead of this game with good design pedagogy. This is from a 1994 South Australian Technology Education curriculum guide:

> Students demonstrate technological capability by:
>
> - being enterprising, innovative, willing to take considered risks and by exercising *critical judgement in developing their ideas*;
> - *demonstrating questioning and critical attitudes to appropriate technological development and application, past, present and future*;
> - building a personal set of intellectual tools through experience with a wide range of technological tasks in different contexts;
> - developing skills in the use of a range of tools and equipment and developing knowledge about their purposeful uses;
> - *designing* and *making* a variety of types of technological products and *appraising* the outcomes. (DECS 1994:9. My emphasis)

5 The South Australian Curriculum, Standards and Accountability (SACSA) Framework

SACSA is a Birth to Year 12 curriculum and has served South Australia for over fifteen years. Full detail cannot be given here but more is available: on its antecedents in DECS (1995) and AEC (1994a, b); on the full curriculum policy at DETE (2001a, b); and, on D&T, in Keirl (2000a, 2001a). In taking account of global and local contexts, SACSA articulates:

- A need for curriculum and pedagogies to be dynamic
- The centrality of an ethical dimension for curricula in changing times
- Recognition that communities and societies no longer exist in isolation
- Its contribution towards an international educational community (Delors 1996)
- Constructivism as its pedagogical theoretical underpinning
- Itself as a *curriculum framework*, that is, not prescriptive but respectful of teachers' professional judgement and local, community-based interpretation
- Seven equity perspectives including those of Aboriginal and Torres Strait Islander peoples
- Eight Learning Areas (reflecting those nationally agreed), one being *Technology* (subsequently, *Design and Technology)*

Constructivist learning is reinforced through five Essential Learnings (ELs). Applying the postmodern pluralised *knowledges* and *learnings*, the ELs are understandings, capabilities and dispositions to be developed through all Learning Areas. 'They are resources which are drawn upon throughout life and enable people to productively engage with changing times as thoughtful, active, responsive and committed local, national and global citizens' (DETE 2001b:7; Keirl 2001a; Spry 2015). The five, with their threads of *power*, *criticality* and *action*, were *communication*, *futures*, *identity*, *interdependence* and *thinking*.

D&T's Learning Area design was overseen by an 18-member Technology Experts Working Group (TEWG) whose role was not only to meet the curriculum design requirements of SACSA but also to accommodate the Learning Area's special challenges, including: early years-primary-secondary progression; multiple 'subjects' and their competing knowledges; emergent technologies; and, continuing transitions from traditional 'technical' towards 'design-rich' pedagogies.

5.1 Technological Literacy in SACSA

The following conceptualisation of technological literacy emerged. The influences of Habermas and critical literacy theory are clear:

> Technological literacy can be viewed as having three dimensions, all of which are equally valid and important. All students benefit from all dimensions of technological literacy and must not be constrained in their learning to one aspect alone. The three dimensions are:

- the *operational*, through which students develop skills and competencies at a technical level to use materials and equipment in order to make products and systems (*they learn to use and do*);
- the *cultural*, through which students contextualise their learning in the world of designed and made products, processes and systems. They recognise the interdependence of technologies with people....and they apply their technical learning in practical ways to realise designs and solve practical problems (*they learn through technology*); and,
- the *critical*, through which students are empowered to take a full and critical role as autonomous citizens in technological societies. They are able to make refined judgements about the worth of the intentions and consequences of technological products, processes and systems on themselves and others... (they learn about, and *to be with*, technology). (DET 2001a, b. My italics)

5.2 Shifting Curriculum Orientation from Content to Process and Issues of Identity

Pre-SACSA, Technology Education was organised around four strands: one 'process' strand and three 'content' strands. The process strand of *designing*, *making* and *appraising* (DMA) had been central to *all* technology activity, while three content strands of *information*, *materials* and *systems* were engaged according to the content being taught. Each strand had its own 'strand organisers'. For DMA these were *investigate*, *devise*, *produce* and *evaluate* (IDPE). Pedagogically, both acronyms 'DMA' and 'IDPE' had become props for inappropriate teaching, each inviting linear, step-by-step approaches; implying singular rather than multiple processes; and inhibiting creative and critical-transformative design pedagogy.

While the primary sector was already celebrating the power of D&T to integrate the whole curriculum, the secondary sector nurtured continuing resistance to anything that seemed to threaten multiple, established (often traditional/technical) 'subjects'. This resistance centred on two assumptions: first, that 'subjectification' is necessary because of *content differences* and, second, that each subject must have its identity maintained.

However, by common agreement, D&T was at heart a *doing* field, and this provided the vehicle for a number of key developments, not least that content variations could affirm their home under common *process* while still maintaining their integrity. By focussing on *process* in redesigning the strands, primary cross-curricular integration was further strengthened and a (potentially) unifying curriculum umbrella was provided for the secondary players. As a result, only three strands, each a *verb (an action word)*, were used to emphasise D&T process and action. Since the TEWG sought to embrace ethical and future perspectives, design, particularly in its senses as 'choice' and as 'intention', was confirmed as a powerful vehicle for this. Design was seen as both central to technological activity and as a rich contributor to general education. Both past best practice and perceived curriculum futures saw *designing* as a worthy strand. *Making*, although stereotyped towards certain types of manufacturing activity, was confirmed as another fundamental of the field.

6 The Emergence of Critiquing

It was recognised that much valuable D&T learning emerges from the deconstruction, physical and otherwise, of designed and made products, processes and systems. Students gain much from finding new ways to question, and make new meanings about, their designed worlds. Such considerations, along with those concerning the technological literacy formulation, brought about the innovative strand: *critiquing*. The three-strand arrangement of critiquing, designing and making (CDM) was seen as wholly interdependent to maintain theoretical rigour and to symbolise the necessary holism of quality D&T education. Notably, the 'C' of critiquing didn't simply replace the 'A' of appraising to become 'DMC'. Rather, an intentional *naming disruption* was aimed at breaking the linear sequencing of the DMA kind.

D&T learning in the age 3–5 phase is also expressed through CDM:

- Children examine, identify and critique processes, products and systems.
- Children use their imagination to generate ideas and participate in the processes of design.
- Children use materials, equipment and processes to design and develop products and systems.

Of D&T's eight curriculum aims, four especially call upon *critiquing*, when students are to develop:

- Ethical, critical, enterprising and future dispositions towards their own and other people's designed and made products, processes and systems
- Capacities to identify and critique the values underlying the intentions, design, manufacture and consequences of any technology
- Capacities to consider and respond to the needs of diverse cultures in relation to developing technologies
- Capacities to apply their design and technology learning to other Learning Areas, to life in the wider community, virtual community, and in accessing further education and training (DETE 2001a, b)

This chapter cannot present the detail of the articulation of the SACSA D&T innovations, that is, the relationships amongst key ideas, standards, outcomes and more. Nor can it include the interdependent and developmental progression of CDM across the years. Full detail and useful tables can be accessed at DETE (2001a, b). However, the six *critiquing outcomes* indicating the progression of learning from early years to age 17 are, in age order, that the student:

- Makes judgements about the significance of different characteristics of products, processes and systems made by themselves and others
- Identifies a range of ways in which the design of everyday products, processes and systems is related to those who use them
- Describes the significance to diverse groups of people of the various criteria used in the design of particular products, processes and systems

- Explains the decisions and choices made in designed and manufactured products, processes and systems and identifies alternative possibilities
- Examines critically the competing values embodied in designed products, processes and systems; clarifies relationships amongst people, products and quality of life; and presents ethical analyses of various possible technological futures.

What, then, might be summarised as key achievements of the SACSA D&T development in terms of critical theory, critical pedagogy and critiquing? These suggest themselves:

- A properly theorised *critical technological literacy* that meets multiple curriculum challenges and actor interests in a holistic and integrated way.
- The articulation of that literacy via three *interdependent* strands all of which are *verbs* (action words). D&T was the only Learning Area to achieve this strategy (privileging *processes* common to all technologies). All other Learning Area strands were *content-focussed* nouns.
- *Design* was established as the central general education concept of the Learning Area and became part of its name.
- *Critiquing* was a true curriculum innovation – a world first for Design and Technology.
- The capacity to readily adapt to such postmodern curriculum arrangements as the five Essential Learnings and their threads of *power*, *criticality* and *action.*
- *No single material*, *technology* or *subject was named* (or valorised over another).
- Because of this, the use of verbs *better embraced technological change* per se – the curriculum was, itself, sustainable and dynamic.
- A *disruptive name ordering* of the three CDM strands was designed to erode the DMA/IDPE mantras.

It should be apparent from these achievements that a critical approach was itself key to SACSA Design and Technology's own design process. By this standard, at least, the curriculum had lived up to its own claims.

7 Subsequent Theorisation and Developments

Critiquing, in D&T curriculum, grew out of turn-of-the-century applications of critical theory articulated as a critical technological literacy for all students. This approach was to serve 'subjects', the Learning Area and general education alike. However, the intended curriculum is one thing and the enacted curriculum as practice is another. While no curriculum is frozen in time, curricula such as SACSA (as a *framework*) are sustainable because of their capacity for (re)interpretation, revision and critique (see, e.g. Spry 2015). All curricula should be open to interrogation and free to evolve. Were they not, in the extreme, they could amount to indoctrination not education.

Over the 15 years since SACSA's launch, there has been further theorisation and application of *critiquing* as D&T curriculum dimension. I close with a selection of six subsequent articulations of critiquing-related curriculum which illustrate the case for its inclusion as an interwoven (not add-on) component of quality D&T education.

7.1 Notes on the Critiquing-Design Relationship

When, in 2002, a colleague questioned the inclusion of critiquing (replacing 'appraising' in DMA), a seed was sown that grew into reflections on the interplay of critiquing and designing in D&T. Clearly there is overlap, there is interplay and there are differences. As happens, a conference paper helped clarify some thinking. It was suggested that designing amounts to arriving at a 'best defensible compromise' and these conclusions were presented on critiquing:

> Critiquing is a skill or a disposition not a methodology. Because it is not concerned with an end point it is used functionally. It needs practice and experience. Critiquing is an excellent tool for arriving at a best defensible compromise (BDC).
>
> Critiquing is responsive to something that exists or has happened – whether an idea in the mind or a physical product. Critiquing is about questioning rather than answering. Its practice helps clarify ill-defined problems through reformulation and reassessment. Critiquing uses many possible differing questions not one best question. Critiquing is reactive – after the fact.
>
> Whilst critiquing is an invaluable tool that enriches designing, it is also something more in itself. Its practice serves democratic purpose and has social value in strengthening democratic society. For D&T education it helps clarify needs-wants issues, values issues, highlights the contestable, exposes the multiple effects of technologies and becomes a mirror for productive thought and action. Its use focuses not on persons but issues, problems, designs, circumstances and supports values resolution.
>
> Critiquing is deconstructive but not destructive. In itself, it has limited problem-solving capacity but it does have excellent problem-finding or fallacy-exposing capacities. Critiquing acts as quality assurance throughout checking and rechecking validity, integrity, worth, accuracy, and fairness. Critiquing may involve looking in the mirror, reflecting alone or together, or placing in the window for public scrutiny.
>
> Critiquing does not have components to be arranged into lockstep sequences – other than understanding the audience for the critique e.g. self, team, assessor, public. Critiquing may lead to a sharper interrogation of assessment criteria and rationales. Critiquing can be used on one's own methodology of practice – on time management, design procedure chosen or research options taken.
>
> Critiquing aids selection of thinking styles. Thus sophisticated critiquing is a form of metacognition… Critiquing may involve discomfort but that is an aspect of critical purpose. The 'discomfort' of self-critiquing is not a matter of positive or negative criticism. It is a phase of the journey to a BDC. Critiquing as experience-building is the interplay of personal experience and knowledge with others' experience (community, research, opinion

> etc). The greater the critiquing experience(s) the greater the critical disposition. Critical friendship is an asset. Like risk-taking in creativity and designing, risk-taking in critiquing requires safety nets. Critiquing is done against a frame of reference which may be personal experience, some agreed or public criteria, or a design brief.
>
> Imagination should not be critiqued. (Keirl 2004:95–96)

The paper suggests that critiquing and designing both:

- Develop socially valuable attributes in students
- Develop thinking styles and confidence
- Are valid components of D&T curriculum for all students
- Are valid components of general democratic education for all students
- Reject fact learning or rote learning
- Are necessary for arriving at a best defensible compromise
- Are undervalued in organisations (Keirl 2004:96)

Meanwhile, these distinct differences were identified:

- Critiquing happens *after* an idea, event, argument or product. Designing *brings into being* these circumstances.
- Designing is proactive, critiquing reactive.
- Critiquing is focussed while designing is holistic.
- Designing always wants imagination to come out and play, but critiquing must never knock on imagination's door.

So far as their working (or living) arrangement is concerned, critiquing is a tool which serves the design enterprise. In fact, good designing demands good critiquing (Keirl 2004:96–97).

7.2 *CDM in an Alternative Arrangement of SACSA*

A second articulation returns to SACSA's five *Essential Learnings* whose role I have outlined above. (I would nominate *ethics, critiquing* and *design* as candidates too.) Such postmodern curricula designs attempt to erode traditional subject silos and avert what Hargreaves (1994) called the *balkanisation* of the curriculum. I have discussed (Keirl 2002a) whether the SACSA matrix that positions the eight *Learning Areas* to be interwoven with five *Essential Learnings* might not be switched – thus privileging the ELs as the primary curriculum organiser. Such modelling shows that the critiquing-designing-making rapport could be successfully and robustly articulated through the ELs. However, any version of D&T without such a rapport (e.g. devoid of explicit and interwoven *critiquing*) would not have the rigour to do so.

7.3 Critiquing, Discomfort and Democratic Citizenship

Critique, as noun and verb, is well established in fields such as the arts, and the act of criticising or passing judgement is as applicable to Design and Technology education as it is to life. Critiquing can be learned, but it can also be thought of as torment, as opposition or as being supportive or empathetic (Walton 1992). How one lives and acts is a matter of one's values; and accommodation of value judgements is the very stuff of active and defensible critique. As Watkins (1978) says:

> There are many bad reasons for placing value judgements outside of the boundaries of genuine critical thought, most of them having to do with the idea that criticism should be neutral and descriptive, that it should say what a poem is or what it means before assessing its value and significance... (The) stance of contemplative neutrality is itself indicative of social and class attitude toward human knowledge. (Watkins 1978:213)

In the above, 'technology' can be substituted for 'poem' and the orthodoxy of technology as neutral is exposed. Walton (1992) describes critical discussion as adversarial, and he points to the function of 'critical doubt' in argument which is not '...having a neutral point of view. It is the bracketing or suspending of the point of view you already have, in order to express doubts and questions. But such a suspension does not imply a neutral attitude' (Walton 1992:267). He articulates the significance for the critic of becoming '...truly conscious of his (sic) own act of thinking, the critical act itself assumes a kind of dialectical reciprocity'. Thus, there may never be 'an end' in the practice of critiquing, merely synthesis towards new beginnings. Rather than maintaining the straitjacketing effect of any singular, linear, step-by-step 'design process', doubt, dialecticism and craftiness become D&T curriculum critiquing assets.

With a healthy critical disposition, one's value judgements are ever under scrutiny by the uncertain self who, not without discomfort, can bring about new possibilities in both being in and acting on the world. As Saul (1995) has argued: 'Criticism is perhaps the citizen's primary weapon in the exercise of her legitimacy. That is why, in the corporatist society, conformism, loyalty and silence are so admired and rewarded; why criticism is so punished and marginalised' (Saul 1995:169–170).

Critiquing, properly engaged, must accommodate *discomfort* (Keirl 2000b, 2004, 2007a, b, 2010). Herein lies at once both an educational asset and a pedagogical challenge. Saul (1995) argues the case for valuing and respecting uncertainty as a partner of critique in democracy, and he acknowledges discomfort and in relation to political consciousness: '(T)he virtue of uncertainty is not a comfortable idea, but then a citizen-democracy is built upon participation, which is the very expression of permanent discomfort. The corporatist system depends upon the citizen's desire for inner comfort' (Saul 1995:195). More recently, I have described *discomfort* as one of three 'curriculum characteristics' of sustainable-democratic curriculum (Keirl 2015a):

> D&T education gives students, through design, real opportunities to be engaged, to participate, and to be creators of their own knowledge but... (they) are (also) practising critique and gaining voice as would-be democratic citizens. This is a strength of quality D&T as a compulsory curriculum component. For D&T's intrinsic richness and its general educational role, critiquing must be taken as purposeful and as a democratic tool of debate, values-weighing, social questioning of technocratic cultures of dependence, and so on. (Keirl 2007a:310)

7.4 *Using Critiquing with Ethics to Explore Technologies*

In developing a case for ethical technological literacy (Keirl 2006), several challenges were to be addressed. The meta-issue of trying to educate about Technology (big T) is its definition-defying pervasiveness, multistability (Ihde 2002), multivalence (Sclove 1995), 'invisibility' (Keirl 2015c) and more. In an attempt to help 'see' technologies (small t), I have suggested that any technology might be witnessed through five phases, *intention*, *design*, *realisation*, *use* and *consequences*, which are not discrete but are co-dependent (Keirl 2009). The phases and their co-dependence are an attempt at exploring T/technology(ies) in ways to respect and support holism. Ethical critiquing (or critiquing in general) of a technology *at each of its phases* is very revealing as circumstances change considerably:

> To achieve the democratics of practice needed to *know* life (in all forms and global sites) with *Technology* at least two other discourses are enabled through the framing – the ethical, addressing a spectrum from values-weighing to big questions like 'How should we live?; and, critiquing – of one's own and others' design decision-making and technological products, processes and systems. Students with well-developed ethical and critiquing dispositions will be well placed to play a role in democratic life. (Keirl 2009:44)

Consideration of others is key to ethical living, and we can explore the meta-picture by considering the four realms in which we coexist (Keirl 2010): with *other people*, with *other species*, with *technologies* and with *the planet*. As individuals we are constructs of our interactions with these four realms. However, learning about ethical-democratic life boils down to understanding that we have choices and that these also shape who we are. Critiquing plays its role here too:

> Design thinking and choice education for learners in terms of 'me, an individual, and I... as I choose to be and who I choose to spend my time with, and how I choose to present myself to others' is rich in opportunity, *so long as* the concept of self is recognised as having consequences for others. The pedagogical repertoire of a D&T teacher can encourage reflection on the self-others-environments relationships and consequences. Using contexts such as design, consumption, and consumerism, with engagement of values, opinions and engagement through active questioning rather than passive acceptance, enables reflection, critique and design activism and develops a higher level of awareness and articulation of choice and consequence. (Keirl and McLaren 2013:1622)

7.5 *Critiquing as Thinking Tool*

A fifth articulation is that of critiquing as a 'thinking tool' – both for its practical value for the learner engaging design issues and for their broader engagements with the world at large – serving both design intelligence and democratic purpose (Keirl 2010). It can play a metacognitive role in the selection of thinking styles so necessary to a designerly repertoire (imagining, analysing, researching, synthesising, advocating are examples). In sum, critiquing:

> ...is a way of thinking, acting and being. Critiquing is the purposeful, practical and metaphorical deconstruction and analysis of any product, process or system in order to expose the values and intentions behind designs, the unanticipated applications of technologies, and the relationships between people and technologies. As when designing, new meanings and knowledge emerge from critiquing and new realisations emerge for seeing, judging and living in the designed world. (Keirl 2015a:169)

7.6 *The Emergent Australian National Curriculum*

At the time of writing, a new national curriculum for Australia is emerging (ACARA 2015a, b). All states and territories have contributed to this. Critique (and its relations) makes some welcome but not well-articulated appearances. Meanwhile, the curriculum's first aim (of five) is that students 'investigate, design, plan, manage, create and evaluate solutions' (ACARA 2015b), that is, "IDPE-plus", a new six-step linear sequence with a saddening potential for inhibiting critical design pedagogies.

8 Conclusion

This chapter has set out something of the motivations, theory, context, history and subsequent developments around the D&T curriculum innovation of *critiquing*.

In the closing section I have attempted to show that there is much to be done both to develop the real potential of critiquing in the D&T curriculum. A start has been made but, I argue, tokenism and vanilla treatments will not do. Those who think that offering a few pluses and/or minuses for this or that technology, design or issue counts as education are doing little of educational value. Critiquing, as presented in this chapter and as part of the South Australian curriculum, was a direct response to the complex and contestable political nature both of technologies and of education.

Critiquing and criticism can be conducted anywhere in a rich curriculum, but, in Design and Technology, their purposes and practice need proper integration with design pedagogies in order to achieve ethical-critical technological literacy. Critiquing cannot (and ought not to) be considered a part of D&T education if it is not addressing inequity, injustice, sustainability or any other ethical issue that arises in any of our four realms of coexistence. The interconnectedness (holism) of critiquing with designing and creating cannot be understated. Nor can its role in technological literacy for democratic life be marginalised or diluted by tokenism, reductionism or superficiality.

Acknowledgement I wish to formally acknowledge the dedication, hard work and arguments that every member of the Technology Education Working Group contributed to what was a multi-faceted curriculum design challenge. Their combined effort created a unique and robust D&T curriculum. Such colleagues, along with all educational researchers, practitioners and teachers who are devoted to achieving better futures for all, stand out against those who would reduce schools to factory production lines and children to products.

References

Apple, M. W. (1979). *Ideology and curriculum*. London: Routledge and Kegan Paul.

Apple, M. W. (2001). *Educating the "Right" Way: Markets, standards, God and inequality*. New York: Routledge Falmer.

Aronowitz, S., & Giroux, H. A. (1993). *Education still under siege* (2nd ed.). Westport: Bergin & Garvey.

Australian Curriculum, Assessment and Reporting Authority (ACARA). (2015a). *Development of the Australian Curriculum*. http://www.acara.edu.au/curriculum/curriculum_design_and_development.html

Australian Curriculum, Assessment and Reporting Authority (ACARA). (2015b). *Australian Curriculum: Technologies*. URL: http://www.australiancurriculum.edu.au/technologies/introduction

Australian Education Council (AEC). (1989, April 14–16). *Sixtieth Australian Education Council* (Hobart Declaration on Schooling: Common and Agreed Goals for Schooling in Australia, No 7).

Australian Education Council (AEC). (1994a). *A statement on technology for Australian schools*. Carlton: Curriculum Corporation.

Australian Education Council (AEC). (1994b). *Technology – A curriculum profile for Australian schools*. Carlton: Curriculum Corporation.

Blackmore, J. (2002, September). Speaking out for critical professionalism and education. *Curriculum Perspectives: The Journal of the Australian Curriculum Studies Association, 22*(3), 60–62.

Boomer, G. (1989/1999a). Democracy, bureaucracy and the classroom. In B. Green (Ed.), Designs on learning: Essays on curriculum and teaching by Garth Boomer (pp. 101–112). Canberra: Australian Curriculum Studies Association.

Boomer, G. (1999b). Education and the media – Makers or mirrors? Dilemmas in the development of Australian culture. In B. Green (Ed.), *Designs on learning: Essays on curriculum and teaching by Garth Boomer* (pp. 71–81). Canberra: Australian Curriculum Studies Association.

Buchanan, I. (2010). *A dictionary of critical theory*. Oxford: Oxford University Press.

Canby, H. S. (1924/1967). *Definitions: Essays in contemporary criticism*. Port Washington: Kennikat.

Comber, B. (1992). Critical literacy: A selective review and discussion of recent literature. *South Australian Educational Leader, 3*, 1.

Comber, B. (1994). Critical literacy: An introduction to Australian debates and perspectives. *Journal of Curriculum Studies, 26*(6), 655–668.

Comber, B., Green, B., Lingard, B., & Luke, A. (1998). Literacy debates and public education: A question of 'Crisis'? In A. Reid (Ed.), *Going public; education policy and public education in Australia*. Canberra: Australian Curriculum Studies Association.

Darder, A., Baltodano, M. P., & Torres, R. D. (2009a). Critical pedagogy: An introduction. In A. Darder, M. P. Baltodano, & R. D. Torres (Eds.), *The critical pedagogy reader* (2nd ed., pp. 1–20). London: Routledge.

Darder, A., Baltodano, M. P., & Torres, R. D. (Eds.). (2009b). In *The critical pedagogy reader* (2nd ed.). London: Routledge.

Delors, J. (1996). *Learning: The treasure within, Report to UNESCO of the International Commission on Education for the Twenty-first Century*. Paris: UNESCO.

Department for Education and Children's Services (DECS). (1994). *Introducing technology education R-7: A guide for teachers*. Adelaide: DECS.

Department for Education and Children's Services (DECS). (1995). *Foundation areas of learning – A curriculum framework for early childhood settings*. Adelaide: DECS.

Department of Education, Training and Employment (DETE), (2001a). *South Australian Curriculum Standards and Accountability Framework* (SACSA). URL: http://www.sacsa.sa.edu.au

Department of Education, Training and Employment (DETE). (2001b). *South Australian Curriculum standards and accountability framework, Hardcopy version*. Adelaide: DETE Publishing.

Emmitt, M., & Pollock, J. (1991). *Language and learning*. Oxford: Oxford University Press.

Feenberg, A. (1991). *Critical theory of technology*. Oxford: Oxford University Press.

Feenberg, A. (1999). *Questioning technology*. London: Routledge.

Feenberg, A. (2010). *Between reason and experience and experience: Essays in modernity and technology*. Cambridge, MA: MIT Press.

Freire, P. (1972). *Pedagogy of the oppressed*. London: Penguin.

Freire, P. (2001). *Pedagogy of freedom: Ethics, democracy, and civic courage*. Lanham: Rowman & Littlefield.

Fry, T. (1992). Green hands against dead knowledge. In N. Ioannou (Ed.), *Craft in society, an anthology of perspectives*. South Fremantle: Fremantle Arts Centre Press.

Fry, T. (1995). Sacred design 1: A re-creational theory. In R. Buchanan & V. Margolin (Eds.), *Discovering design: Explorations in design studies*. Chicago: University of Chicago Press.

Giroux, H. A. (1983). *Theory and resistance in education: A pedagogy for the opposition*. London: Heinemann Educational Books.

Goodson, I. F., & Walker, R. (1991). *Biography, identity and schooling: Episodes in educational research*. London: The Falmer Press.

Habermas, J. (1971). *Knowledge and human interests*. Boston: Beacon.

Hargreaves, A. (1994). *Changing teachers, changing times: Teachers' work and culture in the postmodern age*. London: Cassell.

Ihde, D. (2002). *Bodies in technology*. Minneapolis: University of Minnesota Press.

Johnson, B., & Reid, A. (Eds.). (1999). *Contesting the curriculum*. Katoomba: Social Science Press.

Keirl, S (1996, October 25–26). *Critical technology education: Is it?* Unpublished keynote address to the Technology Teaching and Resource-Based Learning (TTRBL) Conference. Adelaide: Westminster School.

Keirl, S. (1997a). Critical practice in design and technology education: Yarning or weaving? *Design and Education, 7*(1), 3–13.

Keirl, S. (1997b). Technology educators and their 'Curriculum Drama': why the profession must be more than mere performers'. *South Australian Educational Leader, 8*(3), 1–12.

Keirl, S. (1999a/2001). As if democracy mattered . . . design, technology and citizenship or 'Living with the temperamental elephant'. In E. W. L. Norman & P. H. Roberts (Eds.), *Design and technology educational research and curriculum development: The emerging international research agenda*. Loughborough: Loughborough University.

Keirl, S., (1999b), 'Determining technology education: Knowing the orthodox, the interests, and the potential. In B. Johnson & A. Reid (Eds.), *Contesting the curriculum* (pp. 74–89). Sydney: Social Science Press.

Keirl, S. (1999c). The fruits of technological literacy: Wild varieties or crops of mass production. In C. Benson & W. Till (Eds.), *Proceedings of second international primary design and technology conference, CRIPT*. Birmingham: University of Central England.

Keirl, S. (2000a). An episode in technology curriculum refinement: It's only another design brief.... In P. H. Roberts & E. W. L. Norman (Eds.), *Proceedings of the international conference on design and technology educational research and curriculum development*. Loughborough: Loughborough University.

Keirl, S. (2000b). Critiquing as a dimension of technological literacy. In *Proceedings of 1st biennial international conference on technology education research 2000*. Technology Education Research Unit, Griffith University, Qld.

Keirl, S. (2001a) Design and technology and the five 'Essential Learnings' of a new curriculum framework. In E. W. L. Norman & P. H. Roberts (Eds.), *Proceedings of the international conference on design and technology educational research and curriculum development*. Loughborough: Loughborough University.

Keirl, S., (2001b, May 18). Parts, hearts and starts in technology education: What should it be? What could it be?. Keynote address to Technology Education Association of Victoria. *Technotes Journal*, *14*(2), 2001.

Keirl, S. (2002a, December 5–7). Against the provincialism of customary existence: Issues arising from the interplay of 'essential learnings', design and technology and general education. In H. Middleton, M. Pavlova & D. Roebuck (Eds.), *Learning in technology education: Challenges for the 21st century*. Proceedings of the 2nd biennial international conference on technology education research, Parkroyal Surfers Paradise, Gold Coast, Queensland, Australia. Centre for Technology Education Research, Griffith University, QLD.

Keirl, S. (2002b, July 2–5). Emergent technologies and their potential in the shaping of design and technology curriculum. In E. W. L. Norman (Ed.), *Proceedings of 1st design and technology association international research conference*, Coventry, UK. Wellesbourne: Design and Technology Association.

Keirl, S. (2002c, July 10–13). Opportunities for technology education in the context of globalisation. In M. Pavlova & M. Gurevich (Eds.), *Proceedings of 1st biennial international conference on technology education*, Nizhny Novgorod, Russia.

Keirl, S. (2004, December 9–11). Critiquing and designing as keys of technological literacy: Matters arising from the meeting. In H. Middleton, M. Pavlova & D. Roebuck (Eds.), *Learning for innovation in technology education: Proceedings of the 3rd biennial international conference on technology education research* (Vol. 2, pp. 91–98). Surfers Paradise, Australia.

Keirl, S. (2006). Ethical technological literacy as democratic curriculum keystone. In J. R. Dakers (Ed.), *Defining technological literacy: Towards an epistemological framework* (pp. 81–102). Basingstoke: Palgrave Macmillan.

Keirl, S. (2007a). Critiquing in a democratics of design and technology education. In J. R. Dakers, W. J. Dow, & M. J. Vries (Eds.), *Teaching and learning technological literacy in the classroom: Proceedings of PATT 18 – international conference on design and technology educational research* (pp. 306–312). Glasgow: Faculty of Education, University of Glasgow.

Keirl, S. (2007b). Discomforting the orthodox: four debates used to raise curriculum awareness and promote critical thinking in design and technology teacher education. In E. W. L. Norman & D. Spendlove (Eds.), *Linking learning: Proceedings of the design and technology association international research conference 2007*. Wellesbourne: Design and Technology Association.

Keirl, S. (2007c). The politics of technology curriculum. In D. Barlex (Ed.), *Design and technology – For the next generation: A collection of provocative pieces, written by experts in their field, to stimulate reflection and curriculum innovation*. London: Nuffield Foundation.

Keirl, S. (2007d, September) 'Within-it/without-it' and the search for ethical technological literacy. In *Curriculum Perspectives*, *27*(3), 77–80. Australian Curriculum Studies Association, ACT, Deakin West.

Keirl, S. (2009). Seeing technology through five phases: A theoretical framing to articulate holism, ethics and critique in, and for, technological literacy. *Design and Technology Education*, *14*(3), 37–46. URL: http://jil.lboro.ac.uk/ojs/index.php/DATE/article/view/1274/1239

Keirl, S. (2010, June 17–19). Critiquing and designing as thinking tools for technology education for sustainable co-existence. In R Hansen & S. Petrina (Eds.), *Proceedings of the technological learning and thinking: Culture, design, sustainability, human ingenuity conference*, Vancouver, BC (pp. 531–540). URL: http://m1.cust.educ.ubc.ca/conference/index.php/TLT/2010/paper/view/57/5

Keirl, S. (2015a). Against Neoliberalism; for sustainable-democratic curriculum; through design and technology education. In K. Stables & S. Keirl (Eds.), *Environment, ethics and cultures: Design and technology education's contribution to sustainable global futures* (pp. 153–174). Rotterdam: Sense.

Keirl, S. (2015b). Global ethics, sustainability, and design and technology education. In K. Stables & S. Keirl (Eds.), *Environment, ethics and cultures: Design and technology education's contribution to sustainable global futures* (pp. 33–52). Rotterdam: Sense.

Keirl, S. (2015c). 'Seeing' and 'interpreting' the human-technology phenomenon. In P. J. Williams, A. Jones, & C. Buntting (Eds.), *The future of technology education* (pp. 13–34). Dordrecht: Springer.

Keirl, S., & McLaren, S. V.. (2013, May 14–17). Students as choice-makers: Developing altered consciousness as an aspect of design and global citizenship literacy. In J. B. Reitan, P. Lloyd, E. Bohemia, L. M. Nielsen, I. Digranes & I. Lutnæs (Eds.), *Design learning for tomorrow: Design education from Kindergarten to PhD: Proceedings from the 2nd international conference for design education researchers* (vols. 1–4), (Design Research Society/CUMULUS the International Association of Universities and Colleges of Art, Design and Media, Oslo, Norway) (vol. 3, pp. 1611–1625). Oslo: ABM-media/Oslo and Akershus University College of Applied Sciences.

Kellner, D. (2003). Critical theory. In R. Curren (Ed.), A companion to the philosophy of education 161–175, Blackwell, Oxford.

Kemmis, S. (1991/2005). Emancipatory action research and postmodernisms. In C. Marsh (Ed.), *Curriculum controversies: Point and counterpoint 1980–2005* (pp. 308–318). Australian Curriculum Studies Association, Deakin West, Australian Capital Territory.

Kincheloe, J. L. (2008/2010). *Knowledge and critical pedagogy: An introduction*. Springer, Springer.com

Lankshear, C. (1998). Frameworks and workframes: Literacy policies and new orders. *Unicorn, 24*(2), 43–58.

Layton, D. (Ed.). (1994). A school subject in the making? The search for fundamentals. In D. Layton (Ed.), Innovations in science and technology education (Vol. V, pp. 11–28). Paris: UNESCO.

Luke, A. (1992). Literacy and work in 'New Times'. *Open Letter, 3*(1), 3–15.

McLaren, P. (1989/2009). Critical pedagogy: A look at the major concepts. In A. Darder, M. P. Baltodano & R. D. Torres (Eds.), *The critical pedagogy reader*, (2nd Edn, pp. 61–83). London: Routledge.

Ministerial Council on Education, Employment, Training and Youth Affairs (MCEETYA). (1999). *National goals for schooling in the twenty-first century*. URL: http://www.curriculum.edu.au/mceetya

Ministerial Council on Education, Employment, Training and Youth Affairs (MCEETYA). (2008). Melbourne declaration on educational goals for young Australians. MCEETYA Secretariat, Carlton. URL: http://www.curriculum.edu.au/verve/_resources/national_declaration_on_the_educational_goals_for_young_australians.pdf

Morris, V. C. (1966/1990). *Existentialism in education: What it means*. Prospect Heights: Waveland Press.

Morrison, K. (2001). Jürgen Habermas. In J. A. Palmer (Ed.), *Fifty modern thinkers on education from piaget to the present* (pp. 215–224). London: Routledge.

Petrina, S. (2000a). The political ecology of fesign and technology education: An inquiry into methods. *International Journal of Technology and Design Education, 10*, 207–237.

Petrina, S. (2000b). The politics of technological literacy. *International Journal of Technology and Design Education, 10*, 181–206.
Petrina, S. (2003, Spring). Human rights and politically incorrect thinking versus technically speaking. *Journal of Technology Education, 14*(2), 70–74.
Pinar, W. F. (Ed.). (2003). *International handbook of curriculum research*. Mahwah: Lawrence Erlbaum Associates.
Pinar, W. F. (2004/2008). *What is curriculum theory?* Routledge, New York.
Pinar, W. F. (2007). *Intellectual advancement through disciplinarity: Verticality and horizontality in curriculum studies*. Rotterdam: Sense.
Postman, N. (1999/2000). *Building a bridge to the eighteenth century: How the past can improve our future*. Carlton North: Scribe Publications.
Postman, N., & Weingartner, C. (1969/1971). *Teaching as a subversive activity*. Harmondsworth: Penguin.
Reid, A. (2004/2005). Challenging the dominant grammars of an undemocratic curriculum. In C. Marsh (Ed.), Curriculum controversies: Point and counterpoint 1980–2005 (97–105). Australian Curriculum Studies Association, Deakin West, Australian Capital Territory.
Saul, J. R. (1995). *The unconscious civilisation*. Toronto: Anansi.
Sclove, R. E. (1995). *Democracy and technology*. New York: The Guilford Press.
Smith, D. G. (2003). Curriculum and teaching Lwface globalization. In W. F. Pinar (Ed.), *International handbook of curriculum research* (pp. 35–51). Mahwah: Lawrence Erlbaum Associates.
Smith, D. L., & Lovat, T. J. (1991). *Curriculum: Action on reflection*. Sydney: Social Science Press.
Smith, D. L., & Lovat, T. J. (2006). *Curriculum: Action on reflection* (4th ed.). South Melbourne: Thomson/Social Science Press.
Smyth, J. (2011). *Critical pedagogy for social justice*. New York: Continuum.
Smyth, J., Dow, A., Hattam, R., Reid, A., & Shacklock, G. (2000). *Teachers' work in a globalizing economy*. London: Falmer Press.
Spry, L. (2015). Opening up the four walls: Reflections on two South Australian ESD projects. In K. Stables & S. Keirl (Eds.), *Environment, ethics and cultures: Design and technology education's contribution to sustainable global futures* (pp. 285–298). Rotterdam: Sense.
The New London Group. (1996, Spring). A pedagogy of multiliteracies: Designing social futures. *Harvard Educational Review, 66*(1), 60–92.
van Manen, M. (1990). *Researching lived experience: Human science for an action sensitive pedagogy*. London: State University of New York Press.
Walton, D. (1992). *The place of emotion in argument*. Pennsylvania: The Pennsylvania State University Press.
Watkins, E. (1978). *The critical act: Criticism and community*. London: Yale University Press.

Critique as a Disposition

P John Williams

Abstract The appeal of perceiving critique as a disposition lies in its essentiality of action; the ability to critique is an important prerequisite, but when an individual is disposed to act as a result of cognitive critique, then the outcome has effect. This notion aligns well with concepts of technological literacy as the goal of design and technology education. It has been argued that this form of literacy must have a capability element to it, and if the critique element of this capability is dispositional, then technological literate students will be active in their relationships with technology.

Keywords Habits of mind • Design and technology • Dispositions • Values

1 Introduction

This chapter proposes that one of the foci for design and technology education should be the development of critique as a disposition. The literature on dispositions is grounded in the fields of philosophy and psychology. Dispositions have been defined as patterns of behaviour that are exhibited intentionally and frequently, representing habits of mind. Therefore, dispositions are concerned with not only what a student can do but what a student is disposed to do, so addressing the often prevalent gap between abilities and actions. The essentiality of action in a disposition aligns with the manifold notions of activity within design and technology education. Design and technology education is not a passive activity; conceptually it involves the construction of new knowledge, and practically it involves movement and action and construction. It therefore goes beyond the possibly conceptual, although activity-based, notions of technological literacy and ensures an active end point.

Critique is approached and applied in a range of different ways in the chapters of this book, and the definitions consequently vary. As a starting point for this chapter,

PJ. Williams (✉)
Science and Maths Education Centre (SMEC), Curtin University, Perth, WA, Australia
e-mail: pjohn.williams@curtin.edu.au

PJ. Williams, K. Stables (eds.), *Critique in Design and Technology Education*, Contemporary Issues in Technology Education, DOI 10.1007/978-981-10-3106-9_8

I will refer to David's discussion in chapter "The Identification and Location of Critical Thinking and Critiquing in Design and Technology Education" where he refers to the achievement of critique as 'discernment in thinking, based on standards', achieved through processes of critical thinking, and the more general definition Belinda uses in chapter "Modelling as Form of Critique" as 'a disciplinary habit of mind', which is well aligned with the intent of this chapter.

2 A Rationale for Emphasizing Dispositions

Discussions about dispositions range from something that is fixed and unchangeable (Diez 2006) to something that is malleable (Will 2006), from behaviour that indicates a disposition (Katz and Raths 1985) to the notion that dispositions are conceptually distinct from behaviour and 'involve awareness, inclination, and reflection on behaviours and thinking – not just the behaviours or the thinking themselves' (Schussler 2006, p. 257). There is no consensus about the definition of dispositions, although certain terms and definitions have more currency in particular communities. Some of the terms associated with discussions of dispositions include tendencies, values, habits of mind, attitudes and behaviours; the consistent conceptual overlay of these terms is action – that a disposition is not something static, or merely an attitude, but has an essentially behavioural outcome. In addition, it refers to not just what a person can do but what they are disposed to do. Katz (1993) provides a definition of dispositions as patterns of behaviour that are exhibited frequently and intentionally in the absence of coercion, thus representing a habit of mind.

Aspects of dispositions are verbs, that is, not something to be acquired but an automatic response to a circumstance. A student becomes disposed to act in a certain way and, in an insecure or uncertain situation, feels secure in providing a response because the disposition provides the security.

According to Sockett (2009), dispositions seem to have three implications:

- First, a disposition is a proclivity to act. For example, being careful, not merely to 'be' careful; the action of being is integral.
- Second, a disposition to act implies awareness of what one is doing. For example, knowing that in the questions that are being asked, a form of critique is evident.
- Third, acting with awareness implies that a person acts with intention. That is, this specific act is intended as a careful act. To intend to do something is to be aware that this (and not something else) is what one is doing.

Another common perspective is that dispositions have two components – inclination and ability, which are the components of behaviour. An inclination is the person's tendency towards a certain behaviour. For example, a person with an inclination towards critique will tend to be critical when confronted with a situation in which he or she can respond in that way. Ability refers to the capability to engage with the disposition. For example, a person with the ability to critique will know how to question with purpose, isolate elements and perceive patterns and consequences.

3 Dispositions and Values

Projects that engage students in sequences of carefully structured experiences are important to ensure academic progress throughout schooling. Attention to personal development is at least equally important and arguably more important. This is essentially related to character and values, and whether teachers agree that they are engaged in values education or not, the fact is that technology is imbued with values. Consequently, it is necessary to be explicit about this and open it up for discussion; values can then be critiqued and raised to the conscious level. This does not imply the dominance of any one set of values over any other. Technological determinists will have a different set of values than humanists, as will futurists and behavioural scientists. Huitt (1997) suggested a set of values that most educators would accept as important: autonomy, benevolence, compassion, courage, courtesy, honesty, integrity, responsibility, trustworthiness and truthfulness. But the point of emphasis here is that relevant values must first be recognized as being embedded within design and technology education before they can be examined as to their appropriateness.

In his paper, *Dispositions as Virtues*, Sockett (2009) argues that it is incomplete to consider educational dispositions in the absence of moral character dispositions. He holds that education is essentially a moral activity, and, particularly in the empirical tendencies of assessment and notions of skilled teaching emerging from practice, unless the moral is considered, the outcome will be unacceptable. This warning would seem to apply particularly to design and technology education, where its practical focus and instrumental tendencies are not conducive to the consideration of moral dispositions.

Sockett's 2009 literature review suggested a range of perspectives on dispositions – pedagogical, institutional, philosophical and psychological. For instance:

- From a pedagogical perspective, dispositions can be viewed within reflective practice as part of intellectual character and within moral communities of practice.
- From a philosophical perspective, attempts are made to examine meaning and use, as well as the different perspectives offered through moral philosophy.
- Finally, though some psychological perspectives refer to cognitive content, the volume of work on personality, with its strong and authoritative place in psychology, is an additional perspective.

Therefore, dispositions are actions resulting from awareness and intent and are always the result of judgment. 'Our actions thus stem from our cognitive appraisals of situations where we act intentionally within which acts our dispositions are manifest' (Sockett 2009, p. 295).

Research by Stooksberry et al. (2009) proposes that teachers need to be aware of their dispositions so that they can be purposeful in their thinking and actions. There seem to be two main categories of such dispositions: character dispositions, which relate to self-knowledge, the virtues of the will (persistence, perseverance and heed) and the kind of person that an individual is. Secondly are dispositions of intellect that may include accuracy, fairness and impartiality in making judgments and open-mindedness. A thinking disposition is a tendency towards intellectual activity that guides cognitive behaviour. Sockett (2009) refers to integrity, trustworthiness, persistence, fairness, tolerance and civility as relevant to the profession of teaching. Misco (2007) in *Preparing Graduates for Moral Life* refers to dispositions of respect for the dignity of others, sensitivity to cultural norms, and advocacy of equity and access as central to the goal of a democratic education for secondary school students.

No educator would deny that these are worthy dispositions to develop in students, and the list could be quite extensive. However, for this chapter, the focus is on fostering critique as a disposition.

4 Assessing Dispositions

The relatively recent spate of research about dispositions, emanating mainly from the USA, is at least partly the result of the inclusion of this notion in professional teaching standards. The National Board for Professional Teaching Standards, the National Council for the Accreditation of Teacher Education (NCATE 2008) and the Interstate New Teacher Assessment and Support Consortium all mention dispositions as being essential elements of teacher preparation and teacher quality. NCATE explicitly includes 'professional dispositions' as one of its standards, with the expectation that teacher trainees are assessed in their achievement of this standard. The Accreditation Board for Engineering and Technology (ABET) also alludes to dispositions in its reference to professional and ethical responsibility.

Possibly as a result of the inclusion of dispositions in the NCATE standards for teachers, empirical research specifically addressing dispositions, unfortunately, has centred on the assessment of dispositions (see Hillman et al. 2006; Pedro and Miller 2005; Wilkerson and Lang 2005). The assessment of dispositions is fraught with pitfalls, particularly those with moral and ethical dimensions; however, their inclusion in professional teacher development standards highlights their significance.

Thornton (2006) reviewed a number of models that have been developed to assess dispositions in the context of teacher accreditation. These included:

- Dispositions related directly to behaviours in the school setting, which tend to be comprised of checklists, rating scales and rubrics and look more like pedagogical practices or teaching behaviours than dispositions.

- Dispositions developed around professional characteristics such as attendance, work ethic, preparation, punctuality and appropriate dress, which are really minimal dispositions and fall short of capturing the nature of true dispositions.
- Dispositions determined by reflective self-assessment, an attempt to address the complexities and psychological nature of dispositions by requiring a written response to a human relations incident. This is dependent on an individual's ability to self-report and express their metacognitive understanding in writing.
- Dispositions that focus on moral and ethical dimensions, often directed towards diversity and inclusivity.

5 Technological Literacy

Technology education has a history of addressing personal and professional goals, albeit at times in a narrow vocational and instrumental manner. The notion that a fulfilling life in a technological society requires a certain skill set that students gain through practical activities in a school technology workshop environment has been the foundation of many technology programmes. Likewise, the role of technology education as a career awareness experience leading to later prevocational and vocational mastery of competencies has been an oft argued rationale.

The traditional competency-based approach to technical education was too narrow to be classified as literacy. The more recent recognition, through the application of design, that a broad range of cognitive skills exists that could be developed and nurtured through application to a practical context, provided the basis for promoting the notion that this constitutes a unique type of literacy – technological.

Arguably the most significant goal of technology education programmes is technological literacy, generally constituted of an ability/use dimension, a knowledge and understanding dimension and an awareness or appreciation of the relationships between technology, society and the environment (International Technology Education Association 2000; Ministry of Education, New Zealand 2006; Ministry of Education, South Africa 2002; Department of Education Training and Employment, South Australia 2001; Pearson and Garmine 2006). Curricula then go on to elaborate on the specific abilities or outcomes related to these dimensions that are to be achieved in order to reach a school-based level of technological literacy.

Literacy is an essentially dynamic construct that one is always developing and never achieving. This dynamism is elaborated by Waks (2006) in tracing the developments of the concept of technological literacy from its genesis in the 1970s to a contemporary context. He maintains that initial conceptions of technological literacy are no longer valid because of (a) increased localized ethnic and linguistic diversity, (b) economic and technical convergence into internationally networked systems and (c) the need to move beyond the limitations of schooling into less structured 'post-curricular' designs. A fluid construct of technological literacy will accommodate such changes.

Kahn and Kellner (2006) argue for a link between proliferating high technologies and the need for a reconstruction of technoliteracy. Contemporary technoliteracies can '. . . further radical democratic understandings and transformations of our lives, as well as [provide] a democratic reconstruction of education. . . . Technoliteracies must be deployed and promoted that allow for popular interventions into the ongoing and often undemocratic economic and technological revolutions taking place . . . ' (p. 258).

Williams (2009) also calls for a revision of technological literacy and proposes technological multiliteracy as an alternative construct. The proposition being offered is that technological literacy is multiliterate and the parallel drawn is with developments in the general literacy movement. Historically, general literacy was based on a monodimensional construct, but given social and technological developments, a broadening of the construct to multiliteracy provided the platform for a more relevant, useful and ultimately democratic approach. Similarly with technological literacy, reframing the traditional approach to technological literacy as a multiliteracy construct highlights its breadth, incorporates contemporary developments such as multiple modes of communication and empowers students to play a more democratic role in their own development through the potential of, for example, Web 2.0.

The popularization of a multiliteracy approach within education is developed as a response to the multiple modes of communication and increasing cultural and linguistic diversity faced by students (Cazden et al. 1996). Many synergies exist between technological literacy and the notion of multiliteracies (within the general literacy discourse) in developing relevant and engaging pedagogies that promote the critical engagement necessary for students to contribute to and achieve their full potential.

The problem remains, however, that technological multiliteracy as a goal deals with student potential and provides no guarantee that the potential will be realized in a context that demands involvement or a response. A student may perform well in class, develop insightful portfolios or achieve a high score on a test of technology, but not possess the disposition to apply this knowledge and capability consistently and with discrimination to new and real situations. Technological literacy as competency attainment (Dakers 2006, p. 257) fails to provide an impelling rationale for action and is therefore inadequate to that extent. Kimbell and Stables (2008) argue for a focus on capability as a way to ensure Design and Technology involves 'active, interventionist doing and being' (p. 23), as opposed to just understanding and using technology. Dispositional behaviour explains and provides a framework for the desired action and goes beyond simply framing the capabilities or competencies required for this action.

6 Critique as a Disposition

> The positive habits of mind we prize are those which orient decision makers and problem solvers toward using reflective judgment – that is, critical thinking, particularly when working on novel, high stakes, complex, questions in contexts of risk and uncertainty. (Insight Assessment 2013)

With regard to the terminology of critiquing and critical thinking, the tendency in this chapter will be to use the terms interchangeably. As Keirl (2007) quotes Paul (1995) 'Critical thinkers critique in order to redesign, remodel and make better' (p. 526). I do however avoid the use of the term 'critical theory'; although there are many parallels between this terminologies, critical theory as a concept has a range of assumptions, applications and connotations which may tend to confuse this discussion and, as a discourse, brings with it terminology that may not be necessary in a discussion of critical thinking.

Critique is a common process in many areas of design education and related disciplines in which students explain their work usually to colleagues and then respond to questions and comments. Students often find the explanation difficult because much design knowledge is tacit by nature and also because there is an expectation that discussion goes beyond product and process to the designer's cognition which underpins the design (Dannels et al. 2008). This may not necessarily reflect a dispositional approach to critique, as it is a teacher-directed activity, but the activity would come easier to a student who had a disposition towards critique.

So it is helpful to separate the disposition to critique from the ability to critique. A person can be taught the range of skills necessary to be critical and may also be able to recognize when it is appropriate to apply those skills, but then may choose not to, whereas a person who is disposed to critique will naturally choose that approach as their default reaction to a situation.

An additional rationale for taking a dispositional approach to the encouragement and use of critique is, as Keirl (2007) describes, that it can be seen as a way of doing or being, not just a way of thinking. An approach to developing critical thinking skills in students, if it is seen merely as a way of thinking, could become somewhat mechanistic in terms of the pedagogies employed. Critical thinking is more than the successful use of the right skill in an appropriate context. It is also an attitude or disposition to recognize when a skill is needed and the willingness to exert the mental effort needed to apply it. Sears and Parsons (1991) call these dispositions the ethic of a critical thinker.

A Foucauldian approach to critique is sympathetic to its perspective as a disposition as it is characterized by an 'ethos' and is closely related to the notion of experience as a transformative force.

> When I speak of critique I do not mean a work of destruction, of refusal and denial, but rather an investigative work that consists in suspending as far as possible the normative system which one refers to in order to test and evaluate it. (Foucault 1984, p. 68; Lemke (2011) translation)

Foucault characterizes critique through three aspects: the activity of problematization, the art of voluntary insubordination and the audacity to expose one's own status as a subject.

- Problematization: Critique does not relate to a lack of knowledge, but exists in reaction to the limits truth regimes impose on autonomy and democracy, recognizing that there is more than one truth. Problematizing and recognizing the forces, connections and strategies that combine to establish what counts as self-evident and truth is the starting point of critique. There is a relational and collective dimension to this critique referred to by Foucault as 'collective practice' (Foucault 2000, p. 244), which resonates with a sociocultural classroom approach to the development of critique as a disposition. While problematization can itself be the object of analysis, it is also part of the activity of critique, the process in which the critique engages.
- Voluntary insubordination: Critique implies insubordination to the existing normative and institutional systems in exploring ways to reinterpret or transform it. This will result in deviation, dissent and diversity in relation to generally accepted norms.
- The audacity of exposure: Critique is risky in that it exposes the individual's ontological status, which may fall outside the established norms. It is however necessary to make visible 'what we are' (Foucault 1997, p. 319) in order to move beyond that state. This is not a negative process, but an essential part of the transformation entailed in the processes of critique.

The American Philosophical Association (APA) articulates a consensus developed through a 2-year study by the APA with regard to the dispositional dimension of critical thinking. This consensus captures what some have called the 'critical spirit' – a style, a set of attitudes that define a personal disposition to prize and to use critical thinking in one's personal, professional and civic affairs:

> The ideal critical thinker is habitually inquisitive, well informed, trustful of reason, open-minded, flexible, fair-minded in evaluation, honest in facing personal biases, prudent in making judgments, willing to reconsider, clear about issues, orderly in complex matters, diligent in seeking relevant information, reasonable in the selection of criteria, focused in inquiry, and persistent in seeking results which are as precise as the subject and the circumstances of inquiry permit. (American Philosophical Association 1990, p. 3)

The participants in this study saw the need to nurture a disposition towards critical thinking as fundamental to developing good critical thinkers. They cited motivation theory as the theoretical grounds for the assumption that a disposition towards critical thinking would encourage an individual to achieve mastery over critical thinking skills, being motivated to bridge the gap between what is attained and what is valued.

This study was the basis for the development of the California Critical Thinking Disposition Inventory (CCTDI), which consists of seven subscales that statistically represent the factors that contribute to an overall disposition to use critical thinking to form judgements. In examining the constituents of a disposition towards critique, it is useful to analyse the seven subscales (adapted from Insight Assessment 2013).

Truthseeking is the habit of always desiring the best possible understanding of any given situation; it is following reasons and evidence where ever they may lead, even if they lead one to question cherished beliefs. This is sympathetic to Foucault's notion of 'audacity of exposure', the opposite of which is bias which ignores good reasons and relevant evidence in order not to have to face difficult ideas.

Open-mindedness is the tendency to allow others to voice views with which one may not agree. Open-minded people act with tolerance towards the opinions of others, knowing that often we all hold beliefs which make sense only from our own perspectives. Open-mindedness is important in a sociocultural complex where people approach issues from different religious, political, social, family, cultural and personal backgrounds. The opposite of open-mindedness is intolerance.

Analyticity is the tendency to be alert to what happens next. This is the habit of striving to anticipate both the good and the bad potential consequences or outcomes of situations, choices, proposals and plans. The opposite of analyticity is being heedless of consequences, not attending to what happens next when one makes choices or accepts ideas uncritically.

Systematicity is the tendency or habit of striving to approach problems in a disciplined, orderly and systematic way. The habit of being disorganized is the opposite tendency. The person who is strong in systematicity may not know of a given approach, or may not be skilled at using a given strategy of problem-solving, but that person has the desire and tendency to try to approach questions and issues in an organized and orderly way.

Confidence in Reasoning is the habitual tendency to trust reflective thinking to solve problems and to make decisions. A family, team, office, community or society can be trustful of reasoned judgment as the means of solving problems and reaching goals. The opposite habit is mistrust of reasoning, often manifested as aversion to the use of careful reason and reflection when making decisions or deciding what to believe or do.

Inquisitiveness is intellectual curiosity. It is the tendency to want to know things, even if they are not immediately or obviously useful. It is being curious and eager to acquire new knowledge and to learn the explanations for things even when the applications of that new learning are not immediately apparent. The opposite of inquisitiveness is indifference.

Maturity of Judgment is the habit of seeing the complexity of issues and yet striving to make timely decisions. A person with maturity of judgment understands that multiple solutions may be acceptable while yet appreciating the need to reach closure at times even in the absence of complete knowledge. The opposite, cognitive immaturity, is imprudent, black-and-white thinking, failing to make timely decisions, stubbornly refusing to change when reasons and evidence would indicate one is mistaken or revising opinions willy-nilly without good reason for doing so.

Many other authors have identified the constituents of critical thinking. For example, according to Halpern (1998), a critical thinker exhibits the following dispositions or attitudes:

- Willingness to engage in and persist at a complex task
- Habitual use of plans and the suppression of impulsive activity
- Flexibility or open-mindedness
- Willingness to abandon nonproductive strategies in an attempt to self-correct
- An awareness of the social realities that need to be overcome (such as the need to seek consensus or compromise) so that thoughts can become actions

Perkins et al. (1993) also characterized good thinking dispositions – the ones that normally describe productive intellectual behaviour – as consisting of seven broad but key intellectual tendencies. The following list describes these seven dispositions. Ideally, good thinking includes all of these dispositions exhibited appropriately at different times depending on the thinking situation. While other dispositions may contribute to good thinking, they believe these seven to be central, and efforts to teach thinking ought to cultivate them.

1. The disposition to be broad and adventurous: the tendency to be open-minded, to explore alternative views; an alertness to narrow thinking; the ability to generate multiple options
2. The disposition towards sustained intellectual curiosity: the tendency to wonder, probe, and find problems; a zest for inquiry; an alertness for anomalies; the ability to observe closely and formulate questions
3. The disposition to clarify and seek understanding: a desire to understand clearly and to seek connections and explanations; an alertness to unclarity and need for focus; an ability to build conceptualizations
4. The disposition to be planful and strategic: the drive to set goals, make and execute plans and envision outcomes; alertness to lack of direction; the ability to formulate goals and plans
5. The disposition to be intellectually careful: the urge for precision, organization and thoroughness; an alertness to possible error or inaccuracy; the ability to process information precisely
6. The disposition to seek and evaluate reasons: the tendency to question the given, to demand justification; an alertness to the need for evidence; the ability to weigh and assess reasons
7. The disposition to be metacognitive: the tendency to be aware of and monitor the flow of one's own thinking; alertness to complex thinking situations; the ability to exercise control of mental processes and to be reflective

Thinking dispositions are referred to by Costa and Kallick (2000) as habits of mind: the characteristics of what intelligent people do when they are confronted with problems and the resolutions to which are not immediately apparent. They propose 16 habits of mind, not because this is all there is but because these are the ones they have developed so far. They are persisting; managing impulsivity; listening with understanding and empathy; thinking flexibly; thinking about thinking; striving for accuracy; questioning and posing problems; applying past knowledge to new situations; thinking and communicating with clarity and precision; gathering data through all senses; creating, imagining and innovating; responding with wonderment and awe; taking responsible risks; finding humour; thinking independently; and remaining open to continuous learning.

The application of these habits of mind to a disciplinary context is evident in engineering. As engineering professions in many developed countries endeavour to address decreasing university enrolments, some are examining the nature of teaching and learning in school as a way to strengthen the pathway into tertiary engineering studies. For example, the Centre for Real-World Learning has developed and validated a model of practical learning for the Royal Academy of Engineering which consists of six engineering habits of mind (EHoM). These are organized in the following way:

- Systems thinking. Seeing whole systems and parts and how they connect, pattern-sniffing, recognizing interdependencies and synthesizing
- Problem-finding. Clarifying needs, checking existing solutions, investigating contexts and verifying
- Visualizing. Being able to move from abstract to concrete, manipulating materials and mental rehearsal of physical space and of practical design solutions
- Improving. Relentlessly trying to make things better by experimenting, designing, sketching, guessing, conjecturing, thought-experimenting and prototyping
- Creative problem-solving. Applying techniques from different traditions, generating ideas and solutions with others, generous but rigorous critiquing and seeing engineering as a 'team sport'
- Adapting. Testing, analysing, reflecting, rethinking and changing both in a physical sense and mentally (Lucas et al. 2014)

It is notable that 'generous but rigorous critiquing' is identified as a component of creative problem-solving. The report suggestions indicate an affinity with design and technology teaching that '"messy" approaches such as project based and problem-based learning are actively promoted as methods for building engineering habits of mind. These are "signature" engineering pedagogies' (p. 19).

Critical thinking has long been recognized as a core skill in design and technology (Keirl 2007). It is a part of the process of 'purposeful self-regulatory judgement' (Kim et al. 2014, p. 78). It is truly a disposition in that it is not applied selectively in certain contexts, but is a frame of mind that imbues all aspects of designing in technology. It is applied to tangible products and less tangible processes; it is applied to others and to the self. It is never without purpose, not being critical for the sake of just being critical, but it evolves from a rationale, often related to making progress in working towards design solutions to problems by determining the next stage of the creative journey.

The disposition of being critical may be in opposition or may be supportive, it may be objectionable or it may be confirmatory. But either way, it is purposeful and ultimately constructive. These are not discrete dispositions and are therefore not invoked in isolation. For example, being imaginative involves taking risks, and as Keirl (2007, p. 311) points out, critiquing is a form of metacognition. Further indicating their interdependency, Rutland and Spendlove (2007) suggest that components of creativity include flexible thinking, risk-taking and being imaginative.

7 Teaching the Disposition of Critique

The assumption of this discussion is that critique as a disposition can be taught. Dewey (1922) differentiated teachable dispositions from innate characteristics or temperament in emphasizing the importance of acquiring and developing dispositions. However, they cannot be taught directly. Early research has clearly demonstrated the failure of didactic methods (Hartshorne and May 1928) and other direct strategies (Narvaez et al. 1998) to achieve such ends. Rather than attempt to develop dispositions through transmission or instruction, learning experiences must be carefully crafted to foster the development of desirable dispositions. When students have consistent exposure to these learning experiences, dispositions develop as autonomic habits.

Harpaz (2007) terms this indirect teaching of thinking dispositions as a 'pattern of cultivation' (p. 1849). He differentiates it from the 'pattern of impartation', which involves the direct teaching of knowledge. In the pattern of cultivation, knowledge plays a marginal role. 'Dispositions are cultivated indirectly, not by the transmission of knowledge but by a comprehensive culture of thinking that fosters in various ways thinking dispositions' (p. 1850). The fact that a pattern of cultivation is indifferent to knowledge suits design and technology education, in which the knowledge of the discipline is ill-defined and only contextualized by the nature of the design problem being dealt with. This frees the 'cultivation of dispositions' from any set knowledge, and reinforces what Fenstermacher and Soltis (1986) refer to as a liberationist approach to teaching. Cultivating dispositions of critique is liberating in that it frees individuals from unwanted thinking traits and reinstates the individual into a controlling position through the application of good thinking.

Dispositions are a composite of many skills, attitudes, past experiences and influences. Melding all this into a pattern of behaviour and then making judgments about the application to a situation is the workings of a disposition that teachers must consider. All teachers know that each student in their class responds according to the baggage they bring with them: the skills and attitudes derived from home, peer and media influences. With a focus on critique as a disposition, the teacher's role is to encourage the application of all this baggage to new situations in a consistently intelligent and constructive manner.

The multifaceted nature of design and technology lends itself to the nurture of integrated dispositions. By structuring opportunities for reflection and deliberation, students are able to develop defensible arguments based on evidence, develop listening skills that are open-minded and predict consequences of decisions based on sound epistemologies.

8 Teaching for Dispositions in Design and Technology Education

A sociocultural approach to teaching may be more appropriate to the teaching of critical dispositions, where attention is paid by the teacher to the establishment of an environmental context in which a culture of thinking critically is fostered, expected and established. Characteristics of such a classroom would include an openness to healthy scepticism, a presentation of a range of options to any problem, the structuring of social opportunities for interaction, the modelling of the disposition by the teacher and the development of strategies of analysis.

In considering dispositions as classroom goals, the question is: How can teachers take a responsive approach to dispositions? They are not competencies that a student masters or does not, but are comprised of attributes that are often present to some extent in all students, and the teacher's goal is to develop them and increase the likelihood that the student will respond to any situation in a predictably consistent way.

Traditional education outcomes focus on what students know and what they can recall. Dispositions deal with how students behave when they don't know the answer to a problem. What do they do when they are confronted with a problem for which there is no immediate and apparent answer? Teaching for dispositions has the same goal as requiring students to produce, for example, a process portfolio; in fact, one may provide evidence of the other. A process portfolio indicates how students react to an open-ended problem by providing a record of their thinking in working towards a solution; similarly, with dispositions, the focus lies in enhancing students' creation of knowledge, not simply their recall of knowledge.

One goal of good thinking is to have students develop a critical approach to their work: their research, their enquiry, their critique and their collaborative work with others. For example, researching to acquire information is basic and is a skill that must be taught, but it is certainly more important to educate students to evaluate and then apply this information in an intelligent way to the problem at hand than to simply acquire the information.

The understanding of dispositions as essentially behavioural is significant to design and technology education. In this learning area in which practice is central, a student cannot be passively creative or passively critical; it is the action that expresses these characteristics that enables teacher judgments about progress to be made, and consistent thoughtfully applied actions indicate the development of a disposition. In addition, and sympathetically with design and technology education, judgments must be made about the appropriateness of certain dispositions in the given context. A student may be disposed to be critical, but may encounter a situation where it is not clear if this criticism is appropriate.

If the desirable dispositions of design and technology education are to be taught and are something different from personality traits, then they have a cognitive core. The student who has a disposition to be critical makes the judgment to be so after analysing the context and making a deduction of possible responses. It is not a feeling the student has to be critical, but a cognitive and analytical process resulting in the demonstration of the disposition. Therefore, fostering dispositions is about developing student understanding and insight.

Critical thinking as a dispositional outcome of technology education is acquired through institutional and interpersonal social contact. At an institutional level, school culture can support certain dispositions by encouraging democratic involvement by students in school governance. At a more personal level, dispositions in engineering, design and technology can be encouraged through guided learning, including cognitive apprenticeships, reciprocal teaching and expert scaffolding. The teacher can also utilize peer groups to develop good thinking dispositions by establishing an environment for rigorous thinking and thus create social demands for the sought after dispositions.

What can a teacher do to establish an environment that fosters the development of critical dispositions? Claxton and Carr (2004) discuss four aspects of the classroom environment that are relevant to the fostering of dispositions. The first is a *prohibiting* environment in which the pedagogy employed by the teacher makes it difficult for an individual to be uniquely responsive. For example, in the not uncommon situation in design and technology where all students are practicing skills after a teacher has demonstrated them and they are all working on the same project, there is little opportunity for students to respond in unique ways.

Conversely, an *inviting* environment is one in which student responses are encouraged and in which it is clear that individual responses are valued and not denigrated. Further, a *potentiating* environment not only invites the expression of dispositions but encourages students and 'stretches' them to test their responses. Learning is a shared activity between teachers and students, which may exist in the context of students identifying their own design problems and managing their own processes in solving these problems.

If a teacher uses a limited range of pedagogies, the *affordance* thereby provided for students to react in ways that enable them to advance their own learning is also limited and will only appeal to those students who have a complimentary learning style. In order to enable in all students the development of desirable dispositions, a broad range of pedagogies must be employed by teachers.

It is relevant to consider what might be the nature of progress in working towards dispositions in design and technology. It may be that robustness, that is, the strength of disposition, is one indicator of progress. In this case, the critique disposition is robust enough to be evident, even in the face of an unsupportive context or forceful pressure to respond in a certain way. For example, a technological issue for which the prevailing social attitude is obvious and demeans any alternative response, and so most people conform; but a certain disposition evokes an alternative response (refer to Foucault's 'audacity of exposure').

Breadth may also be a measure of progress whereby a student develops enough confidence in a disposition to apply it to a broad range of technological contexts. As a disposition is applied to a broadening range of contexts, and this application is rewarded and is complementary to the students' world view, they will be emboldened to cement the disposition as an appropriate response to an increasingly broad range of contexts. Transferability is the mechanism for achieving breadth by providing the opportunity for students to apply their dispositions in a range of contexts, to test them, to refine them and to strengthen them.

9 Generalizability of the Dispositions

The goal of education designed to help students become better thinkers is transferability to out-of-the-classroom situations. With this goal in mind, the ideal learning assessment would occur naturally in the course of one's life, in multiple settings, and would provide comparable measures before, during and long after the activities designed to develop critique.

The extent to which the aspects of critical thinking dispositions are generalizable is an issue. It is clear that transferability of knowledge and skills between contexts, even within the domain of technology, is not straightforward and cannot be assumed to take place without support. Glaser (1984) was one of the first to recognize the discipline-bound nature of knowledge and skills. This has developed into a range of more recent research on the situated nature of cognition (Hennessy 1993) and the consequent problematic notion of transferability (Georghiades 2000).

Those proposing that knowledge and skills are discipline bound (Glaser 1984), and more recent situated cognition literature, indicate the importance of the context in grounding learning and the difficulties individuals have in generalizing their learning (or in this case, applying their disposition) to different contexts. It may be the case that the generality of a disposition will, in application, build upon quite specific contextualized abilities.

10 Conclusion

If the aim is to develop students' ability to deal with design and technological issues at a personal and social level intelligently and confidently, then a school classroom culture can foster dispositions of critique. According to this conception, action to encourage the desired dispositions must address both components: inclination and ability, which requires teachers to provide students with opportunities to set goals and make plans for themselves in meaningful contexts.

This focus on critical thinking dispositions provides educators with the opportunity to extend the goals of education beyond the instrumentalism of technology education or the superficiality of design, which is evident in some education. With essentially behavioural outcomes, thinking dispositions are built on skills, competencies and the potential of technological literacy to ensure that critical thinking is applied in an appropriate and considered manner to opportunistic contexts. The careful structuring of classroom activities in sequences that are designed to elicit dispositions to critique is a fundamental design and technology teaching activity.

References

Cazden, C., Cope, B., Fairclough, N., & Gee, J. (1996). A pedagogy of multiliteracies: Designing social futures. *Harvard Educational Review, 66*(1), 60–92.

Claxton, G., & Carr, M. (2004). A framework for teaching learning: The dynamics of disposition. *Early Years, 24*(1), 87–97.

Costa, A. L., & Kallick, B. (2000). Assessing the habits of mind. In A. L. Costa & B. Kallick (Eds.), *Assessing and reporting on habits of mind* (pp. 29–53). Alexandria: Association for Supervision and Curriculum Development (ASCD).

Dakers, J. (2006). *Defining technological literacy*. New York: Palgrave Macmillan.

Department of Education Training and Employment, South Australia. (2001). South Australian curriculum standards and accountability framework. Retrieved September 17, 2007 from http://www.sacsa.sa.edu.au/index_fsrc.asp?t=LA

Dewey, J. (1922). *Human nature and conduct*. New York: Henry Holt & Company.

Diez, M. E. (2006). Assessing dispositions: Five principles to guide practice. In H. Sockett (Ed.), *Teacher dispositions: Building a teacher education framework of moral standards* (pp. 49–68). Washington, DC: American Association of Colleges for Teacher Education.

Fenstermacher, G., & Soltis, J. (1986). *Approaches to teaching*. New York: Teachers College Press.

Foucault, M. (1984, July 6–12). 'Dupouvoir' (Interview with P. Boncenne in 1978)'. *L'Express, 1722*, 56–68.

Foucault, M. (1997). What is enlightenment? In *Ethics, subjectivity and truth. Essential works of Michel Foucault* (pp. 303–319). New York: The New Press.

Foucault, M. (2000). Interview with Michel Foucault. In *Power: Essential Works of Michel Foucault* (Vol. 3, pp. 239–297). New York: The New Press.

Dannels, D., Gaffney, A., & Martin, K. (2008). Beyond content, deeper than delivery: What critique feedback reveals about communication expectations in design education. *International Journal for the Scholarship of Teaching and Learning, 2*(2). Article 12. doi: http://dx.doi.org/10.20429/ijsotl.2008.020212

Georghiades, P. (2000). Beyond conceptual change learning in science education: Focusing on transfer, durability and metacognition. *Educational Research, 42*(2), 119–139.

Glaser, R. (1984). Education and thinking: The role of knowledge. *American Psychologist, 39*, 93–104.

Halpern, D. (1998). Teaching critical thinking for transfer across the domains. *American Psychologist, 53*(4), 49–455.

Harpaz, Y. (2007). Approaches to teaching thinking: toward a conceptual mapping of the field. *Teachers College Record, 109*(8), 1845–1874.

Hartshorne, H., & May, M. (1928). *Studies in the nature of character: Vol. 1. Studies in deceit*. New York: Macmillan.

Hennessy, S. (1993). Situated cognition and cognitive apprenticeship: Implications for classroom learning. *Studies in Science Education, 22*(1), 1–41.

Hillman, S., Rothermel, D., & Scarano, G. (2006). The assessment of preservice teachers' dispositions. *Teacher Educator, 41*(4), 234–250.

Huitt, W. (1997, April 18). *The SCANS report revisited*. Paper presented at the fifth annual Gulf South Business and Vocational Education Conference, Valdosta State University, Valdosta, GA. Retrieved December 1997, from http://chiron.valdosta.edu/whuitt/student/scanspap.html

Insight Assessment. (2013). *Critical thinking dispositions*. http://www.insightassessment.com/FAQ/FAQs-Test-CT-Attributes-Dispositions/CT-Dispositions-Habits-of-Mind-Attributes#sthash.NPvhhkRy.dpuf

International Technology Education Association (ITEA). (2000). *Standards for technological literacy*. Virginia: Author.

Kahn, R., & Kellner, D. (2006). Reconstructing technoliteracy: A multiple literacy approach. In J. Dakers (Ed.), *Defining technological literacy* (pp. 253–274). New York: Palgrave Macmillan.

Katz, L. (1993). *Dispositions: Definitions and implications for early childhood practice*. ERIC #211. Retrieved from http://ceep.crc.uiuc.edu/eecearchive/books/disposit.html

Katz, L., & Raths, J. (1985). Dispositions as goals for teacher education. *Teaching and Teacher Education, 1*(4), 301–307.

Keirl, S. (2007). Critiquing in a democratic of design and technology education. In J. Dakers, W. Dow & M. J. de Vries (Eds.), *Proceedings the pupils attitude towards technology (PATT-18) conference* (pp. 306–312). Glasgow: Faculty of Education, University of Glasgow. Retrieved from http://www.iteea.org/Conference/PATT/PATT18/fullprog-21a[1].pdf

Kim, D., Moon, S., Kim, E., Kim, Y., & Lee, S. (2014). Nursing students' critical thinking disposition according to academic level and satisfaction with nursing. *Nurse Education Today, 34*, 78–82.

Kimbell, R., & Stables, K. (2008). *Researching design learning*. London: Springer.

Lemke, T. (2011). Critique and experience in Foucault. *Theory, Culture & Society, 28*(4), 26–48.

Lucas, B., Hanson, J., & Claxton, G. (2014). *Thinking like an engineer*. London: Royal Academy of Engineering.

Ministry of Education, New Zealand. (2006). Technology in the New Zealand curriculum. Retrieved September 17, 2007 from http://www.tki.org.nz/r/nzcurriculum/draft-curriculum/technology_e.php

Ministry of Education, South Africa. (2002). *Curriculum 2005*. Pretoria: Ministry of Education.

Misco, T. (2007). Did I forget about the dispositions? Preparing high school graduates for moral life. *The Clearing House, 80*(6), 267–270.

Narvaez, D., Bentley, J., Gleason, T., & Samuels, S. (1998). Moral theme comprehension in third grade, fifth grade and college students. *Reading Psychology, 19*(2), 217–241.

National Council for Accreditation of Teacher Education (NCATE). (2008). *Professional standards accreditation of teacher preparation institutions*. Washington, DC: NCATE Retrieved from http://www.ncate.org/documents/standards/NCATE%20Standards%202008.pdf.

Paul, R. W. (1995). *Critical thinking: How to prepare students for a rapidly changing world*. Cheltenham: Hawker Brownlow Education.

Pearson, G., & Garmine, E. (2006). *Tech tally*. Washington, DC: The National Academies Press.

Pedro, J., & Miller, R. (2005). Assessing pre-service teacher teachers' dispositions. Paper presented at the 57th annual meeting of the American Association of Colleges for Teacher Education, Washington, DC.

Perkins, D., Jay, E., & Tishman, S. (1993). Beyond abilities: A dispositional theory of thinking. *The Merrill-Palmer Quarterly, 39*(1), 1–21.

Rutland, M., & Spendlove, D. (2007). Creativity in design and technology. In D. Barlex (Ed.), *Design and technology – For the next generation* (pp. 140–153). Shropshire: Cliffeco Publishing.

Schussler, D. L. (2006). Defining dispositions: Wading through murky waters. *The Teacher Educator, 41*(4), 251–268.

Sears, A., & Parsons, J. (1991). Toward critical thinking as an ethic. *Theory and Research in Social Education, 19*, 45–46.

Sockett, H. (2009). Dispositions as virtues: The complexity of the construct. *Journal of Teacher Education, 60*, 291–303.

Stooksberry, L., Schussler, D., & Bercaw, L. (2009). Conceptualizing dispositions: Intellectual, cultural, and moral domains of teaching. *Teachers and Teaching: Theory and Practice, 15*(6), 719–736.

Thornton, H. (2006). Dispositions in action: do dispositions make a difference in practice? *Teacher Education Quarterly, 33*(2), 53–68.

Waks, L. (2006). Rethinking technological literacy. In J. Dakers (Ed.), *Defining technological literacy* (pp. 275–296). New York: Palgrave Macmillan.

Wilkerson, J., & Lang, W. (2005). *Measuring dispositions with practicality, utility, validity, and reliability in mind.* Paper presented at the 57th annual meeting of the American Association of Colleges for Teacher Education, Washington, DC.

Will, G. (2006, January 16). Ed schools vs. education. *Newsweek* (U.S. ed.), 147, 98.

Williams, P. J. (2009). Technological literacy: A multliteracies approach for democracy. *International Journal of Technology and Design Education, 19*(3), 237–254.

Empathy as an Aspect of Critical Thought and Action in Design and Technology

Bill Nicholl

Abstract User-centred approaches to design stress the importance of the designer understanding the needs and experiences of the user when designing products (Sanders E, Dandavate U, Designing for experiencing: new tools. In Overbeeke CJ, Hekkert P, (eds) Proceedings of the first international conference on design and emotion, 3–5 November 1999, Delft University of Technology, Delft, pp 87–92, 1999). How designers and others involved in designing have understood these needs has evolved since Taylor's seminal work in the early 1900s. One emerging and influential user-centred approach to design over the last decade has been inclusive design. Researchers working in this field have developed ways of working or 'signature pedagogies' that allow them to think critically and empathise with users, to understand their needs from their perspective and to use this understanding to critically inform their own actions when designing, as well as educating others in the practices of inclusive design. I will discuss these signature pedagogies, arguing that they are crucial for developing critical thinking dispositions and engendering empathy when designing and educating others. I will then discuss how the signature pedagogies of inclusive design were successfully introduced into high schools in a number of countries.

Keywords Empathy • Critical thinking • User-centred design • Signature pedagogies • Designing Our Tomorrow (DOT)

1 Introduction

Formal approaches to meeting the needs of users as part of the processes of designing have been around for over a hundred years with Taylor's methodological approaches to understanding how people worked to improve efficiency (see Baumgart and Neuhausre 2009) and Henry Dreyfuss' pioneering work (Dreyfuss 1955) on anthropometrics in the design of household products being two early

B. Nicholl (✉)
University of Cambridge, Cambridge, UK
e-mail: ban22@cam.ac.uk

PJ. Williams, K. Stables (eds.), *Critique in Design and Technology Education*, Contemporary Issues in Technology Education, DOI 10.1007/978-981-10-3106-9_9

examples. User-centred approaches have stressed the importance of the designer understanding the needs and experiences of the user when designing products (Sanders and Dandavate 1999). How designers and others involved in designing have understood these needs has evolved since Taylor's seminal work. One emerging and influential user-centred approach to design over the last decade has been inclusive design. Researchers working in this field have developed ways of working or 'signature pedagogies' that allow them to think critically and empathise with users, to understand their needs from their perspective and to use this understanding to critically inform their own actions when designing, as well as educating others in the practices of inclusive design. I will discuss these signature pedagogies, arguing that they are crucial for developing critical thinking dispositions and engendering empathy when designing and educating others. I will then discuss how the signature pedagogies of inclusive design were successfully introduced into high schools in a number of countries. First however, I will discuss the relationship between empathy, critical thinking and design.

2 Conceptualising Empathy

Empathy is a relatively recent construct that has its roots in philosophy and psychology. The origins of empathy as a construct can be traced back to 1873 when Visher used the term 'Einfühlung' which is German for 'feeling into' (Hickman 2013). Later, Titchener adopted the same word, Einfühlung, but used it to align with notions from aesthetics and defined its meaning as being 'to project yourself into what you observe' (Baron-Cohen and Wheelwright 2004, citing Tichener 1909, p. xx). More recently, 'empathy has been termed an ability, an attitude, a feeling, an interpersonal process, a trait, a state, a sensitivity, and a perceptiveness' (Sutherland 1993, p. 309). Kunyk and Olsen analysed the concept of empathy and found five different uses of the term, namely, empathy as human trait, empathy as a professional state, empathy as a communication process, empathy as caring and empathy as a special relationship (Kunyk and Olson 2001). This, they conclude, suggests that empathy as a construct has not yet fully matured.

The philosopher Maxine Greene suggests that empathy is 'the capacity to see through other's eyes, to grasp the world as it looks and sounds and feels from the vantage point of another' (Green 2001, p. 102). Conceptualisations of empathy from the philosophical literature are congruent with the psychological literature. Although acknowledging empathy as being multidirectional, there seems to be a consensus among psychologists that empathy has two broad strands, namely, emotion and cognition (Lawrence et al. 2004). The emotional strand of empathy refers to 'an emotional response to . . . emotional responses of others' (Lawrence et al. 2004, p. 911). Emotional empathy has also been labelled 'affective' empathy where the emphasis is on the 'appropriateness of the viewer's emotional response' (Baron-Cohen and Wheelwright 2004, p. 164). Affective empathetic responses can be further classified as 'parallel', for example, 'feeling fear at another's fright' or reactive responses, which go beyond mirroring the observed state and include a feeling of sympathy or compassion (Lawrence et al. 2004, p. 911 citing Davis 1994). In this view, sympathy is seen as a component of affective empathy (Baron-Cohen and Wheelwright 2004).

The cognitive strand of empathy refers to 'the intellectual/imaginative apprehension of another's mental state' (Lawrence et al. 2004, p. 911) and emphasises the observer's 'understanding and/or predicting what someone else might think, feel, or do' (Baron-Cohen and Wheelwright 2004, p. 165). The emphasis here is on 'taking the role or perspective of another person' (Baron-Cohen and Wheelwright 2004) and is based on the cognitive processes of role-taking and perspective taking (see, e.g. Mead 1934; Piaget 1932). Although discussed separately, some writers suggest that both affective and cognitive components are strongly interrelated (Kouprie and Visser 2009, citing Damasio 1994). Rogers conceptualises empathy as a process where one enters the world of another where one must 'be sensitive, moment to moment, to the changing felt meanings which flow in this other person ... communicating your sensing of his/her world as you look with fresh and unfrightened eyes ... checking with him/her as to the accuracy of your sensings, and being guided by the responses you receive ... you help the person ... move forward in the experiencing' (Rogers 1975, p. 4). Both affective and cognitive aspects of empathy, as well as empathy as a process, are important in design, and this will be discussed in due course. I would now like to turn to discuss empathy and critical thinking.

3 Empathy and Critical Thinking

In this section I draw on, and extend, the discussion presented by Williams in chapter "Critique as a Disposition". In particular, I would like to extend his discussion to explore the dispositional dimension to critical thinking in relation to the current discussion on empathy. Ennis defines critical thinking as 'reasonable reflective thinking focused on deciding what to believe or do. The emphasis is on reasonableness, reflection, and the process of making decisions' (Ennis 1996, p. 166). Halpern too stresses critical thinking as a process which requires one to be reflective, show sensitivity to the particular context one is working in, and be able to monitor one's progress throughout the process, involving 'judgement, analysis, and synthesis' necessary for solving ill-defined problems (Halpern 1998, p. 451). Monitoring and reflecting on one's own actions throughout this process is known as metacognition (Flavell 1987) and is characteristic of the dispositional dimension to critical thinking cited in the literature (e.g. Perkins et al. 1993). Each of these definitions describes critical thinking as a process which is summarised by Scriven and Paul:

> intellectually disciplined *process* of actively and skilfully realising, conceptualizing, applying, analyzing, synthesizing, and/or evaluating information gathered from, or generated by, observation, experience, reflection, reasoning, or communication, as a guide to belief and action. (Scriven and Paul 2003, emphasis in italics added)

Furthermore, each of these authors also stresses critical thinking as being a disposition. Katz (1993) defines dispositions 'as patterns of behaviour that are exhibited frequently and intentionally in the absence of coercion, thus representing habits of mind' (Williams citing Katz, chapter "Critique as a Disposition"). Williams suggests that dispositions have two components, namely, ability and inclination. An inclination is a person's tendency towards a certain behaviour, and ability refers to the capability to engage with the disposition. I shall return to critical thinking as a process later; for now however I would like to make the link between critical thinking and empathy. One of the thinking dispositions characterised by Costa and Kallick (2000) makes explicit reference to empathy, namely, 'listening to others with understanding and empathy' (Costa and Kallick 2000, p. 4). Empathy is clearly evident in the work of Ennis (1996) who has identified three broad critical thinking dispositions, one of which is 'to care about the dignity and worth of every person' which includes the dispositions to 'discover and listen to others' views and reasons; take into account others' feelings and level of understanding; be concerned about others' (Ennis 1996, 171–172). Finally, the disposition to be 'open-minded', which is commonly cited in the critical thinking literature (e.g. Perkins et al. 1993; Halpern 1998), would suggest that it is important to listen to, and understand what someone else might be thinking, how they might be feeling or what they might do in certain situations. This is particularly important when the 'observer' is different to the person being observed, in terms of their age or gender, religious or political beliefs or social background. Empathy, or being empathic, is therefore an essential part of what Williams refers to as the critical spirit, which he describes as 'a set of attitudes that define a personal disposition to prize and to use critical thinking in one's personal, professional and civic affairs' (Williams chapter "Critique as a Disposition"). The set of attitudes that contribute to an overall disposition to use critical thinking to form judgements are truthseeking, open-mindedness, analyticity, systematicity, confidence in reasoning, inquisitiveness and maturity of judgement (Williams chapter "Critique as a Disposition"). Conceptualised in this way, empathy is embodied within an overall disposition to think critically. In other words, being empathic is essential to critical thinking, and this will be discussed further in relation to design.

4 Empathy, Critical Thinking and User-Centred Design

At about the same time, as psychologists and philosophers were debating and honing their conceptualisations of empathy, researchers working in the field of user-centred design began to critique their own practices, and this revealed a number of interesting findings. For example, some researchers realised the tools traditionally used to investigate users, such as questionnaires, were inadequate in 'creating a rich, empathetic understanding of the users' desired experiences' (Battarbee and Koskinen 2005, p. 6). Other research found that designers 'designed for themselves', and this is due, in part, to limitations of time, budget and logistical requirements faced by designers when solving problems (Cardoso and Clarkson 2012, p. 1; Coleman et al. 2003). Whether designers designed for themselves or used tools that were inadequate for engendering empathy, the result was often products that were difficult, frustrating or even dangerous to use (Porter and Porter 1999). Findings such as these raised serious questions about whether designers had the *ability* or the *inclination*, that is, had the disposition to think critically and, in particular, to emphasise with people outside their own empathic horizons when designing (McGinley and Dong 2011; McDonagh-Philp and Denton 1999). The interest in the role of empathy within user-centred design led to the term 'empathic design' in the late 1990s (Koskinen et al. 2003). This is where 'designers attempt to get closer to the lives and experiences of (putative, potential or future) users, in order to increase the likelihood that the product or service designed meets user's needs' (Kouprie and Visser 2009, pp. 437–438). A number of research projects investigated the designer's engagement in critical thinking and in becoming 'more sensitive to users, be able to understand them, their situation, and feelings: to be more empathic' (Kouprie and Visser 2009, p. 438). In order to get closer to the lives and experiences of users, understanding their situation as well as their feelings requires an empathic approach to design which is part of an overall disposition to think critically. Designers working in the emerging field of inclusive design have developed such an approach, which addresses the issues of a global ageing population. Empathy as an aspect of critical thought and action within the field of inclusive design is illustrated in work they have published recently. Visual ability is crucial when using products, for example, reading signs in public places and recognising icons. They found that the data readily available on visual abilities focused on a narrow set of measures. If these data were used to guide the actions of designers, they could potentially exclude a large proportion of the population. They conducted a survey examining a wider range of human capabilities and characteristics, including ones on vision. One of the recommendations resulting from this study was that text size needs to be 17–18 % larger for 'comfortable viewing' and meet the needs of a wider range of users (Goodman-Deane et al. 2016, p. 150). This illustrates how empathy as an aspect of critical thinking can guide designer's actions, and this will be discussed further.

The importance of user-centred approaches has grown in line with dramatic demographic changes. For example, it is estimated that 2 billion people will be over 60 in 2050, compared with only 200 million in 1950. The implications of an ageing population range from threatening the solvency of social security systems (pensions and public health) around the world (United Nations 2009) to products that are difficult, frustrating or dangerous to use. Research has shown that the people most likely to be affected by products that are difficult or frustrating to use are those who suffer from some form of capability loss, especially those associated with, although not exclusive to, the effects of ageing, such as depreciating vision or limited dexterity in one's hands due to conditions such as arthritis (Keates and Clarkson 2003). User-centred approaches to design, and in particular inclusive design, have a key role to play in helping us to understand and address the problems faced by people with capability loss who can be excluded from using products and services safely and with dignity. At the heart of inclusive design is the need for the designer to think critically and be disposed to see things from the perspective of the user, to understand what someone else might be thinking, how they might be feeling or what they might do in a certain situation. Empathy or being empathic, as part of an overall disposition to think critically, is central to that understanding (Kouprie and Visser 2009).

Researchers and practising designers working in the field of inclusive design have developed a way of working that embodies this critical spirit, which comprises four interrelated 'principles': explore, create, evaluate and manage. This is called the 'design wheel' or process they use which is represented graphically below (see Fig. 1). The inclusive design wheel and principles serve two purposes, one of which I would like to discuss now and the other later. Firstly, inclusive designers use the wheel and principles to guide their critical thought and actions when they are solving design problems commercially, that is, to say, in the real world. There are many similar representations of design described in the design literature, and Cross suggests they comprise three core elements, namely, 'analysis, synthesis and evaluation' (Cross 2011, p. 27). Furthermore, these conceptualisations of design emphasise the iterative nature of designing, which requires the designer to monitor and reflect on his/her thinking when attempting to solve complex, ill-defined problems (Lawson 2001; Schon 1983). These conceptualisations describing design as an iterative process use a similar language to the psychologists and philosophers working in the field of critical thinking discussed previously. The critical thinking dispositions and how they relate to empathy, using the work of Kouprie and Visser (2009) and the inclusive design principles, are summarised in Fig. 1.

I would now like to discuss in a little more detail some of the empathic tools that have been developed to guide designers' thoughts and actions. Hosking et al. (2015) suggest there are two broad types of empathy tools: direct and indirect. Direct contact is where the designer explores by engaging first-hand with potential users via techniques such as observing users in their own context and user focus groups, where designers talk with end users early in the design process. Focus groups can also be used to generate ideas (create) or to feedback (evaluate) on ideas and prototypes that have been developed as part of an empathic, critical and iterative approach to design (Kouprie and Visser 2009; Dong et al. 2009). Indirect contact techniques are used when direct contact is not possible and serve the same purpose. Indirect techniques include simulation or role-playing techniques including 'product handling', 'experience prototyping', 'bodystorming' and 'informance' (Buchanau and Fulton-Suri 2000). Of particular interest here are simulation tools such as

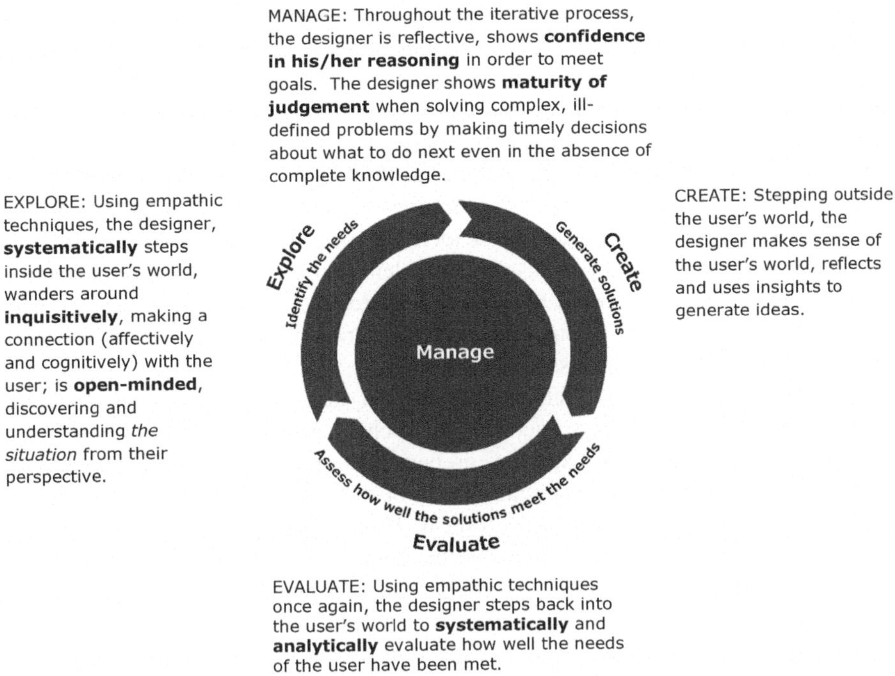

Fig. 1 Elaboration of inclusive design wheel showing relationship with critical thinking dispositions (Design wheel © University of Cambridge)

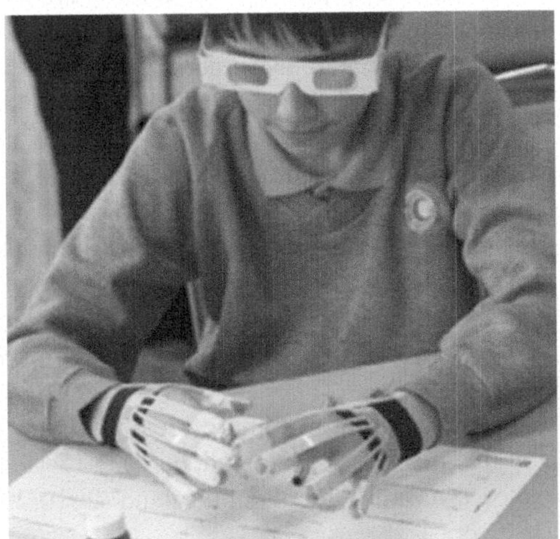

Fig. 2 Student (11 years old) using simulation gloves and glasses to role-play an older person with capability loss

glasses that simulate how one's vision depreciates with age and gloves which simulate the effects of arthritis (see Fig. 2). Simulation tools enable the designer to experience some of the effects of capability loss as they allow the designer to 'step into parts of the user's experience by simulating the user's condition' (Kouprie and

Visser 2009, p. 440). This perspective taking, via role-play, allows the designer to 'identify and evaluate the nature (the kind of capability) and magnitude (the level of capability) of the capability demands imposed upon the user … it is necessary to be able to assess the features of the product to identify those that present difficulties to the user and, ideally why they present difficulty' (Keates and Clarkson 2003, p. 109). Consequently, simulation tools foster critical thinking when solving problems in a particular situation. This critical exploration phase can then be used to inform idea generation (create phase), and these ideas can be evaluated, critically, using the simulation tools, to see if the needs of users identified as part of the explore phase have been met. Managing what to do next requires the designer to reflect on and monitor his/her 'thinking' throughout multiple iterations when solving ill-defined problems (metacognition discussed previously).

Another indirect technique is task analysis. In a scenario-based task analysis, the designer 'imagines' they are someone else, such as someone suffering capability loss, undertaking a task, for example, preparing the table for a family meal. The designer would identify, systematically, each of the steps it takes to gather cutlery and layout the table in preparation for a meal. This systematic approach would then be used analytically, to identify and assess the demands each step places on the user, for example, picking up a knife from the table can be demanding for users who suffer from arthritis as it requires a pinch grip. Task analysis then requires the designer to be systematic, analytical, open-minded and truthseeking, all of which are attitudes that capture the critical spirit. Thinking critically in this way opens up opportunities to be creative as the designer can generate ideas (create) by focusing on reducing the demands for this task step, for example, generating ideas for a knife so that it is easier to grip/pick up.These ideas are systematically evaluated, and providing the demands in other steps has not increased; the new concept for a knife is more inclusively designed. Designers/researchers working in the field are referring to the process of empathy embodied throughout the process of designing (Kouprie and Visser 2009). The inclusive designer steps into the user's life in order to connect with them, both cognitively and affectively, in order to get a critical and 'deep understanding' of their life, and steps out of the user's life, in order to take on the 'role of the designer and makes sense of the user's world … to reflect [and] deploy the new insights for ideation' which are then evaluated (Kouprie and Visser 2009, pp. 444–445). Consequently, empathy at the heart of a critical thinking process can lead to new insights which can lead to more creative solutions (McDonagh and Thomas 2011).

The disposition to think critically, with a particular emphasis on empathy, and how this might motivate students (12–15 years) and lead to creative solutions was explored in high schools in England (2010/2011), Ireland (2012) and India (2015). This work is discussed next.

5 Empathy as Critical Thought and Action in High School Design and Technology

In discussing empathy as part of critical thinking in high schools brings me back to the purpose of the design principles mentioned earlier. The second purpose of the process and associated principles is to educate graduate and experienced designers into the practices of inclusive design. Budding inclusive designers are introduced to the simulation tools and use them by interacting with product outcomes they have recently designed. This early immersion in the use of simulation tools gets the designers to reflect, critically, on these outcomes, and how large parts of the population have been excluded. The principles and tools are then introduced, via the design wheel, and the designers reflect, critically, on the design processes they went through when designing their product. This critical reflection is then used to guide the actions of budding inclusive designers. In order to do this, a problem they are currently working on is used, with the hope that they go through a critical, empathic, iterative process that can lead to producing more inclusively designed products. In this way, the wheel and associated empathy tools and techniques became the 'signature pedagogies', and this is important as they can shape how professionals behave as Shulman states:

> Signature pedagogies make a difference. They form habits of the mind, habits of the heart and habits of the hand. As Eriksson observed in the context of nurseries, signature pedagogies prefigure the culture of professional work and provide the early socialisation into the practices and values of the field. Whether in a lecture hall or lab, in a design studio or clinical setting, the way we teach will shape how professionals behave .... (Shulman 2005, p. 59)

This approach is aligned with sociocultural theories of learning that places an emphasis on 'contexts and social practices - and sees these as important "cultural resources" that are available to the learner from that setting' (Pollard 2002, p. 148). The interdependence between social and individual processes in the co-construction of knowledge can be traced back to the work of Vygotsky and has given rise to a number of interptretations including the situated cognition approach by Lave and Wenger (1991) and cognitive apprenticeships by Brown et al. (1989). Furthermore, contextualising learning experiences, and by implication teaching practices (principles and tools discussed here), within meaningful, real-world problems and practices, that is, the work of designers, has a long tradition and is consistent with the philosophical approach of Dewey who advocated 'meaningful school activity that extends experiences and practices of the adult world' (Dewey 1938, p. 3).

Using these signature pedagogies and real-life design problems as part of socialising designers into the practices of inclusive design is something we have investigated with high school students. Students had to find a design problem, based around a context, namely, 'dining', and were introduced to the simulation tools and guided through the design wheel and principles over a period of 12 1-h lessons. We were also mindful, however, that educating practicing or graduate designers is different to educating novice designers in high schools. For this reason, the signature

Fig. 3 Task analysis sort exercise

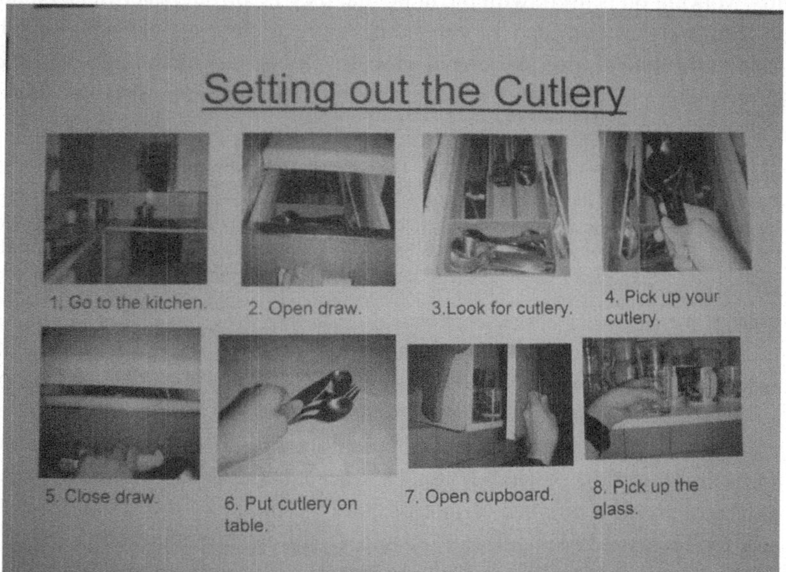

Fig. 4 Example of a student's task step analysis showing task steps for preparing a meal (This page shows 8 steps of the 24 task steps identified (female 12 years))

pedagogies described above were complimented with other tools and strategies to guide and scaffold students' thinking and actions in solving the problem we gave them. This included showing them how to analyse problems, modelling examples of good practice (Wood et al. 1976). For example, in order to teach task analysis, we modelled how to do this via a card sort exercise. In groups of four, students were asked to sort the steps for a task not related to dining (see Fig. 3). Students could then do a task analysis for the problem they were working on, dining (see an example in Fig. 4). This is an example of task-related scaffolding, which is crucial for maintaining the ambiguity of ill-defined problems like design problems, without reducing the task to a series of 'closed' steps or procedures to follow (Galton 2007). Other activities were varied, and many involved students working in groups, taking a dialogic approach to teaching and learning (Alexander 2008). Consequently, the signature pedagogies of inclusive design discussed earlier were important in establishing an environmental context where a culture of critical thinking could be fostered. The culture of critical thinking was further enhanced by strategies that helped students develop strategies of analysis. Further strategies will be discussed as part of the findings.

In terms of framing the research, in each country, we interviewed and surveyed students about their recent experiences in design and technology (D&T) prior to introducing the Designing Our Tomorrow (DOT) activity. We interviewed and surveyed the students again at the end of the project. Some of this work has been published (Nicholl et al. 2013, 2014). What follows are extracts from the student interview data, thematically analysed around three broad themes: students' reflections about the empathic activity, their motivations associated with this and their perceptions on how the empathic approach relates to their own creativity, in particular their generation of ideas. This was supported with survey data and field notes including examples of students' work and lesson observations. The findings, with some discussion, are presented next.

Given that students did not have any experiences of designing for other people prior to DOT (see Nicholl et al. 2013), students were able to appreciate the importance of seeing things from other people's perspective when designing, someone who was different to them as this excerpt reveals:

> If you want to make a product for them you've got to know how they use things and how they think, because they're different to us, we're not elderly people. So we have different things that we like and stuff, and how we react. (Female/13 years/England)

Furthermore, perspective taking and physically experiencing the simulation tools were commented on by students as being important in fostering and feeling empathy as these excerpts reveal:

> I think it's helpful to design for other people, because you kind of feel what their everyday life is like. With old people, like we had to use a finger restrictor, how … how they kind of like move their fingers is tough for them. (Male/13 years/England)

> If you look at it from a person of visual impairment and arthritis it really is difficult 'cos you actually get to simulate how tricky like simple things are like even just making cup of tea which I am sure pretty much all of you like and want to do at some point during the day and even the simple things like that … or putting salt onto your dinner it's just interesting to see and put yourself in that situation rather than putting it from a perspective … you actually get to do and experience. (Female/14 years/England)

> But the thing is, you've actually learnt what it's like to be visually impaired and how it would be like to not be able to pick something up and how frustrating it would be. (Male/12 years/England)

The importance of the simulators in facilitating the development of empathy, both affective (feeling) and cognitive understanding of users' needs, was understood by students. for example, 96 % of students strongly agreed/agreed, with the item 'Using special finger restrictors/glasses to change how easy it is to move my hands and to see really helped me understand people' s (e.g. elderly people and young children and people with poor eyesight) differing abilities'. The level of empathy exhibited by many students was exceptional. The following excerpt is used to illustrate this point. This 14-year-old girl was asked if the simulation tools helped her understand the problems faced by older people:

> yeah massively ... 'cos like erm we had ... we had little medicine pots and we thought you know ... you put your glasses on you know ... but you really, really can't and like tube maps ... I could no way could I find ... and especially if you have never seen a tube map before ..erm which you didn't quite understand before you would say "oh yeah, you might not be able to see it very well" but you didn't understand the depth of it ..and especially with arthritis ... it is so much harder than you think just little things like when you think shaking a salt pot ... that's simple, but it's not, there's things like taking it out and putting new salt in it, which you didn't realise was so hard, without using the gloves you wouldn't understand. (Female/14 years/England)

In the excerpt above, a 14-year-old girl clearly understands that having no previous experience of using a product, in this case an underground map, would make that product more visually demanding to use than if one was familiar with the map prior to one's vision declining with age. This illustrates a high level of understanding of capability loss in relation to prior knowledge, as this can make a product easier to use. Furthermore, in terms of dexterity, this girl also understood that the most demanding step in using a salt shaker was not necessarily picking the salt shaker up and dosing one's food, but in actually refilling the shaker when it becomes empty, which is particularly demanding in dexterous terms as it requires a 'pinch' grip in order to remove the cap. This can be very difficult for a person who suffers from arthritis. This shows a very sophisticated level of critical insight with respect to systematically 'imagining' the steps and locating where the peak demands might be for a person with capability loss. In turn, this leads to the design problem being reformulated which maximises opportunities to generate ideas that are judged creative, as novel starting points can lead to more innovative solutions (Getzels and Csikszentmihalyi 1976). In another example, a student identified 24 task steps a person has to go through in order to layout the cutlery on a table for a family dinner. This sequence of 24 steps was systematically captured on her smart phone (see Fig. 4). The examples discussed here illustrate that perspective taking, if done systematically and analytically, can lead to critical insights. Students told us that experiencing empathic design tools and techniques challenged students' assumptions about the nature of everyday products such as using cutlery as these excerpts reveal:

> if the first thing you asked me was how many people in the world can use cutlery? I'd say everybody could, but then kind of like if you look at that it kind of puts it into proportion, and shows you. (Male/15 years/Ireland)

> then you kind of get into it, so you know a little bit about it [an aging person]. And you kind of say, oh, I didn't think that many people had this problem or whatever. And then you kind of have that set in your head for the rest of it, for the rest of the project. (Male/15 years/Ireland)

Challenging one's own assumptions was a strategy we introduced as part of the cultural activity and is another important constituent of critical thinking as one must be open-minded and flexible. Furthermore, challenging one's own assumptions is crucial for creativity (Csikszentmihalyi 1999). Having experienced what it is like being an ageing person, the student is able to understand their feelings and understand the implications this has when they interact with the made world; students were required to generate ideas using creative thinking strategies informed by the literature (Ward et al. 1997; Cross 1997; Nicholl et al. 2008). We asked students whether the empathy tools and creative strategies they experienced helped them with generating creative ideas. Students told us that the empathy tools helped inform their ideas as this excerpt reveals:

> Our designs were a lot better because of it ... because we wouldn't ... have maybe thought the things we did if we hadn't understood how they quite felt and how simple it was like even if was like a touch or a grip thing ... we were saying oh it's light we can do it...but it's hard but we made our designs so much better from using it [empathy tools]. (Female/14 years/England)

> We were very creative. I never knew that it was inside me. ... and I want it to be like that every day. (Female/15 years/India)

This was supported by items in the questionnaire. For example, 96 % of students strongly agreed/agreed with the item 'My understanding of people's (e.g. elderly people and young children and people with poor eyesight) various abilities helped me come up with my design ideas'.

We wanted to get indications of how students embraced the DOT activity, which they found very engaging. A number of students stressed how much they valued being given the opportunity of tackling a real-life problem such as designing for an ageing population and how much they enjoyed using the empathic tools as these excerpts reveal:

> Well, I liked [the project] because we get to experience, like with the finger restrictors, experience what other people have with hand disabilities and stuff ... To how to like open a bottle or something like that. (Male/15 years/Ireland)

> We are not just making something for the sake of making it. Like it is going to apply to somebody in life ... It is not just something that the teacher wants us to do, like it is actually going to apply to somebody. And if it works, maybe we can take it a step further, and try and make someone's life easier. (Male/15 years/Ireland)

> Well, to design something that you have no experience with before, like everyday items that you deal with, maybe talking to older people, young parents or something to see what they use, so you might see before and get the chance to design it ... instead of just taking everything as your own, you know, consider other people in the design. (Male/15 years/Ireland).

Survey items on engagement supported the interview data, for example, 86 % of students in Ireland agreed that they liked ‘having a real-life problem to solve’, 89 % said they felt happiest when working on a project they felt ownership of and 100 % found the resources interesting and helpful. Making activities meaningful and relevant is crucial not only to motivation (Anderman and Maehr 1994) but also to developing critical thinking skills (Halpern 1998) and learning (McCormick 2004). Finally, we asked students about the empathic approach and whether this had any impact beyond the D&T classroom experiences. The following excerpt reveals how the DOT experience extended beyond the classroom:

> Thinking about other people that are disabled, and cutlery. And then you think that when you’re having your tea and then you’re cutting it up, and you’re thinking that people can’t actually do that. I’m quite lucky actually. (Female/12 years/England)

> Erm Well like. I went to my grandma’s the other day and she was … she’s got arthritis and she wears glasses… her vision isn’t that bad but she is very arthritic and before I used to say… “Ahh bless her” where know I … I… that is actually really hard like … you just didn’t quite understand that before but now, now seeing it you … you really do feel for them and it did change it a lot. (Female/14 years/England)

It was both heartening and encouraging to see and hear students talking about the experience of DOT in contexts other than the classroom. Perhaps one of the most profound insights from all of our work to date came from an Indian student whose excerpt below summarises just how powerful and liberating teaching for dispositions within a D&T context can be:

> … before this workshop we had chapters in our textbooks, and we would have to write essays about ageing, and looking after old people. Looking at like ... trying to think of what an old person feels like, what it is like to age, and to feel helpless. But I think that was just a little bit sympathy, and maybe pity. With what happened yesterday was empathising with them. Feeling the way they feel, and that ... that’s not the same as looking at them from a different point of view, and looking at their problem. Feeling their problem is different from looking at their problem. And what we did yesterday really had a powerful effect. (Female/15 years/India)

6 Closing Thoughts

> The designer sets off to explore. To discover something new, rather than to reach somewhere already known, or to return with yet another example of the already familiar. (Cross 2011, p. 8)

Given the quote above by Cross about the role of the designer as well as the discussions on user-centred design outlined in this chapter, design and technology would seem to be well placed to foster critical thinking. As Paul asserts, ‘Critical thinkers critique in order to redesign, remodel and make better’, and this captures precisely what inclusive designers do (Paul 1995, p. 526). In their respective chapters, however, both Stables and Williams raise concerns about learners of D&T being set tasks that are formulaic, leading to learners’ outcomes being identical or with design being superficially addressed. Indeed, these features were typical of what we saw in our research prior to introducing inclusive design, where the focus was on the teacher teaching technical knowledge directly to students via activities that were procedural leading to practical outcomes that were identical (see Nicholl et al. 2013). This has major implications for the teaching of critical thinking dispositions as the 'image of the teaching and learning' is one where the teacher 'transmits knowledge and skills' and where the child listens to the adult (Pollard 2002, p. 152). Thinking dispositions cannot be taught directly in this way, but must be cultivated, indirectly:

> Dispositions are cultivated indirectly, not by transmission of knowledge but by a comprehensive culture of thinking that foster various ways of thinking dispositions. (Harpaz 2007, p. 1852, citing Passmore, 1967)

This is where the signature pedagogies of inclusive design are key. D&T student activities should be based on authentic and messy problems typically faced by designers, for example, the problems associated with capability loss and how this affects older people when eating. At the heart of the cultural activity is the iterative design process of explore, create, evaluate and manage (Hosking et al. 2010). This process becomes the 'organising pedagogical principle' that embodies the critical spirit, which is crucial for solving design problems (Lucas et al. 2014, p. 14). Direct and indirect empathy tools and, in particular, the use of role-play and perspective taking (Mead 1934) can be used iteratively and make up some of the 'signature pedagogies' that help 'form habits of the mind' as they 'provide the early socialisation into the practices and values of the field' (Shulman 2005, p. 59). McCormick citing the work of Schoenfeld in mathematics education agrees when he states that learning in D&T 'is not a matter of mastering a body of knowledge' but 'to understand the nature of these areas [e.g. design and technology] they [pupils] need to experience what it is like to engage in mathematical (or any other subject) activity' (McCormick 2004, p. 23).

This culture of thinking is more aligned with sociocultural theories of learning that place an emphasis on 'contexts and social practices-and sees these as important "cultural resources" that are available to the learner from that setting' (Pollard 2002, p. 148). Here the 'image of the child is active' and socially interacts with teachers and peers via 'challenges [that] can clarify thinking and extend meaningful understanding' (Pollard 2002, p. 152). The important role of the teacher in establishing a classroom environment in which 'a culture of critical thinking is fostered, expected and established' is stressed (Williams chapter "Critique as a Disposition"). The teacher is the ultimate cultural resource, mediating students through this ambiguous, ill-defined activity. Teaching for thinking dispositions, therefore, requires the teacher to 'embody in... personality and behaviour the disposition toward which he wishes to educate' (Harpaz 2007, p. 1852). This means a fundamental shift in current teaching practices, where the focus on the teaching of technical knowledge to the teaching of thinking dispositions in a way that can 'influence the values, dispositions, and characters of those who learn' (Shulman 2005, pp. 57–58).

The data presented here tells an optimistic story. Given appropriate learning experience(s) or cultivating activities, students can think critically, quickly develop feelings for and understand people beyond the characteristics of their own age group and broaden their own 'empathic horizon' (McDonagh-Philp and Denton 1999). Furthermore, they are willing and able to use this empathic understanding critically, exploring users' needs, identifying the demands products place on the user and generating solutions to meet the needs of an ageing population. In turn, this provides opportunities for students to think critically about their ideas, whether their ideas meet the needs of an ageing population. I am not suggesting that the students who participated in these studies will naturally choose this critical and empathic approach when designing problems. They have only had one experience of this type of cultivating activity. Williams reminds us, however, if students are consistently exposed to these types of learning experiences, then dispositions can be cultivated and developed into habits of mind. The findings discussed in this chapter should offer some encouragement to educators and teachers of D&T who value and want to develop empathy as part of critical thinking. Through them, their students have so much to gain, as Gallo states:

> Empathic role taking fosters imagination by providing opportunities for immersive, holistic, spontaneous, and novel responses to problems that are engaging and complex. In so doing, it exercises and nurtures intrinsic motivation for tasks requiring imagination, a tolerance for complexity and ambiguity, as well as self-esteem and courage. (Gallo 1982, p. 114).

Acknowledgement I would like to acknowledge and thank my work colleagues, Ian Hosking, Julia Flutter and Katie Klavenes for their valuable contributions they have made in the wider work that has informed this chapter.

Bibliography

Alexander, R. (2008). *Towards dialogic teaching: Rethinking classroom talk* (4th ed.). Cambridge, MA: Dialogos.

Anderman, E., & Maehr, L. (1994). Motivation and schooling in the middle grades. *Review of Educational Research, 64*(2), 287–309.

Baron-Cohen, S., & Wheelwright, S. (2004). The empathy quotient: An investigation of adults with asperger syndrome or high functioning autism, and normal sex differences. *Journal of Autism and Development Disorders, 34*(2), 163–175.

Battarbee, K., & Koskinen, I. (2005). Co-experience: User experience as interaction. *CoDesign, 1*, 5–15.

Baumgart, A., & Neuhausre, A. (2009). Scientific management in the operating room. *Quality Safety Health Care*. doi:10.1136/qshc.2009.032409.

Brown, J., Collins, A., & Duguid, P. (1989). Situated cognition and the culture of learning. *Educational Researcher, 18*(1), 32–42.

Buchanau, M., & Fulton-Suri, J. (2000). Experience prototyping. In D. Boyarski & W. Kellogg (Eds.), *Proceedings of the conference on designing interactive systems: Processes, methods, and techniques* (pp. 424–433). New York: ACM Press.

Cardoso, C., & Clarkson, P. (2012). Simulation in user-centred design: Helping designers to empathise with atypical users. *Journal of Engineering Design, 23*(1), 1–22.

Coleman, R., Lebbon, C., & Myserson, J. (2003). Design and empathy. In P. J. Clarkson, R. Coleman, S. Keates, & C. Lebbon (Eds.), *Inclusive design: Design for the whole population* (pp. 478–499). London: Springer.

Costa, A. L., & Kallick, B. (2000). Assessing the habits of mind. In A. L. Costa & B. Kallick (Eds.), Assessing and reporting on habits of mind (pp. 29–53). Alexandria: Association for Supervision and Curriculum Development (ASCD).

Cross, N. (1997). Descriptive models of creative design: Application to an example. *Design Studies, 18*(4), 427–440.

Cross, N. (2011). *Design thinking*. London: Bloomsbury.

Csikszentmihalyi, M. (1999). Implications of a systems perspective for the study of creativity. In R. J. Sternberg (Ed.), *Handbook of creativity* (pp. 313–335). Cambridge: Cambridge University Press.

Damasio, A. (1994). *Descartes' error: Emotion, reason, and the human brain*. New York: Gosset/Putmam Press.

Davis, M. (1994). *Empathy: A social psychological approach*. Dubuque: Brown & Benchmark.

Dewey, J. (1938). *Education and experience. Kappa Delta Pi lecture series*. New York: Macmillan.

Dong, H., Nickpour, F. & McGinley, C. (2009). Designing ergonomic data tools for designers. In DS 58-8: Proceedings of ICED 09. The 17th International Conference on Engineering Design, Design Information and Knowledge 8: 53–64.

Dreyfuss, H. (1955). *Designing for people*. Re-released in paperback by Allworth Press (2004).

Ennis, R. (1996). Critical thinking dispositions: Their nature and assessability. *Informal Logic, 18*(2&3), 165–182.

Flavell, J. H. (1987). Speculations about the nature and development of metacognition. In F. E. Weinert & R. H. Kluwe (Eds.), *Metacognition, motivation and understanding* (pp. 21–29). Hillside: Lawrence Erlbaum Associates.

Gallo, D. (1982). Educating for empathy, reason and imagination. *Journal of Creative Behavior, 23*(2), 98–115.

Galton, M. (2007). *Teaching and learning in the primary classroom*. London: Sage.

Getzels, J., & Csikszentmihalyi, M. (1976). *The creative vision*. New York: Wiley.

Glaser, R. (1984). Education and thinking: The role of knowledge. *American Psychologist, 39*(2), 93–104.

Goodman-Deane, J., Waller, S., Latham, K., Price, H., Tenneti, R., & Clarkson, P. (2016). Differences in vision performance in different scenarios and implications for design. *Applied Ergonomics, 55*, 149–155.

Green, M. (2001). *Variations on a blue guitar*. New York: Teacher College Press.

Halpern, D. (1998). Teaching critical thinking for transfer across domains: Dispositions, skills, structure training, and metacognitive monitoring. *American Psychologist, 53*(4), 449–455.

Harpaz, Y. (2007). Approaches to teaching thinking: Toward a conceptual mapping of the field. *Teachers College Record, 109*(8), 1845–1874.

Hickman, R. (2013). Empathy and art education. In B. White & T. Costantino (Eds.), *Aesthetics, empathy and education* (pp. 235–246). New York: Peter Lang.

Hosking, I., Waller, S., & Clarkson, J. (2010). It is normal to be different: Applying inclusive design in industry. *Interacting with Computers, 22*(6), 496–501.

Hosking, I., Cornish, K., Bradley, M., & Clarkson, P. (2015). Empathic engineering: Helping deliver dignity through design. *Journal of Medical Engineering & Technology, 39*(7), 388–394.

Katz, L. (1993). *Dispositions: Definitions and implications for early childhood practice*. ERIC #211. Retrieved from http://ceep.crc.uiuc.edu/eecearchive/books/disposit.html

Keates, S., & Clarkson, P. (2003). *Countering design exclusion: An introduction to inclusive design*. London: Springer.

Koskinen, I., Batterbee, K., & Mattelmaki, T. (2003). *Empathic design, user experience in product design*. Helsinki: IT Press.

Kouprie, M., & Visser, F. (2009). A framework for empathy in design: Stepping into and out of the user's life. *Journal of Engineering Design, 20*(5), 437–448.

KunyK, D., & Olson, J. (2001). Clarification of conceptualisations of empathy. *Journal of Advanced Nursing, 35*(3), 317–325.
Lave, J., & Wenger, E. (1991). *Situated learning: Legitimate peripheral participation*. Cambridge: Cambridge University Press: Cambridge.
Lawrence, E., Shaw, P., Baker, D., Baron-Cohen, S., & David, A. (2004). Measuring empathy: Reliability and validity of the empathy quotient. *Psychological Medicine, 34*, 911–924.
Lawson, B. (2001). *How designers think: The design process demystified* (3rd ed.). Boston: Architectural Press.
Lucas, B., Hanson, J., & Claxton, G. (2014). *Thinking like and engineer*. London: Royal Academy of Engineering.
McCormick, R. (2004). Issues of learning and knowledge in technology education. *International Journal of Technology and Design Education, 14*, 21–44.
McDonagh, D., & Thomas, J. (2011). Design + empathy=intuitive design outcomes. *The Design Journal, 14*(2), 147–150.
McDonagh-Philp, D., & Denton, H. (1999). Using focus groups to support the designer in the evaluation of existing products: A case study. *The Design Journal, 2*(2), 20–21.
McGinley, C., & Dong, H. (2011). Designing with information and empathy: Delivering human information to designers. *The Design Journal, 14*(2), 187–206.
Mead, G. H. (1934). *Mind, self and society*. Chicago: University of Chicago Press.
Nicholl, B., McLellan, R., & Kotob, W. (2008). *Understanding creativity for creative understanding, Research report*. Cambridge: Cambridge University.
Nicholl, B., Hosking, I., Elton, E., Lee, Y., Bell, J., & Clarkson, P. (2013). Inclusive design in the Key Stage 3 classroom: An investigation of teachers' understanding and implementation of user-centred design principles in design and technology. *International Journal of Technology and Design Education, 23*(4), 921–938.
Nicholl, B., Flutter, J., Hosking, I., & Clarkson, J. (2014). Joining up the DOTs: Authentic teaching and learning in Design and Technology education. *Cambridge Journal of Education, 43*(4), 435–450.
Passmoor, J. (1967). On teaching to be critical. In R. S. Peters (Ed.), *The concept of education*. London: Routledge & Kegan Paul.
Paul, R. (1995). *Critical thinking: How to prepare students for a rapidly changing world*. Cheltenham: Hawker Brownlow Education.
Perkins, D., Jay, E., & Tishman, S. (1993). Beyond abilities: A dispositional theory of thinking. *The Merrill-Palmer Quarterly, 39*(1), 1–21.
Piaget, J. (1932). *The moral judgement of the child*. London: Kegan Paul, Trench, Trubner.
Pollard, A. (2002). *Reflective teaching in schools*. London: Continuum.
Porter, C., & Porter, J. (1999). Designing for usability: Input of ergonomics information at an appropriate point, and appropriate form, in the design process. In W. Green & P. Jordan (Eds.), *Human factors in product design: Current practice and future trends* (pp. 26–36). London: Taylor & Francis.
Rogers, C. (1975). Empathic: An unappreciated way of being. *The Counseling Psychologist, 5*(2), 2–10.
Sanders, E., & Dandavate, U. (1999). Designing for experiencing: New tools. In C. J. Overbeeke & P. Hekkert (Eds.), *Proceedings of the first international conference on design and emotion* (pp. 87–92). Delft: Delft University of Technology.
Schon, D. (1983). *The reflective practitioner: How professionals think in action*. New York: Basic Books.
Scriven, M. & Paul, R. (2003). *Defining critical thinking*. Available online at www.criticalthinking.org. Accessed Mar 3rd 2015.
Shulman, L. (2005). Pedagogies. *Liberal Education, 91*(2), 18–25.
Sutherland, J. (1993). The nature and evolution of phenomenological empathy in nursing: An historical treatment. *Archives of Psychiatric Nursing, 7*, 369–376.
Tichener, E. (1909). *Elementary psychology of the thought processes*. New York: Macmillan.
United Nations. (2009). *World population aging: 1950–2050*. New York: United Nations.

Vygotsky, L. (1978). *Mind in society*. Cambridge: Harvard University Press.
Ward, T., Smith, S., & Vaid, J. (1997). Conceptual structures and processes in creative thought. In T. Ward, S. Smith, & J. Vaid (Eds.), *Creative thought: An investigation of conceptual structures and processes* (pp. 1–17). Washington, DC: American Psychologist Association.
Wood, D., Bruner, J., & Ross, G. (1976). The role of tutoring in problem solving. *Journal of Child Psychology and Psychiatry, 17*(2), 89–100.

Critiquing Teaching: Developing Critique Through Critical Reflection and Reflexive Practice

Susan V. McLaren

Abstract This chapter will explore critique of teaching, with reference to reflection, critical reflection, reflexive practice, professional inquiry and learning. The aim is to illustrate why active engagement, with critique of teaching, practice and beyond, facilitates professional learning and professional development. In addition to the overview of *why* developing a mind-set for reflexive practice, critical reflection and critique of teaching is of value in terms of improving practice, the chapter explores models and strategies to support the *how* and *when* of these processes. It is through critique that teachers can ensure they are professional, fluid and informed in their responses as, and when, scenarios and contexts demand and be true to their personal ethics.

Critique is effortful, uncomfortable and disruptive. Teachers must want to involve themselves in the hard work of critique and see some results for their efforts; otherwise, why bother?

The chapter comprises three sections:

1. Exploration of conceptual frameworks of critique
2. The value of critique in developing design and technology (D&T) education practice
3. Models and methods of critical reflection to scaffold critiquing D&T teaching

Keywords Critique • Teacher-as-designer • Inquiry-as-stance • Learning journals • Creative growth

1 Exploration of a Conceptual Framework of Critique

The importance of reflection has been stressed for many years as a driver for the continued professionalisation of teaching (c.f. Dewey 1933; Schön 1983; Bandura 1993; Calderhead 1989; Hargreaves 1998; Boud et al. 1985, 2006). However, the ubiquitous promotion of reflection for professional learning and growth has

---

S.V. McLaren (✉)
University of Edinburgh, Edinburgh, UK
e-mail: susan.v.mclaren@ed.ac.uk

P.J. Williams, K. Stables (eds.), *Critique in Design and Technology Education*, Contemporary Issues in Technology Education, DOI 10.1007/978-981-10-3106-9_10

often been presented, or adopted, too simplistically, and been assumed to be an instrumental, technical and individualistic tool or a 'recipe to follow' so as to be *seen to be* reflective. This has resulted in superficial approaches which offer little useful learning (Boud and Walker 1998). Perhaps it is time to reconceptualise reflection and develop more meaningful critique within the context of design and technology (D&T) teaching.

The aim for D&T teachers, as for any other subject discipline specialist, is to develop as a critical thinker, a professional who has the autonomy, motivation and high self-efficacy necessary to operate with wisdom and skill for the benefit of their learners (Bandura 1993). D&T teachers enter the profession with personal assumptions and beliefs, fears and desires developed through experience over many years prior to any professional indoctrination and/or education. They may not be fully cognisant of these nor aware of how such assumptions and beliefs shape their thinking and their practice as a teacher (Rogers 2002). In order to be in a position to engage in critique of teaching, a teacher needs to accept that their personal values can be challenged by others and by the systems in which they teach. Teachers do not teach in a vacuum. The context in which they are located is often one created by political ideology which may be underpinned by conflicting ideas about the purposes of education overall. Teachers work in collaboration with colleagues within the complex and shifting interrelationships of a school as a system, a structure, a community and curricula. All of which may be subject to change. In addition, the context is populated with learners who are unpredictable and create situations which are unique and variable.

LaBoskey (1993), as others (cf. Arnold et al. 2012; Mezirow 1990) suggest that the impetus for taking time to study the constructs and power structures of society, and analyse how these impact on educational policies, curriculum, assessment, accountability and pedagogical choices, is not linked to a particular professional life phase of a teacher but considered more as a professional frame of mind. This frame of mind demands that D&T teachers develop the skills to critique *what* they are doing, and *why* they are doing *what* they are doing, within the specific context they are operating by raising and asking questions. This helps to determine how their own beliefs, ideologies and assumptions impact on ways they enact educational policy, curriculum and assessment and adopt teaching methods and materials with agency.

To sum up, the critique of D&T teaching comprises:

- Critically examining interrelationships between pedagogy, curricula and assessment
- Adopting a mind-set of exploration and continual growth
- Exploring personal beliefs about teaching and learning processes (including emotional attachment to the discipline they teach)
- Transforming initial responses and tacit understandings into reflective action
- Praxis, i.e. taking creative risks to go beyond reflection to reflexive informed action

2 Conceptual Framework for Critique for Design and Technology Teaching

For a D&T specialist, the concept of critique involving aspects of professional life, such as managing complexity, coping with (and creating) uncertainty, challenging assumptions, embracing creative risk, taking responsibility for innovation and progressing action, maps well with the methodologies and attitudes of designerly thinking (Cross 2006) which is at the core of D&T learning experiences as determined by the majority of curriculum guidelines across the globe. Designerly thinking involves the learners working iteratively and creatively with cognisance of constraints or unforeseen challenges and issues within any given system, scenario or context to model an appropriate resolution or proposal for the client and user. Indeed, the construct of systems thinking and designerly learning places D&T in an interconnected and complex web and creates parallels with teaching and learning. Even when taken outwith political ideology, or a national/state definition of the purposes of education, the context in which D&T teachers are situated is one of continual change, surprise and challenge as engendered by the general underpinning construct of D&T education. This implies D&T teaching is dynamic, interactive and subject to scrutiny by the hierarchy of governance, subject to various external influences and technological cultural shifts, and by the teachers and learners themselves.

The processes of critique relate to what could also be considered as a framework for designing where designerly thinking involves:

- Seeking out issues and 'spark finding' (Kimbell 2002)
- Asking critical probing questions
- Challenging assumptions and personal beliefs
- Sourcing, identifying and critically analysing evidence that supports or conflicts
- Generating multiple alterative solutions and appraising each
- Taking intellectual risks
- Working iteratively to develop and arrive at a resolution that may be considered the best/elegant fit to offer a proposal or conclusion framed in consideration of consequences

Such characteristics work well to begin to describe D&T teaching where teachers encourage learners to suspend judgment, be willing and open to exploration, to deal with uncertainty, develop their technological creativity and take responsibility for their own learning and design decisions. Using designing as a framework offers opportunities to engage creatively in a critique of D&T teaching with the intention of gaining insights on educational issues, specific dilemmas and personal and professional pedagogical content knowledge and take informed action on the basis of interpretation of findings, discourse and dialogue with the intention to enact change. What is considered to be the purpose(s) of schooling and education in the specific context the D&T teacher is teaching? What informs the basis of the planning and enactment of the learning experience? Who is granted (and who grants?) the autonomy to choose and plan appropriate D&T teaching approaches and learning experiences for learners? What knowledge should be taught, and what knowledge is not to be taught? Who decides? On what basis?

Boud et al. (1985) suggest that critique involves reflective thinking, emotions, feelings and cognition in a complex personal process with the intention of future action. Dewey (1910) as cited by LaBoskey (1993: 30) promoted an attitude of open-mindedness, responsibility and wholeheartedness as being integral to reflective action. Although not fully defining critique, accepting such principles as a framework for what D&T purports to involve lends itself to a framework of critique of D&T teaching in itself.

3 The Value of Engaging with Critique

This section argues the reasons for developing critique as a frame of mind and the value it offers D&T education practitioners. It serves to help teachers develop an understanding of the way they operate as a professional and what guides and frames their responses, actions, choices and decisions. Such enhanced understanding enables a deeper appreciation of what it is that makes teachers the teachers they are and informs them of how they can become the teachers they could be. Critique offers opportunities to develop an informed, personal and collegiate repertoire of actions through which professional teaching practices are enriched. The process of critique develops a proactive appraisal of experiences (Rogers 2002). It is less about immediate guilt or self-blame when, for example, learning is not achieved as had been anticipated, and more about developing a more secure understanding of what can be done to modify, enhance and/or address the aspects of practice which are spotlighted by the process. This, in turn, results in a renewed vigour and agency to address the issues at play.

Teachers are encouraged, from their induction as student teachers through to accomplished practitioners, to ask critical questions of educational policies and practice. Beliefs, values and practices need to be regularly reappraised as society and the needs of learners change and as a teacher's understanding develops. This notion of teachers engaging in critical reflection and adopting reflexive practice is central to ideas of responsiveness and relevance of the teaching profession. Eisner (1985) describes the value of critique as going beyond the skilfulness and effectiveness of being a teacher. He argues it helps practitioners, through meaningful professional learning, move towards 'connoisseurship' and to develop an 'artistry of teaching'. Eisner urges teachers to develop the ability to appreciate the different dimensions of their observations and experiences and to explore how they relate to each other and examine how these dimensions connect with their own values and commitments. Subsequently, by adopting the role of a critic and employing criticism to scrutinise all the various interrelationships of a complex system such as education, 'as experienced' in thc wider context, a teacher can make them explicit and engage in discussion with others to construct meaning or challenge existing paradigms. Larrivee agrees and suggests (2000: 294) 'Unless teachers engage in critical reflection and ongoing discovery they stay trapped in unexamined judgments, interpretations, assumptions, and expectations'. Critical reflection is not concerned with the how-to of action, but more keenly with the *why*, the reasons for and the consequences of what we do (Arnold et al. 2012). The more one develops knowledge and understandings of the ways in which environmental, social, cultural, political and economic systems function, the better one can appreciate, and be more curious about, how such systems interconnect with, and impact on, professional practice. Framing the questions to ask in order to challenge and develop personal and professional knowledge and understandings in interconnected systems takes effort, but this is a necessary part of the continuum of constructive, critical enquiry. The balance of the interrelationship between curriculum, teaching and assessment is sensitive to distortion if any one aspect dominates. The value of critique lies in examining the dilemmas, conflicts, puzzles and lines of enquiry that present themselves and proactively create opportunities for deeper, wider or further investigation. As a result, new models, alternative frames and different ways of thinking may be prompted.

However, the cognitive effort and energy required to engage with critique is great, and so some value must be recognised in return. Such critical inquiry and self-reflection enable teachers to acknowledge their strengths, to identify aspects for further development and to seek continuous improvement. In order to develop the skills and mind-set to engage purposefully in critique, teachers need to be willing to examine and challenge their personal and professional belief systems and the impacts and consequences of any enactment of these through their teaching. The processes of critique are not always comfortable; hence clarity in the value is paramount. In summary, critique has the potential to:

- Add and create meaning from what is already known and experienced through professional practice
- Challenge one's own actions and understandings, in order to develop/change practice
- Free one's own practice, in an informed and deliberate manner, from any externally imposed assumptions and beliefs
- Support teachers to engage in collegiate critical dialogue
- Facilitate processes of iterative enquiry related to teaching for purposes of transformation of teaching, learning and assessment practices
- Enable teachers to view themselves 'as transformative intellectuals' (Giroux & McLaren 1996)

The value of engaging with critique is teased out further in this chapter through discussion of creative growth and adopting inquiry-as-stance. This is followed by some illustrative models and exemplar scaffolding methods through which D&T practitioners can embark on the processes of critiquing D&T teaching.

3.1 Critique as Creative Growth

Designerly thinking involves turning the act of looking into seeing more specifically and critically through different lenses (Cross 2006). This helps the designer to appreciate the needs, desires and values of the various users and the range of stakeholders for whom they are designing. It is important, therefore, that any underlying assumptions, of all those involved, are uncovered and made explicit. Sterling (2009) and Schön (1987) discuss what can be achieved by reframing problems and illustrate the value gained in shifting from problem-stating and solution-finding strategies to creating question-framed problems. This enables the underlying causes, constructs and influences to be challenged and exposed and accepts that there are relational and complex factors to be addressed. Thus sensitivities and capabilities are required to interpret and explore a wide range of resolutions and empathise more authentically with those whose values and beliefs vary.

D&T learners are introduced to strategies to appraise existing environments, artefacts and systems. The relationship of the design outcome to the context, the user, the intended function and purpose is scrutinised and questioned, and informed judgments are formed and articulated. It follows, then, that the skills and strategies which the learners develop can be transferred to critiquing the efforts of their D&T teacher and their teaching. D&T teachers who value creative growth are those that have a keen appreciation of how their learners are responding to the D&T experience. Brookfield (1995) advises that by welcoming the learners' voice, inviting critique and listening to their perspective, opportunities arise to model genuine critical enquiry for the learners, thus stimulating them to do likewise. Learners in the D&T workshop, lab and studio can learn from their teacher's critical reflective approach and begin to be inducted into the model of critique to develop their own capabilities and growth mind-sets. Brookfield argues that this does much to alter the traditional teacher-learner relationship that is historically based on power and control or management of learner behaviours and attainment and can help to create a more productive partnership in learning.

D&T education strives to develop active, critical citizens and creative contributors who have the capability to adopt a designerly eye, to challenge, disrupt, evaluate and appraise the worlds they encounter. D&T also integrates objective and subjective, visceral and emotional values, with creative alternative and imagined worlds framed by personal, lived experiences and cultures. Transpose this conceit to the skills and values of a D&T teacher to adopt a teacher-as-designer stance, i.e. someone who critiques their personal pedagogical framework and values, who deals with uncertainty and thrives in a more than one solution design space, in order to create that very unique D&T educational experience which centres on creative growth. And yet, a D&T teacher may encounter a professional arena that is entirely contrary to designerly thinking and which offers no space for exercising creative growth. They may find themselves working in an environment that is bound by professional standards and that subscribes to professional competences which are at odds with their own. They may become accountable through statistics arising from examination awarding bodies that prescribe criteria for standardised tests which reward formulaic performance over creativity (Atkinson 2000). This may, in turn, create a sterile environment of operation which conflicts with their view of D&T as an educational experience. Nicholl and McLellan (2008: 588) describe such tensions as 'dual values of creativity and performativity'.

Adopting a 'teacher-as-designer' stance and accepting that creativity is beneficial in the professional practice of education, particularly D&T, Christenson (2001, cited in Thompson and Jan Pascal 2012: 37) suggests that 'Any society that values creativity also needs to enable criticism. If we cannot question the way we are doing things and thinking about things at present, it will not occur to us that they could be thought of or done differently'. This concept of critique and the value of a 'critical approach', common language to the practice of art and design, is an 'important part of promoting creativity and preventing stagnation'. Thompson and Jan Pascal (2012) reiterate that this is entirely consistent with reflective practice and quote Adams (2002: 87) who notes, 'the two do not always go together: Critical practice is not just reflective practice, because the critical practitioner does not take the world for granted and does not automatically accept the world as it is. Reflective practice contributes to critical, transforming practice. ... Critical practice involves reflectiveness but transcends it'.

The richness of designerly thinking is evident when shortcomings of models of reflection are scrutinised. Reflection, when regarded as an individual process and single perspective, personal view of an experience, is limited and can be counter-productive. It is not enough to seek and find a 'technical fix' for a specific issue as identified through a solo-internal reflective process. This in itself will not enable creative growth as a professional nor contribute to any genuine development of a practitioner. It will, perhaps, simply 'sort' a *specific* problem (as defined in a *particular* way) in a *specific* scenario, in a *specific* culture, or environment, if indeed it does achieve that much.

Why then would a D&T teacher not engage in critique? There are many possible answers to this question. LaBoskey (1993) noted in her study that some teachers were secure and confident with their practice and felt as though they had 'got it sussed'. They considered themselves to be effective in their teaching. They obtained results for their learners, and, more importantly, the results were those required by the system, and therefore they had no need to take time to engage in critique. As a D&T specialist teaching in a system where individual performance is judged by 'added value' in terms of grades attained, it may appear to be easier to adopt an approach of acceptance and compliance where teaching to the test may prove to be a mode of survival. LaBoskey (1993) also found that some teachers chose not to engage in critiquing due to a lack of personal confidence and because they felt overwhelmed and distressed by the multiple requirements and demands on their professional and personal time. Teachers can feel insecure or threatened about critiquing their teaching, and that of others, for fear of not getting 'the right answers' or 'not doing it right'. Critique serves as professional learning and demands scrutiny of theories and experiences through a range of lenses, to reveal hitherto unrecognised possibilities which serve to inform future practices. This is deliberately provocative and understandably unsettling for some.

3.2 Critique and Inquiry-as-Stance

Much as critique in the context of designing, which aims to develop further design inquiry for deeper understanding to arrive at alternative design ideas or to make more secure design decisions, critique in the context of teaching has the potential to inspire new lines of inquiry and generate excitement in terms of meaningful and personalised practitioner research. Critique of D&T teaching drives the search for something different, the inquiry into practice, knowledge and understanding; the quest for something richer. 'Inquiry-as-stance' is a term used by Cochran-Smith and Lytle (2009) to describe the ways in which practitioners see and act, and relates to the lenses through which they look. It is a 'way of being' a teacher which partners critique well. Inquiry-as-stance is not time-bound, as a project, or initiative or strategy, but serves more as a construct to frame personal and professional learning and posit a teacher's orientation towards knowledge, intellectual ideas, their relationship to the practices, purposes and systems of schools and schooling. For Cochran-Smith and Lytle (2009), a teacher who adopts inquiry-as-stance is one who acts in ways that are considered beyond the well-rehearsed, reflective routines and one who more readily adopts a questioning approach to what they do in their own practice and critiques the socio-political context in which they, as professional D&T teachers, are teaching. Through rigorous interrogation of planning, teaching and assessment, concerns, issues and questions are raised. This process of examining and 'problematisating' practice is central to inquiry-as-stance (Arnold et al. 2012). This inquiry goes wider than a teacher's own practice and wider than the immediate experiences of teaching. It also deliberately makes current educational policies and systems problematic by examining the underlying assumptions and unpicking the ideology that is embedded in the institutions, doctrines and documents that guide curriculum design and assessment. It contributes to a dialectic cycle of questioning, observing, acting and learning in collaboration and dialogue with other members of the school community, seeking alternative perspectives and making visible the personal, professional and political thinking and decision-making. It acknowledges the struggles that individual practitioners face in their attempts to tease out knowing why, how and what. Inquiry-as-stance has the potential to explore consequences, evidence impact and expose beliefs, assumptions, values and practices. Critique concerns construction (and deconstruction) of knowledge and ways of knowing, through conversation, discourse, collaboration, analysis and interpretation, thus making the tacit visible and the complexity of teaching more explicit. The culture of critique, through such inquiry, is rich and disruptive.

A D&T teacher who develops inquiry-as-stance does not feel obliged to accept the status quo and is willing to challenge assumptions, identify potential consequences and conflicts and appraise alternatives. There is curiosity and purpose in mind. Inquiry-as-stance can be considered much like a creative process through which there is the development of a critical habit of mind and higher-order thinking towards seeking patterns and relationships. As with design thinking, such inquiry deals in the currencies of uncertainty, hypothesis, controversy and dilemma. Craig (2010: 206) suggests that when teachers pose questions, create and undertake enquiry, shaped by the contexts of their own teaching, they have the ability to 'unpack the unintended consequences of public policy'. Fichtman Dana & Yendol-Hoppey (2014: 6) use the term 'wonderings' to describe the initiation of the processes of critical inquiry from which teachers embark as knowledge creators and co-constructors of understandings. Teachers, as Hargreaves (1998) notes, are more likely to be motivated towards the change when the change is driven by teachers themselves, and not by 'outsiders' such as politicians, educational administrators or university researchers. The transformation of D&T teaching is made possible through teachers reconstructing, reforming, renewing, refining and reformulating knowledge in, knowledge of and knowledge for practice, as a natural part of their professional learning (Cochran-Smith and Lytle 1999).

4 Towards Critiquing: Models and Methods of Critical Reflection

The previous sections have discussed what is required for reflection to be transformed more usefully into critical reflection and ultimately serve as critique. The argument for engaging in the processes of critique has been presented in terms of creative growth and value for professional learning. This section illustrates some models to develop approaches to critiquing teaching. Although each is underpinned with similar concepts, there are subtle and, at times, distinct differences between some of them. The different models all involve uncertainty and will provide no definitive answers. Critique is not about problem-solving, and the various models discussed here should not be considered as problem-solving tools seeking a fast 'technical fix'. It is also important to note the distinction between the processes of reflection, critical reflection and critique. Critical reflection takes reflection beyond the analysis of personal experience with a view to solving problems encountered in personal practice, by considering the wider socio-political dimensions in which the experience is located. Critique then develops this further and proactively challenges and questions these dimensions. As such, it is messy and complex and requires acknowledgement of the many shifting variables.

Common to all models is the dialogical process which reveals alternative perspectives. This is considered central for any reflective practice as it creates a dynamic process, motivates professional learning through enquiry and deepens critique. The processes of critique will question assumptions about our own actions, intentions and values and those of others. The models encourage practitioners to examine their personal experience as located in, defined and bounded by political and social structures. They all take cognisance of multiple perspectives. Critique requires comparison and examination with what is already known about the issue and will draw on existing literature, discourse and research. So too then, when engaging with critique, a D&T teacher recognises the limitation of their personal experience, as felt and understood from the subjective and the emotional states, through the lens of a very personal frame of reference. Personal values and beliefs can serve as *part of* the process of critical reflection but not serve *as* critical reflection. Critique is not a process undertaken through one static lens, and this therefore highlights the limitations of some models of reflection.

4.1 Schön's Reflective Practitioner: Contribution Towards Critiquing Teaching

One model of reflection was made explicit by Schön (1983). He explored professional 'ways of knowing', 'reflection-in-action', and 'reflection-on-action' and drew on several professional practices from the world of design education. In brief, 'reflection-in-action' is the immediate, intuitive, tacit, reactive approach of the professional teacher in the classroom, studio or workshop with the learners, or being reactive in meetings with fellow practitioners and colleagues. In D&T practice, as with all teaching, teachers will encounter messy, unplanned situations, and they will 'reflect-in-action' to decide upon alternative approaches, adopt a different 'language', try varied strategies and assess for counterresponse, *in the moment*. Eraut (1995) suggests that, for a classroom teacher, in the limited time frame available, particularly in crowded settings, the need for such rapid decisions results in scant analysis, and therefore the actions that are taken tend to follow convenient institutional protocols, emulate routinised reactions of a more experienced teacher or imitate a recently read evidence-based theory. 'Reflection-*in*-action' is sometimes (wrongly) interpreted as seeking technical fixes through an on-the-spot experiment or restructuring of strategy. The spontaneous, yet conscious, 'knowing-in-action', which accompanies 'reflection-in-action', draws on a repertoire of learned responses from previous experiences in different contexts.

In contrast to the possibly ill-informed immediacy of 'reflection-in-action', 'reflection-*on*-action' demands deeper, more deliberate thought about the unique experience as encountered from different perspectives and is undertaken with the intention of rethinking and constructing new understandings. It is only when the initial situation and the subsequent actions and reactions are discussed and reviewed with a colleague, or considered alone, through a retrospective lens does the 'reflection-*on*-action' enable further questions to critique the phenomena as experienced. This critical approach to reflection involves deliberate reliving and re-rendering: who said and did what, how, when, where, and, importantly, why (Waks 1999). The intention is this process leads to insight(s) about something hitherto not noticed or not understood. It aims to identify details or underlying issues which, for example, were undetected in the 'heat' of the teaching episode. It is effortful and involves finding strategies to further question our own attitudes, thought processes, values, assumptions, prejudices and habitual actions, in order to understand our complex roles in relation to others and in relation to the experience as lived, and system in which the teaching is bounded. With practice, this develops the discipline of reflexivity (often a missing component). Over time, with practice, the reflective process develops from practical pragmatic/technical fixes and praxis towards reflexivity and critique (Carr & Kemmis 1986; Arnold et al. 2012).

4.2 *Brookfield's Lenses: Contribution Towards Critiquing Teaching*

Brookfield (1995) suggests that the aspect omitted from Schön's initial work, or at least given less focus, is that of 'reflection-*for*-action'. Brookfield argues the importance of personal and professional learning through a heightened awareness of planning, foresight and teaching. He proposes a model for embarking on critical reflection with four explicit lenses which, he suggests, offer different perspectives through their specific focus. These enable a teacher to make a deliberate shift from tacit commitments and constructs to becoming a critically reflective teacher and question their way of thinking and deeply held implicit assumptions and how things have come to be as they are. The lenses serve to reveal personal assumptions and frameworks that lead teachers to understand more about their own practice and why they 'operate' as they do. This can serve as a stepping stone to the processes of critiquing.

> Lens one adopts an autobiographical exploration. This can include examination of past personal learning experiences including initial degree disciplines; previous places of work; range of experiences as a teacher; self-evaluations of teaching episodes; feedback received and feedback given; personal goal setting; previous places of work; and profiles of teachers adopted as role models.
>
> Lens two refers to insights from student learners and student voice. This involves taking on board their feedback to the teaching and learning experiences, paying due respect to their interpretation of the teacher portrayed; analysing patterns of responses; reviewing less/ more successful engagement; less/more secure performances; analysing assessment data.
>
> Lens three respects experiences of colleagues and includes dialogue, debates and critical conversations about, for example, planning, implementation, assumptions and subject and pedagogy constructs, conflicts, purposes of education, accountability and performance. This enriches personal frameworks through increased exposure to diverse and/or novel insights from those who experience similar contexts.
>
> Lens four refers to continuous scholarly reading, research, and enquiry. This serves to source a wider realm of voices and theories about, for example, D&T specifically and the contribution of D&T research to the wider educational arena and vice versa, locally, nationally and internationally, providing further topics to examine and challenge.

Brookfield urges teachers to develop their critical reflective capabilities such that they can justify their professional actions and the intended consequences through the development of a critical rationale for practice which he claims is a 'psychological, professional and political necessity. Without it we are tossed about by whatever political or pedagogical winds are blowing at the time. A rationale serves as a methodological and ethical anchor' (Brookfield 2009:11). His model of 'four lenses for critical reflection' in concert with his process of 'hunting assumptions (causal, prescriptive and paradigmatic)' can help unearth the power dynamic that impacts on the purposes and practices of teaching and distorts social justice. 'Critical reflection is inherently ideological. It is also morally grounded. It springs from a concern to create the conditions under which people can learn to love one another; and it alerts them to the forces that prevent this. Being anchored in values of justice, fairness and compassion, critical reflection finds its political representation in the democratic process' (ibid 1995: 26–27).

5 Scaffolds to Support Critique of Teaching

With the priorities for a teacher being determined by the reality of a school day, reviews, reflective journal entries and/or meetings with colleagues to discuss teaching can become superficial and the quality of discourse poor. In a limited time frame, a brief outline 'story' is relayed, a cursory thought is cast towards how to address this 'next time' and a record of 'next step(s)' is noted. It is common to identify a technically orientated goal to fix the 'problem' as doing so avoids asking the bigger or deeper causal question(s) necessary to examine the assumptions and behaviours which shape the initial 'story'. As Valli (1993) suggests, such technical rationality bypasses the more important questions of critique. What is needed is analysis, explicit links and dialogic connections with professional knowledge and theories which help to discover and construct new knowledge or understandings from the experience(s). This is effortful and requires a framework to help guide the process to ensure it holds value and serves purpose beyond the mechanistic.

What follows are some practical approaches to engage with various levels of critique. Whichever model is framing the critique, some scaffolds may be required to serve as prompts and tools to aid the process, working towards greater integration, interrogation and iteration of theory and practice and heightened metacognition. Some of the 'tools' and strategies which have the potential to contribute to the process of critique include learning journals, reflective writing, significant incidents, learning rounds, lesson study, fictitious narratives, alternative views, 'a story in the round', learning dialogues/discourses/reflective dialogical exercises (with peers or mentors) and role plays. Four examples of scaffolds for guided critique are described below: learning journals, critical incidents, fictitious writing and lesson study. These are selected to illustrate approaches suitable for an individual and also for collegiate critiquing.

5.1 Learning Journals

Some teachers may feel that a learning journal is only valid when there is something traumatic and/or dramatic to reveal in their entries. There may be a sense from beginning teachers, especially, that they are writing a learning journal for their tutor or mentor, and the purposes of the journal remain unclear. Brookfield (1995) cautions against the journal becoming a ritualistic and mandated confessional, written for others to read. Teachers may think that if they do not have anything painful or exciting to 'share', or big questions to ask, they will be 'judged' less reflective than those that do (McGarr & McCormack 2014). However, Morrison (1996) suggests that a learning journal may be advantageous for several reasons, for example, to chart experiences and development over a period of time and for this record to offer an overview of the developing dialogues between academic work, professional practice and personal development, for oneself. The learning journal can provide a tool to encourage increased self-awareness through the ability to theorise about the nature of experiences and encounters and make the author more explicitly aware of the choices and decisions they themselves are empowered to make. The process of writing in a learning journal, much as the practice of keeping a design-sketch sourcebook, can generate narratives based on experiences, and these narratives, with other observations and 'headlines' from scholarly readings, can provide a device for enabling teachers to synthesise a variety of different experiences into a coherent whole. The intention is that learning journals provide scaffolding for the teacher to reflect on their own development in the context in which she/he is operating. Learning journals require guidance on how to use them, what raw data to include and prompts regarding the variety of tasks to instigate their use to avoid them being as Bolton (2010: 11) also cautions, 'becoming only confessional'. Morrison (1996: 323) suggests the focus is made explicit in terms of four key headings: personal, professional, academic and evaluative development. Within this overall framework, data could relate to progression and development in terms of:

- Increasing knowledge (including institutional, content and pedagogical content knowledge)
- Increasing ability to articulate and identify issues
- Increasing ability to make issues (their own and those of others) explicit and clearly articulated
- The expansion (in depth and breadth) of their understanding of an issue
- The expansion (in depth and breadth) of their vision and personal construct
- The replacement of one set of beliefs (or theories) with another or confirmation of beliefs
- Attitudinal changes over time
- Changing practices in the institution in which they work
- Changing relationships with colleagues

Learning journals have the capacity to expose contradictions, misconceptions and conflict. A frame to enable this could be to regularly note aspects of the specifics of particular situations and behaviours in order to analyse what a teacher does (behaviours), why a teacher does it (values, belief, assumptions, aspirations) and how a teacher feels (emotional intelligence). Through such efforts, a critiquing process of teaching, planning, implementation, assessment and relationships can begin to identify specific strategies for change. Learning journals tend to work best in conjunction with other strategies, rather than being considered as *the* means to develop critical reflection; otherwise as McGarr and McCormack (2014) note, there is a false comfort in 'doing' reflection, and little learning ensues.

5.2 Significant Incidents

A significant incident is not necessarily a dramatic incident. A critical incident, as a significant incident, does not need to be an exciting enthralling, unusual and/or puzzling experience (Tripp 1993). It can be situation of any duration and scale. It can be unanticipated, and rare, but equally an incident that occurs frequently and be familiar, or even common. They are however incidents that have impact and contribute towards the trajectory of the learning, teaching, planning and/or implementation of an experience and as such are indeed *significant* and offer scope for critique. There will be an incident which can be described and situated in a scenario or a context. There will be an emotional, visceral or tacit response and resultant or subsequent actions which can also be described. The issues or concerns that are noted help to suggest the significance of the event. The descriptions and detail enable analysis to be possible. It is the process of drilling deeper into the incident and viewing it from a range of standpoints that creates the significance and makes the incident *critical* (Mezirow 1990). The approach may go something like this:

- Briefly describe a situation that occurred that affected you as an individual or as a team.
- Why are you describing this incident? Did you experience challenges in meeting it? Did you exhibit strengths? Did you learn something? About yourself? About others?
- Is there an overarching problem here? Are there values at stake?
- What were you feeling at the time of the incident/situation?
- What were your thoughts at the time of the incident/situation? Did you have preconceived ideas or assumptions?
- Has this experience challenged your assumptions, prejudices, biases or beliefs?
- What specific questions have you been able to raise?
- What specific (potential) 'solutions' have you been able to identify? What further questions arise from this?
- Will this experience alter your future behaviours, attitudes, understandings or aspirations? If so, in what ways?

By writing or talking, in response to prompts, the incident becomes a vehicle for critiquing an existing rationale or construct which frames the way the teacher, as a personally constructed professional, acts, views the world and assumes their role within the specific context. The writing (or discussion) should consider alternatives. This may require further reading to seek, but not necessarily accept, ideas from research-based evidence, to support, develop or contest any assumptions that have been revealed. This then frames the existing, exposed ways of understanding and sets these against any new understanding. The understanding gained, and disruption that the altered consciousness causes, is what renders the significance of an incident critical.

5.3 Fictitious Critical Writing

Critical reflection has been shown to be supported by seeing through different lenses (Brookfield 1995). This requires the practitioner to step outside oneself, be curious and unsettled, create discomfort and disrupt the familiar status quo. Bolton (1999, 2010) suggests that writing an all-imagined retrospective view of an experience, or episode, from the learners' perspective, capturing their thoughts and feelings and who said and did what, when and why, can serve as a useful strategy. It may appear that such fictitious writing is creatively a step to far for the purpose of critique, and yet it can serve a valuable contribution to the process. It provides more than a story, albeit a story nonetheless. The result is a story that incorporates the implicit theories of the author whilst also garnering the various points of view of all the actors through the story, words, thoughts and actions. Fictitious writing contributes more to critical reflection than problem-solving and target setting. It is a tool for exploring the *why* things are experienced the way they are and how they are perceived. The writing, for example, could explore the responses of various actors to the annual statistics reporting high-stake assessment results for D&T courses in the teacher's school. The actors in this instance would be the learners and their parents, the teachers in the D&T department and colleagues elsewhere, the school senior management and the government. Such writing exposes additional data, power dynamics and political ideologies which are useful for critique. It can serve as a comparison of incidents, thus revealing patterns, making meaning in the social, political economic and ethical context and system in which the experience/phenomenon is located.

5.4 Lesson Study

Guskey and Passaro (1994) note that teachers, who are high in self-efficacy and engage in critiquing their teaching, are more creative in their job. They tend to intensify their attempts to look for different strategies and methods and are less likely to become complacent and compliant. 'Lesson study' and 'learning rounds' are collaborative approaches to localised practice-based inquiry, which can develop greater ownership of reforms and as such can be useful in terms of critiquing teaching. In brief, lesson study, as described by Yoshida (in Stigler and Hiebert 1998), is most commonly a teacher-led professional learning process and often takes the focus of curriculum development with a view to improving teaching by studying how learners learn. It tends to be worked in small groups of teachers who identify a long-term aim as a line of enquiry and make detailed plans for the study together. Preparation is complex and requires in-depth research into whatever topic is being studied. This tends to involve examination of syllabi/content frameworks, teaching resources, established teaching approaches, reports and related research literature, for example, adoption of roles for a cooperative learning approach to a robotics project; use of flow charts for differentiation in identification of commercial manufacture processes and materials; and techniques for learning creative thinking and idea generation. The teachers will then observe the learners in the classroom, as they are being taught by one of the lesson study group. They collect the data as agreed, and using the insights the observers report, including a learner perspective, the group reflects on what was learned specifically about teaching and learning of the specific topic being taught and more broadly the dynamic between teacher and learner, and learner and learner, the teaching and the content framework and the resources incorporated. It is the richness of the collaborative discussion that provides the insights for the inquiry to progress through iterative cycles. Fernandez (2002) acknowledges, however, that there are many challenges in undertaking lesson study. For example, dialogue with colleagues which focuses on personal shortfalls can undermine confidence (Bandura 1993: 125), and the teachers engaged in lesson study must have mutual trust. All those involved are required to adopt inquiry-as-stance in order to pose researchable questions, specify the type of evidence to be collected and interpret and generalise results through robust and collegiate discussion. Lesson study as an approach cannot claim critique of teaching is inevitable.

Bandura (1986, 1991) looks to developing self-efficacy and agency, and lesson study can contribute towards this. He suggests a teacher's self-efficacy impacts on their willingness to explore alternative pedagogical approaches and deal with uncertainty. Teachers with high self-efficacy are more likely to take intellectual risks rather than 'play safe' and less likely to adopt formulaic teaching, learning and assessment approaches. Critical reflection, through lesson study, can be transformative and stimulate the process of critique through actively encouraging doubt and uncertainty with the explicit purpose of seeking to challenge and disrupt. The key to lesson study is in the posing of questions to challenge the current context and to examine the underlying assumptions and purposes of the status quo.

In summary, this section has demonstrated that there are several approaches to scaffold the processes of critical reflection, and each practitioner, and group of teachers, will select and adapt their own strategies to suit their unique contexts and purpose. There are opportunities for teachers to reveal new, co-constructed knowledge, understanding and/or meaning, which offer new perspectives that can inform subsequent actions, challenge ways of knowing and critique ways of being (Hargreaves 1998).

6 Conclusion

This chapter has explored the value of critical reflection with a view to developing informed practice through professional learning, creative growth and critique which has the potential to transform teaching. It has alluded to the processes of teaching as designing and teacher-as-designer. It has drawn on literature from reflection, critical reflection, praxis and reflexivity to review models and strategies to support critique. As Mezirow (1990) suggests teachers become critically reflective by challenging the established definition of a problem or by revealing the underpinning cause of an issue encountered, perhaps by finding a new ethos that orientates efforts in an alternative direction. This demands that they reassess the way they have arrived at their values, beliefs, ways of knowing, feeling and acting; they examine policies and structures; they revisit values and systems and the relationship between these, particularly in terms of planning and practice and how these impact through informed forethought which results from the insights of critical reflection.

Bandura's research (2003) underlines the importance of the quality and purpose of the dialogue, and the language used, for critique in the education community. Teachers learn how to develop the skills of critical consciousness, self-regulation and self-efficacy. Larrivee (2000) argues for greater *examination* of the broader socio-political level, where practice occurs (questioning, challenging, desire for change), and explicit acknowledgement of *struggle* (inner conflict, surrender, uncertainty, chaos, power) and *perceptual shift* (reconciling, personal discovery, new practice) as essential components to arrive at *transformation* of D&T practice through critique.

Critique is not necessarily a systematic process, and yet it demands an understanding of holistic systems thinking and the interrelated, interconnected aspects therein. It is not about gaining a veneer of accomplishment (Hennessy et al. 1993). The process of critique is not an emotional confession, not a description or defence, nor a self-indulgent examination of self in one moment in time. It is not intended to be a process which carries with it negative connotations nor is it about seeking out 'cause and effect' to signal blame. Having an experience in itself does not lead to quality, meaningful learning nor is improvement always achieved by repeating or continuing the same action or experiences or rituals. Mezirow (1990) urges teachers to shift attention from procedural protocols towards a systematic review and critique of the why they do what they do and recognise the consequences of their practices. This is an iterative and continuous process, more a frame of mind, or ongoing habit, reappraised as their career progresses, as society and the needs of learners change and as understanding develops. Collaborative reflective practice offers collective strength, and when changes are determined collegially with all members of the educational community engaging rigorously in the processes of critique, there is combined strength in the commitment to take action.

This chapter outlined some of the prerequisites which enable D&T educators to be active participants in critiquing their practice and the socio-political context in which they are located. These include designerly thinking, innate curiosity and a willingness to ask critical and deep questions which challenge assumptions. A teacher who critiques teaching requires the professional integrity and will to challenge dangerous ideas and make informed pedagogical decisions, and has the personal motivation to take purposeful, progressive action and collegiately enact constructive disruption.

Finally, critiquing teaching is complex and messy. It requires time and effort. When tackled with intent and underpinned by a well-considered philosophy and understanding, it will serve to integrate theory, practice, context and values to the advantage of all stakeholders and those involved in the design and technology education realm.

References

Arnold, J., Edwards, T., Hooley, N., & Williams, J. (2012). Conceptualising teacher education and research as "Critical Praxis". *Critical Studies in Education, 53*(3), 281–295.

Atkinson, S. (2000). Does the need for high levels of performance curtail the development of creativity in design and technology project work? *International Journal of Technology and Design Education, 10*(3), 255–281.

Bandura, A. (1986). Fearful expectations and avoidant actions as co-effects of perceived self-inefficacy. *American Psychologist, 41*(12), 1389–1391.

Bandura, A. (1991). Human agency: The rhetoric and the reality. *American Psychologist, 46*(2), 157–162.

Bandura, A. (1993). Self efficacy in cognitive development. *Educational Psychologist, 28*(2), 117–148.

Bolton, G. (1999). Reflections through the looking-glass: The story of a course of writing as a reflexive practitioner. *Teaching in Higher Education, 4*(2), 193–212.

Bolton, G. (2010). *Reflective practice: Writing and professional development* (3 ed.). London: Sage Publications.

Boud, D., & Walker, D. (1998). Promoting reflection in professional courses: The challenge of context. *Studies in Higher Education, 23*(2), 191–206.

Boud, D., Keogh, R., & Walker, D. (1985). *Reflection: Turning experience into learning*. London: Kogan Page.

Boud, D., Cressey, P., & Docherty, P. (Eds.). (2006). *Productive reflection at work: Learning for changing organizations*. London: Routledge.

Brookfield, S. (1995). The getting of wisdom: What critically reflective teaching is and why it's important. In S. Brookfield (Ed.), *Becoming a critically reflective teacher*. Jossey-Bass: San Francisco.

Brookfield, S. (2009). Understanding reflection as hunting assumptions. UWI/Guardian Life Premium Open Lecture retrieved 29th Mar 2016 https://sta.uwi.edu/cetl/wshops_events/events/openlectures/documents/UWIGLOpenLecture2009-SBrookfield.pdf

Calderhead, J. (1989). Reflective teaching and teacher education. *Teaching and Teacher Education, 5*(1), 43–51.

Carr, W., & Kemmis, S. (1986). *Becoming critical: Education, knowledge and action research*. Lewes: Falmer.

Cochran-Smith, M., & Lytle, S. (1999). Relationships of knowledge and practice: Teacher learning in communities. *Review of Research in Education, 24*, 249–305.

Cochran-Smith, M., & Lytle, S. (2009). *Inquiry as stance: Practitioner research for the next generation*. New York: Teachers College Press.

Craig, C. J. (2010). Reflective practice in the profession: Teaching. In N. Lyons (Ed.), *Handbook of reflective inquiry: Mapping a way of knowing for the profession Reflective Inquiry*. New York: Springer.

Cross, N. (2006). *Designerly ways of knowing*. London: Springer-Verlag.

Dewey, J. (1933). *How we think: A restatement of the relation of reflective thinking to the educative process*. Lexington: D.C. Heath.

Eisner, E. (1985). Five basic orientations to the curriculum. In E. Eisner (Ed.), *The educational imagination on the design and evaluation of school programs* (pp. 61–86). New York: MacMillan Publishers.

Eraut, M. (1995). Developing professional knowledge within a client-centred orientation. In T. R. Guskey & M. Huberman (Eds.), *Professional development in education* (pp. 227–252). New York: Teachers College Press.

Fernandez, C. (2002). Learning from Japanese approaches to professional development: The case of lesson study. *Journal of Teacher Education, 53*(5), 393–405.

Fichtman Dana, N., & Yendol-Hoppey, D. (2014). *The reflective educator's guide to classroom research: Learning to teach and teaching to learn through practitioner's inquiry*. Thousand Oaks: Corwin Press.

Giroux, H. A., & McLaren, P. (1996). Teacher education and the politics of engagement: The case for democratic schooling. In P. Leistyna, A. Woodrum, & S. A. Sherblom (Eds.), *Breaking free: The transformative power of critical pedagogy* (pp. 301–331). Cambridge, MA: Harvard Educational Review.

Guskey, T. R., & Passaro, P. D. (1994). Teacher efficacy: A study of construct dimensions. *American Educational Research Journal, 31*, 627–643.

Hargreaves, D. (1998). *Creative professionalism: The role of teachers in the knowledge society*. London: Demos.

Hennessy, S., McCormick, R., & Murphy, P. (1993). The myth of general problem-solving capability: Design and technology as an example. *Curriculum Journal, 4*(1), 74–89.

Kimbell, R. (2002). Assessing design innovation: The famous five and the terrible two. *Journal of Design and Technology Education, 7*(3), 172–180.

LaBoskey, V. K. (1993). A conceptual framework for reflection in pre-service teacher education. In J. Calderhead & P. Gates (Eds.), *Conceptualizing reflection in teacher development*. London: The Falmer Press.

Larrivee, B. (2000). Transforming teaching practice: Becoming the critically reflective teacher. *Reflective Practice: International and Multidisciplinary Perspectives, 1*(3), 293–307.

McGarr, O., & McCormack, O. (2014). Reflecting to conform? Exploring Irish student teachers' discourses in reflective practice. *The Journal of Educational Research, 107*(4), 267–280.

Mezirow, J. (1990). How critical reflection triggers transformative learning. In J. Mezirow & Associates (Eds.), Fostering critical reflection in adulthood. San Francisco: Jossey-Bass

Morrison, K. (1996). Developing reflective practice in higher degree students through a learning journal. *Studies in Higher Education, 21*(3), 317–332.

Nicholl, B., McLellan, R. (2008) We're all in this game whether we like it or not to get a number of As to Cs.' design and technology teachers' struggles to implement creativity and performativity policies. *British Educational Research Journal, 34*(5), 585–600

Rogers, C. (2002). Seeing student learning: Teacher change and the role of reflection. *Harvard Educational Review, 72*(2), 230–253.

Schön, D. A. (1983). *The reflective practitioner: how professionals think in action*. New York: Basic Books Inc.

Schön, D. A. (1987). *Educating the reflective practitioner: Toward a new design for teaching and learning in the professions*. San Francisco: Jossey-Bass.

Sterling, S. (2009). Ecological intelligence viewing the world relationally. In I. P. Villiers-Stuart & A. Stibbe (Eds.), *The handbook of sustainability literacy* .Retrieved Mar 2016 http://arts.brighton.ac.uk/stibbe-handbook-of-sustainability

Stigler, J., & Hiebert, J. (1998). Teaching is a cultural activity. *American Educator, 22*(4), 4–11.

Thompson, N., & Jan Pascal, J. (2012). Developing critically reflective practice. *Reflective Practice: International and Multidisciplinary Perspectives, 13*(2), 311–325.

Tripp, D. (1993). *Critical incidents in teaching. Developing professional judgement*. London: Routledge Falmer.

Valli, L. (1993). Reflective teacher education programs: An analysis of case studies. In J. Calderhead & P. Gates (Eds.), *Conceptualizing reflection in teacher development* (pp. 11–22). London: The Falmer Press.

Waks, J. L. (1999). Reflective practice in the design studio and teacher education. *Journal of Curriculum Design, 31*(3), 303–316.

A Critique of Technology Education for All in a Social and Cultural Environment

Jacques Ginestié

Abstract As in many countries, technology education (TE) has been introduced into general education in France, where it was introduced in 1985. The initial curriculum expressed a real will to position it in a social science perspective in terms of the relationship humans have with their technical environment on the one hand and, on the other hand, how this technical environment organises social relations between human beings. Thirty years later, it is clear that this orientation has largely failed and that the teaching of technology today is far removed from the original intention as regards both elementary and secondary education. The purpose of this article is not to trace the history of the last 30 years but to understand why such ambition has failed, through a critical study of choices as well as lack of engagements of educational authorities.

The critique is epistemological in terms of the TE curricula as well as its integration in the school structure. Critique is also sociocultural, including understanding of the world in which humans are living and developing. It is ultimately educational whenever societies organise schools for the transmission and development of knowledge, including the interrelations between school subjects and the efficiency of the teaching-learning process.

Keywords Technical object • Individuation • Socialisation • Education

1 About Epistemology

The connection that we maintain with the environment, particularly the technical environment, enables us to establish relationships with the things that make it up. Understanding these relationships inspires many fields of research in social sciences. The designers of the first curriculum saw this teaching as one that would allow youth to build significant understanding of their world. They adopted a particular focus on applying socio-constructivist approaches and systemic

J. Ginestié (✉)
Aix-Marseille Université, EA 4671 ADEF, ENS de Lyon, 32, rue Eugène Cas, 13248, Marseille Cedex 4, France
e-mail: jacques.ginestie@univ-amu.fr

PJ. Williams, K. Stables (eds.), *Critique in Design and Technology Education*, Contemporary Issues in Technology Education, DOI 10.1007/978-981-10-3106-9_11

interactions. They departed drastically from the previous manual and technical education and with technology as an application of science.

Translating this epistemological ambition into curriculum was the main difficulty encountered by the designers. Translating this ambition into school tasks is not easy, specifically in terms of the acquisition of knowledge and development of competences. In this section, we discuss how humans organise their relationships with the technological world they design, produce and use and how this epistemological approach influences the technology education (TE) curricula.

1.1 From Thing to Object

Education aims to organise the perceptible environment, from the most familiar to the farthest, into an intelligible world. This movement is based on the qualification of undefined things that occupy this environment into objects with which humans can know, act and think. To study this complex process, we refer to the theory of systems and the understanding of complex organisations (De Rosnay 1975; Ellul 2004; Le Moigne 1984; Le Moigne and Morin 2001; Morin 1991, 1992, 1994, 1995).

Understanding our environment depends on the objectification of the things that make it up and the networks of complex interrelations between different things. As Simondon theorises, human activity is largely designed to establish relations with these things, and the nature of the relations defines the nature of the objects we build (Simondon 1989a). This human activity becomes more complex with the increase in the complexity of organisations that help humans to act in their environment by rearranging or modifying it. In this sense, the mode of existence of technical objects, i.e. the production by a human of a relationship with one thing, is a manifestation of our attention to the social organisations by which and for which these objects exist. For now, we will look a little further at this process of creating technical objects understood as the objects produced by humans for a specific need.

A priori, understanding one's environment involves establishing relationships with the different things that inhabit it; the object exists when a subject is building a relationship with it; the nature of the object depends on the nature of the subject-object relationship. Technical objects are mediators of our relationship with the world. There is an inbuilt simultaneity in this relationship between the world of undefined things and the symbolic representations we develop to read and build objects that we handle, design, produce, modify, etc. (Simondon 2004). For example, a chair has existence only when a subject considers it in its environment, either because he uses it, or he wants to make one, or he wants to describe it. In all cases, the relationship he establishes will define a different object with which he will act differently—use, manufacture, describe, etc.—within its environment.

The undefined thing becomes an object when a subject establishes a relationship with it. This relationship represents above all a meaning that will guide the actions of individuals. Thus, the same thing can be at the origin of different objects depending on the nature of the relationship.

1.2 Object as Point of View

Each relationship established with the undefined thing presumes different points of view adopted by the subject. Each point of view defines the nature of each relationship and accordingly the nature of the object. The multiplication of point of view increases the ability to understand an undefined thing in its complexity and therefore in its entirety. Thus, a single point of view reduces the relationship to a cause-effect approach and reduces the construction of the object to a very particular case. This particularism is usually enough to explain a localised phenomenon. Going from explanation to understanding broadens from this localism by expanding the network of subject-object relations, on the one hand, and, on the other hand, by including this network in an organisation of interrelations (Simondon 2005). The diversification of point of view allows one to move from an explanatory mode to an understanding mode. The objectification of things that make up our environment is a process based on the subjectivity of the subject that establishes a special relationship in order to build a specific object. This definition of the subject-object relationship constitutes a major contribution to constructivism: humans are a 'machine' to build relationships that make possible the existence of objects (Simondon 1989b). This mode of existence of objects involves complex mental operations, such as categorisation, prioritisation, enlargement, deepening, development, stabilisation and transfer. Simultaneously, the individual learns to act on and with his/her environment and becomes aware of his/her ability to act.

The different points of view always contextualise the object thus constructed. This process of objectification derived from individual subjectivity allows us to specify the object in its environment and thus the subject characterises this environment. Neither the object nor the subject is isolated from its environment; the system so formed treats equally subject-object interrelationships, object-environment interrelationships and subject-environment interrelationships. For Simondon (1964), this process concerns the object just as much as the subject. It is constituted as an individual acting on and with its environment, which Simondon defines as individuation. Acting, explaining and understanding organise the relationship between a human being and his/her environment.

In line with the fact that the nature of the relationships determines different registers of human activities, the register of techniques defines the technical attribute of a class of objects (Andreucci and Ginestié 2002). An object is technical when it carries within itself a technique, that is to say, it proves it can achieve a predetermined goal (Séris 1994). More simply, techniques can be defined as effective traditional acts (Haudricourt 1988; Mauss 1936, 1948), highlighting the fact that there is no technique without transmission and, so without tradition, and no technique without significant material effect (Sigault 1990). The technical nature of the object implies its recognition as a human construction without any ambiguity about its mode of existence as a response to a human need (Simondon 2014). However, this external definition is no longer operative (Cazenobe 1987) when the materiality, causality and finality criteria cease to be supportive of each other or when we refuse to detach the human background from the material object in which it is embedded (Akrich 1987). This is the case, i.e. for the most familiar objects humans use, like a bicycle, pencil, mobile phone, computer, car, chair, etc.

In France, TE, in its first principles, aimed at articulation between action and understanding. TE was designed to provide for pupils the key for understanding their technical environment, i.e. recognising human activity in and with this environment, understanding the purpose and the organisation and ultimately building their own individual relationship with the world of technical objects (Deforge 1993). The curriculum considered the objects manufactured as products of human organisations and from four privileged points of view:

1. As a user, pupils identify the functions of use (which purpose for which need) and the functions of esteem (why a user would buy this object and not another one that responded to the same use).
2. As a family of objects, pupils recognise the purpose of this family and why and how one particular object is different of the other.
3. As a trader, pupils investigate the distribution of the product and the economic dimension of trade and sale.
4. As a manufacturer, pupils study the fabrication of the object with a special attention to its design by transforming all the functions identified at the three previous stages in technical functions and integrated solutions.

By specifying these points of view, the first French curriculum gives opportunity to pupils to elaborate a wide understanding of the mode of existence of technical objects, supported by the use of formal languages and skills. According to this approach, TE might enlarge understanding of the technological character of the manufactured object; but what are the effects of this teaching on the pupils' understanding? This is explored in the following subsection.

1.3 How Pupils Deal with the Concept of Technical Object

As we said, we could think that TE produces an extension of the concept of technical object. The results of a study conducted in the early 2000s (Andreucci and Ginestié 2002) show the limited understanding that pupils have of the concept of technical objects. It also shows that the extension given to this concept is becoming increasingly narrow in school. However, pupils seem to be able to tell the difference between animal and human fabrications. Furthermore, the nature of the material—natural or synthetic—used to achieve the object is not enough to recognise it as a technical object; the biological reference seems to retain the primacy of the technical character even if it concerns an explicit artificialism (e.g. a plastic flower). Pupils seem relatively aware that the notion of human production relates to the technical nature of the criteria.

The same study shows that pupils between 11 and 15 years old have a tendency to reduce the span of this concept during their schooling; they 'naturalise' objects more easily at the end of middle school than at the beginning. This trend does not apply to the category of the best representations, which remains stable over time; in this category, two objects, computers and video games, reinforce their status. Some objects (flute, artificial lake, wiring diagram, boiler, ruler, handsaw, photography, slingshot, handmade pull, pen, etc.) see their status deteriorate significantly to reach the category of poor or very poor representations. The ordinary technical objects, whose use does not justify specific school learning, become poor representation of the concept, while objects that require intellectual investment become good representations. This result is quite surprising when we consider this fundamental concept: a technical object exists because it is designed and manufactured to meet the need of users. These processes of 'naturalisation' or 'instrumentation' illustrate the instrumental genesis by which a tool becomes an instrument (Rabardel 1995).

In another study (Ginestié 2002), we observe that teachers massively focus their school organisation on design-manufacturing and they widely provide pupils with instructions to follow. They reduce the situations of problem-solving. In fact, the technical nature of the objects does not result from school tasks of design-manufacturing. This choice shows, on the one hand, the epistemological limitations of these founding principles and, on the other hand, the limitation of approaches like 'learning by doing'. The question of the activities' purpose becomes central.

1.4 From Gesture to Word

The link between the action and the meaning of the action is just as essential as the subject-object relationship. The French translation of *understand* is *comprendre*, which means *take with*. This clearly fits into the register of actions on and with the environment: humans act knowingly. This awareness of the action simultaneously creates the foundation of the skilfulness (the action) and the meaning of the gesture (the understanding of action). This simultaneity is a key element of individuation; it fits in a constructivist perspective and understanding appears as a very high level of learning (Simondon 1989b). Establishing relationships with the environment involves close articulation between the way to do something and the purpose (why do it and, above all, why do it in this way). Simondon agrees with anthropologists that action is always aimed towards a goal—of which the subject is aware or not—and requires organisation to be realised. The development of understanding involves knowledge acquisition.

For anthropologists, knowledge is a social construction that conditions the development of an individual; he/she constructs his/her own knowledge to enable him/her to act in and with the social group to which he/she belongs. This individual enlarges his/her capacity to act on and with his/her environment by enlarging the field of knowledge (understanding more things) and by deepening some of them (understanding better and better). Objects, relationships, descriptions, representations and symbols all lead to organising and stating human exchanges with the world of things. The development of understanding is a dynamic process with acute consciousness of the action. There is no real hard evidence of the need to develop high-level language to develop techniques but the two always go together (Leroi-Gourhan 1989). Humans simultaneously manufacture tools and symbols (Latour 1991; Rashed 1997). It is the final orientation of their activities that gives meaning to the practice that organises a praxeology indicating ways of acting (Castoriadis 1999). The elements of social diffusion ensue from these ways of acting collected in the praxeology, and this makes the transmission active because it has been enriched with new ways experienced by others (Séris 1994).

The development of technologies and the evolution of tools and objects are systematically accompanied by a development of language (Leroi-Gourhan 1973). Acting evolves with the modes of symbolic representations that describe it (Castoriadis 1975). The completion of an activity refers to the objects produced by humans, to objects that result from activity and which give rise to symbolic elaborations (Latour 1992). An object does not exist at the fleeting immediate time of its use, but it is registered in a more global scheme of meanings, specifically in the register of the potentialities which organise human activity (Wallon 1979). In fact, this dynamic generates another temporality and a new generic order that superpose the temporality and the natural order (Althusser 1994). The established relationship, which simultaneously generates objects and activities on and with these objects, causes a change in the status of the object that becomes a tool as soon as it is registered in this relationship. Tool and word have empirical existence immediately doubled by a universality; they are tool or word as concrete instances of this tool and this word (Leroi-Gourhan 1992). The process of individuation gives the reality and the appearance of control by the individual user but also the appearance and the reality of the control of tool and word over the individual for whom they pre-exist and that could not be without them (Althusser 1986).

This semiotic mediation broadens the subject-object relationship; the object becomes a socially shared tool, carrier of a tradition of uses and meanings that simultaneously determine the manner of use and the aim of such use.

1.5 Languages as Structuration of Codes and Symbols

The codes and the symbols that describe the objects-tools and define their possible fields of action with and on these tools are organised in specific languages according to a grammar, a syntax and a semiology. Languages symbolically organise this double arrangement of procedural schemes (how to do it) and semiotic schemes (why do it in this way). Language, as symbols organisation, makes sense of the continuity from the thing to the object, then to the tool, by making intelligible relationships. The structure of languages, whether for thinking or communicating, is formal: language puts into words that which represents the abstract tools in a meaningful way for oneself (understanding) or for others (communication). The formalisation of the description of the expected outcome also requires a clear description of how to achieve it. The achievement of the task as intended involves formalisation inevitably accompanied by a rationalisation of the arrangement of the means to implement it. Some of the emerging technical characteristics can be attributed to languages. This is the first formalisation of a prescription from the very specific perspective of the expected outcome.

It is also a normalisation as soon as one invites the dialogue partners to adopt a single perspective. Using a set of rules and descriptions of symbolic significance stems from the construction of meaning when it is concerned with the relationship between the signifier and signified. The development of techniques goes together with the development of technical languages. At the same time, the number of languages increases and extends their universality. For example, standardisation efforts have largely left the workshop to extend standards (i.e. AFNOR, CEE, ISO, etc.). The formalisation, description, precision and standardisation of relationships are of course a matter of abstract codes and symbols that make the objects more real when the subject knows about the languages used.

The mediating role played by technical languages has been widely studied in many works (Bessot and Vérillon 1992; Rabardel and Weill-Fassina 1987; Weill-Barais 1997; Weill-Fassina and Rabardel 1985). Indeed, in these approaches, technical languages appear as structuring factors of human action at the same time as they allow us to structure their thoughts. The aim is to reflect on a particular view of an action process. The logical formalisation activity is related to the language because, on the one hand, it autonomously produces statements and, on the other hand, these logical statements can themselves be heteronomous with other statements (Wittgenstein 1961). The technical languages are tools for the formalisation process of concrete achievements. A design drawing says nothing about the designer, let alone the workers responsible for carrying out the design. On the other hand, it allows their actions, the material they handle, how they manipulate it and the result to be codified. There is an extraction of individual praxis of a praxeology that tends to depersonalise this description in order to generalise to the same class of problems regardless of the actors. This process of depersonalisation and contextualisation induces the level of description of generality of a language. This level determines the language's extent to a more or less large community who shares the same meaning unequivocally.

Our ability to produce symbols and to communicate with and through them makes our system of values very unstable. For example, reference to *natural languages* appears to facilitate the understanding of a very artificial environment and accredits the idea that the natural order predominates artificial orders. This inversion of signs is in flagrant contradiction with the development of human social organisations, based on the domination of the natural order by the growing of techniques and languages that allow description, transmission, development and so on. This *naturalisation* of a language by its mastery is closer to the process of *naturalisation* of technical objects we observed.

The semiotic development supported by TE tends to naturalise the relationship to the object and consequently the object itself. This same process is at work when a user loans intentionality to the object he/she uses.

Fundamentally, for a large number of actors in France, TE is not closely linked to the role of specialised languages and teaching these languages because building relationships with the technical environment appears as a premature specialisation. For these promotors of 'learning by doing', technical languages are too conceptual, abstract and theoretical. In fact, they promote a procedural approach to achieve tasks based on skilfulness and techniques, and they don't engage with a semiotic approach based on meaningful elaboration. This choice, in contradiction with Simondon's individuation theory, reinforces the differences between sciences, as the place where pupils think, and technology, as the place where they apply and make, i.e. pupils study the electricity circuit principle in sciences and they apply it in technology by building an electric torch. This approach does not provide a real understanding of what an electric torch is; Ohm's law doesn't explain this. The French curriculum emphasises the development of scientific knowledge through an experimental approach; at the same time, TE aims at mastering professional skilfulness and meanings on the social division of labour, widely based on manufacture organisation (workshop, methods office, design office, etc.) and skill levels (from worker to engineer). This reference to the 'real' industrial world is widely present. The evolution of TE curricula, also including an introduction to science and technology in primary schools over the past 30 years, swings between these two approaches. If the relationships that an ordinary person has with the world of objects he/she manipulates and uses daily are widely present in the curricula, this approach do not find legitimacy in the teaching practices. All the constituent principles of different curricula include this educational dimension of future citizens but without any consequence in teaching plans. Should this education focus on life lessons in which a technical object is studied from every angle? Should pupils study the mode of existence of this object through tools and successive transformations and the social organisations in which this object exists? The answer is not simple and the reality is probably even more complex.

1.6 From Tool to Instrument

The semiotic networks, supported by specific languages, allow a potential action to be assigned to each object created by a relationship and given the status of a tool. The action is part of a project with a goal that can be explicit or not and is organised in the activity of the subject. The tool thus defines a potential action that organises the activity field by setting a field of possibilities. The status of a tool is not an intrinsic characteristic; its recognition as such is based on familiarity, social utility and potentialities. An object could be recognised as a tool by one person and not by another; i.e. a chair is easily recognised for a large part of humanity as the tool to sit in and not for those sitting Indian style on the ground.

Recognition of tools is an ordinary educational situation. A teacher knows the usefulness, the social significance and the potential of the tools he/she introduces in his/her teaching; he/she purposes tasks to organise the discovery of these tools by pupils, and, in fact, he/she expects that pupils recognise the situation as a possible domain of use of these tools to act with and to develop new knowledge. Tools allow a pupil to improve his/her performance, increase the quality of the solution and aim at greater satisfaction. The link between procedural schemes such as *instrumentation* (how to do it) and semiotic schemes such as *instrumentalisation* (why to do it in this way) defines an artefact that simultaneously includes the action, the meaning of this action and the values that guide the whole. This artefact becomes an instrument that supports the subject's activity to reach his/her goal. Therefore, the mastery of the activity is characterised by knowledge and defined as the organised power of action.

The activity is the motor of learning: it defines the organisation of actions produced by a subject to achieve a task. This logic of organisation characterises the strategy for fulfilling the task completion. This strategy is planned and structured a priori. The procedural and semiotic schemes are highly automated with increasing expertise, thanks to the acquired experience. Expertise includes the forecasting and planning that guides the activity through the execution of actions (which actions, in which arrangement, when and for which expected result) in a permanent appreciation of the difference between planned strategy and results. From this perspective, understanding becomes an instrumented activity as instrumental genesis. This complex mental process characterises expertise; ultimately, the gesture performed, the task carried out, seems very simple, independently of their real complexity. The appropriation of the technique by the individual explains the close link with invention and a creative act.

This irreducible dimension of originality at work in all human activity, even in the most automated skilfulness, guarantees the possibility of renewed questioning about the meaning of this gesture, of this technique and of this work. At the end of his documentary, the director Henri-Georges Clouzot tells the painter Picasso of his wish to see him create a masterpiece in 30 min. The painter replies: *No, not in thirty minutes, Mr. Clouzot, but in seventy-five years and thirty minutes* (Clouzot 1956). He implies that the originality of the object lies not only in the implementation of the technical gesture but in the meaning given to it through the experience gained in a long reflexive journey which in turn changes and determines the technical gesture.

1.7 Understanding Through Expert/Novice Activity

The set of interactions between the task to be carried out, knowledge available and activity deployed to achieve it defines the learning situation. A new task produces new knowledge if the subject cannot achieve it with his/her existing knowledge. Spontaneously, the subject tries to address the task with what he/she knows; the willpower to achieve this task implies a conflict. This inability to produce a socially satisfactory solution creates an imbalance, and the subject seeks ways to produce a solution, searching in his/her social environment for new and available tools. He/she develops new relations with new things and tools for the social group in which he/she lives. At the same time, he/she develops procedural and semiotic schemes;

he/she builds a new instrument for acting, that is to say, an operational instrument for achieving this new task. This new instrument has produced a new structure of his/her knowledge that integrates these new schemes and, in the end, a new knowledge—new knowledge that allows him/her, on the one hand, to enlarge the ability to act on and with his/her environment and, on the other hand, to extend his/her understanding of his/her world (Cheneval-Armand and Ginestié 2009). Learning is the result of a new problem that arises for the subject, far from the reproduction of gestures acquired by mimicry in a linear logic; it is a logic of adaptation of the gesture to the situation.

Many authors (Béguin 2007; Bonnardel 2009; Borgmann 2001; Chevalier et al. 2009; Christiaans and Venselaar 2005; Coles and Norman 2005; Darses 2009; Dorst and Cross 2001; Engeström 2000; Fortus et al. 2005; Kroes 2002; Lebahar 2007; Liu 1998; Mathews and Swainston 1992; Tricot et al. 2006) deepen the analysis and understanding of the activity of experts in situations of problem-solving. They focus on the initial description of the problem as the most important part of the activity, allowing an expert to establish the orientation's base of his activity by organising the planning of actions in time. The solution is the consequence of this initial description and the choices made at this moment. Three phases divide the initial description of the problem: (1) a basic description of the problem, (2) a theoretical description of the problem and (3) an exploratory analysis of the problem to ensure some qualitative properties of the solution. This initial description is difficult to analyse. It is a mental activity, which does not leave particular traces. The routines used by experts are familiar, with a high level of automation. The initial description organises an orientation base of activity; an expert progresses step by step, following this orientation, collecting information that confirms or not the effectiveness of each action, anticipating the results and continuing with the same plan (or modifying it).

In contrast, novices make a partial initial description; they discover a small number of constraints, mostly explicit constraints. They adopt a strategy of immediate transformation of the constraints, one by one, into local solutions without integration in a comprehensive solution. They have a poor idea of the expected result and progress by trial and error. The general solution is a collection of local solutions that are more or less complete (Amigues and Ginestié 1991).

A teacher has to organise the conditions of initial description. Classes are organised as workshops with a large autonomy devoted to groups of two or three pupils. They have at their disposal a description of the job to be done, a technical dossier and the tools and resources to do the job. In many cases, there is a detailed description of the order of actions to execute. By this description, the teacher shortcuts the initial description of the task, indicating the orientation base and the planning of actions. Guided in the task's achievement, pupils perform the task with a good rate of success and obtain a solution which conforms to the expectations. But, because they are centred on a procedural resolution of the task, without 'real' problems to solve, they just enact procedural schemes and they have a very poor construction of semiotic schemes. Finally, they perform the task with a low level of understanding and learning. In this perspective, TE appears to be a subject without any great interest (Ginestié 2008b).

The initial curriculum aims at helping pupils to understand the world of objects produced by humans and to be aware of their ability to use them for acting with and in their environment, but epistemological reflection was not sufficiently deep to give a sufficient ambition to TE. Pedagogical guidance, coupled to few 'real' problem-solving situations, reduces the impact on pupils' learning.

2 About the Sociocultural Inscription of Technology

By focusing on the relationship with the world of technical objects, TE gives prominence to the pupils' individuation. Their sociocultural dimension remains largely implicit. Development of technologies suggests a discourse on technique within a sociocultural tradition where innovations meet specific social pressures. TE, in the first curriculum, claims to understand a technical object within the social organisations for and through which it exists and aims at understanding this not only in terms of structural aspects of technical objects but also as social facts. It is not enough to answer the question 'how does it work?'; we should also ask 'why does this object exist?' and 'why does it exist in this shape?'.

2.1 Object as Social Artefact

The construction of a discourse on technique enlightens human and social sciences such as history, anthropology, ethnology, sociology, economics and ergonomics (Perrin 1988; Sigault 1985; Spengler 1969). Consider Simondon's theses on the promotion of awareness of the meaning of technical objects in their social reality (Simondon 1989a). Objects contain a human reality in order to fight against any blindness towards technology, whether in terms of technophobia or an unconditional acceptance of progress. Awareness of the existence of technical objects is consubstantial to consciousness of the existence of the subject as an individual and as member of a particular social group: *tell me what objects you are using (and how you use them) and I tell you who you are.* If the technical objects appear in a special human context, conversely, the human is also part of a technical universe that existed previously, governing and changing his/her future. Lecourt (1997) illustrates this double game of influence through evolutions of the digital world. Based on the banalisation of these technologies (computers, notepads, smartphones, etc.), this world becomes unattainable for the one who ignored this, not because he lacked a service he has today, but because his future is profoundly modified, whether it involves his objective relationship with time and also all its emotional relationships with others. Techniques are not external to humans; they are derived from life and integrate and set out its standards (Lecourt 1997, 2009). In other words, there is a positive mutual inclusion of human life in the world of technique. This does not exclude appropriation by the user of the object but generates invention and creativity; the sociocultural dimension exists through the tools that human groups use, share, develop and create.

We just underline the power to act that the instrument gives to each human and, by evidence, the role of TE as the way to access to tools that characterise the social group in which the young live. In this sense, the transmission of tools is the process of conservation, including intergenerational, of their achievements; it is also a process of development and expansion as these sharing tools are based on the sum of the knowledge constructed by each individual who composes the group. Conservation is a process of withdrawing into oneself, into one's social group, whereas development is a process of opening up to others.

2.2 Dialectic Socialisation-Individuation

In French, the term knowledge has two significations:

- *Savoir* mainly refers to the individual's potential for acting; it describes the knowledge, the know-how and the values brought by the person. It defines the individual's understanding and represents his/her potential of actions (consciousness within the individual of what it is possible to do because he/she knows).
- *Connaissance*, which mainly refers to the social institution, defines the social heritage and represents a potential of knowledge (awareness of the individual that, in his/her environment, there is a social group who knows).

The acquisition of *savoir* by an individual thus allows him/her access to *connaissance*. The awakening to this double level of consciousness, self-awareness and consciousness of the other, is a strong element of learning process, based on the individuation and socialisation of the person. Knowledge is structured through this dialectic tool-instrument in which *connaissance* is linked to the tool, whereas *savoir* is linked to the instrument. Procedural schemes are easily identifiable; semiotic schemes fall within this interpenetration between the meaning that each gives to a thing and the meaning commonly assigned to this thing by the social community. This interpenetration induces the nature of the relationship by simultaneous combination of both the processes of:

- Socialisation, by using a car, a knife, a chair, an idea, etc.; the individual marks his/her belonging to this community that produces and uses such objects.
- Individuation, by marking his/her familiarity of these objects he/she uses as my car, my knife, my chair, my place, my job, my idea, etc. (Lebahar 1994).

Such dialectic testifies to the complexity of these relationships and aims to open up many other essential debates out of this paper. In sum, the broad sharing of the same objects, the same categorisations of these objects and the same tool-instrument potentialities defines the homogeneity of a social group. Languages, whether general or very specialised, are the semiotic instruments that allow members of a group to share their uses, their potential and their mode of production or evolution. The logic of conservation aims to preserve the homogeneity of the group by standardising these exchanges, codifying and developing them.

Expression like 'good usage or practice' valorises the individual praxis in the light of social praxis shared by the group. Doing as others do while proclaiming one's own personality is an ongoing internal dialogue that generates socio-cognitive conflict, i.e. a tension between the desire to conform to the group and the desire to mark one's difference. Far from reproducing a gesture identically, humans transform and modify the gesture unpredictably. The example of the above-mentioned smartphone, as trivial as it is, reveals that its use is not the same for everyone. Depending on the person, it can become a *negation of its existence*, for those who are de facto ignored (partially or entirely) by the user, or as a means *to expand the spectrum of feelings and reasoning*, an instrument for work or for socialising (Schwartz and Durrive 2009). The group becomes heterogeneous through its openness to other individuals, even if it is only to ensure the intergenerational transmission necessary for its own conservation.

From a systemic point of view, phases of instability and stability alternate. This alternation allows stable states, as an unstable equilibrium position, to be found after destabilisation phases generating evolutions. The slightest disruption—i.e. the affirmation of a new point of view, new idea, new method, more rational, more cost-efficient, more effective, etc.—will undermine the internal organisation. Therefore, the group will reorganise itself to find a new unstable equilibrium position, thus initiating a new stable phase. This process is a permanent dialogue between tools that constitute the heritage of the group and the instruments brought by the individuals. The incoming of new *savoir* brought by an individual unsettles the organisation of the *connaissances* of the group and vice versa. The stakes of knowledge are always stakes of power. Evidently, it is easier to promote instruction based on access to *connaissances* than education based on elaboration of *savoirs*.

2.3 Understanding as Complex Social Activity

Understanding, in the sense built here, is a decisive issue in education and an ambition in 1985 for the TE curriculum. School is responsible for giving meaning to social actions of pupils in their social community, and TE should help them to develop relationships with their environment, which is highly technologised. This idea is not new. For Dewey, technique and innovation offer a better understanding of our environment that science alone cannot make intelligible (Dewey 1916). Educated citizens are aware of the essential terms of their environment. Sharing the democratic control of development in our societies involves educating pupils in the social logic of the technological world. Knowledge sharing is performed in order to share power.

Technology, taking effect in the heart of human activities, breaks barriers that previously separated people; it expands human relationships. It creates the interdependence of interests on a wide scale. It brings with it the belief that mastery of nature for the benefit of humanity is possible and leads humans to look at the future and not the past. Now they look at the future with the firm belief that well-used intelligence can free us from the evils formerly thought inevitable. It is no longer a dream that devastating epidemics can be overcome; it is realistic to expect to overcome poverty. Technology has familiarised us with the idea of development in the gradual and constant improvement of the fate of the human community (Dewey 1916). This way of thinking at the beginning of the last century was present in the ambition of the first TE curriculum; 30 years later, this hymn to the glory of progress has been widely questioned. Without engaging in philosophical or political debate on the values of progress and its consequences, we can see that these 30 years have shown the social role played by teachers and schools faced with social evolution (globalisation of the economy, impact of technological developments on the natural environment, increasing social inequality, etc.). These debates have affected particularly the community of technology's teachers regarding the meaning of their teaching and its relevance. The curriculum has been progressively expurgated of all these points of debates. It has given gradually priority to the realisation of procedural tasks (by guidance of actions) to the detriment of the construction of social meanings.

The development of the guidance of action is not a simple pedagogical convenience; it is also a way to eliminate any significant alternative to the construction of critical sense by pupils. Faced with non-problematic tasks, guided in achieving them with a low autonomy of action, pupils repeat storylines written by the teacher that they perform without great motivation. At the end, they express little interest in TE and a disaffection with this teaching. Several studies conducted during this period show this (Ginestié 2002, 2005, 2008a, c).

3 Education for Developing Socialisation Through Individuation

If we want human communities to continue to improve, it is necessary for education to give young people the intellectual means of invention and innovation (Howes 2008). This formulation sounds like a slogan: education is the way to develop humanity. By linking the fate of human communities with the development of intellectual means, Howes makes explicit the relationship between *connaissance* and *savoir* and between socialisation and individuation. TE is a way to awaken youth to collective knowledge to empower them to invention and innovation, i.e. to carry on in turn the progress of knowledge. For school entry, knowledge is organised and grouped into fields that integrate progression, what needs to be studied and the order in which it is studied, concepts and procedures for the use of these concepts. School subjects are social constructions supposedly representative of the social knowledge they organise, but there is not a direct link between both.

The reconstruction of fields exclusively for teaching is fragile because they do not derive from social references but were organised for teaching. The references have no particular role in the definition of knowledge for school purposes (Cheneval-Armand and Ginestié 2009). Based on a traditional academic dividing, school subjects no longer meet the modern challenges of teaching, education and training (Johsua and Dupin 2003). The logics of structuration of school knowledge have tended to give themselves a coherence of progression that is all their own. The disaffection linked to low social enhancement of technological studies adds to the lack of interest related to the organisation of the teachings mentioned previously.

The role of the teacher, specifically for TE, extensively evolves with the changes of traditional school organisations. The meaning of a school for all changes the balance between instruction and education under the pressure of the acceleration and globalisation of sources of knowledge, through the digital networks. This dynamic promotes a comprehensive approach to the social, economic, cultural and technical environment of the pupils. The school can no longer play the almost exclusive role of transmitting knowledge; the mastery of knowledge slips gradually and resolutely towards the control of access to knowledge resources.

If we follow the thesis of Simondon, learning is the construction of relationships that will allow the pupil to act with instruments he/she constructs (Rabardel and Béguin 2005). This thesis is quite relevant to the evolution of school organisations, on the one hand, and, on the other hand, the logic of learning of individuals. The teaching-learning process is the result of the confrontation of the three separate logics: of the curriculum, of the teacher and of the pupil. Referring to the theories of activity, the school situation is characterised by the task entrusted by the teacher to the pupils but it is not a guarantee of the engagement of pupils to complete the task.

It is not enough merely to give a problem to a pupil and ask him/her to solve it. The teacher must play a decisive role in pupils' efforts to become involved in the task as well as to supervise their activity. The different modes of interactions put in place by the teacher characterise the different kinds of teaching-learning processes. They determine whether or not devolution of the problem occurs and whether or not pupils make progress during its accomplishment. This process is one of the key elements in constructing knowledge and pupils' cognitive progress, notably through discursive episodes. The teacher plays the role of facilitator in building knowledge aims (Roux 2003a, b, c; Trognon et al. 2006). The task must exemplify the importance of the knowledge targeted by teaching. The obstacles must be salient and the learning environment must allow for overcoming them. The task must allow supervision of the pupils' learning activity. Pupils must do things they have never done before; the problem must be original, and the pupils must identify obstacles they need to overcome in order to find the solution within the constraints incorporated in the problem. The pupils use the task-oriented environment; they choose the available resources (or the means of accessing them). In order to overcome each obstacle, pupils plan a chronology of their actions and structure their activity by defining and by anticipating the use of available resources aims (Rabardel 1995; Vérillon and

Andreucci 2006). Such task organisation goes beyond the procedural descriptions usually detailed in the traditional guided learning (Verillon and Rabardel 1995). In fact, the problematic dimensions of a situation must be recognised as such by the pupil. The teacher cannot claim that this task is a problem to solve for him.

The teacher can impose on the pupil the achievement of the task as school duty, but not to solve the problem. It is not easy to design and build tasks, which make the obstacles salient, which make resources available and which organise the conditions to maintain their activity. This supposes the dynamic management of interactions between teacher, pupils and knowledge, which is not in the tradition of French schools; the teacher is tempted, through these interactions, to lead pupils to the solution by imposing his/her own logic.

4 Some Conclusions

How should we understand the current situation of TE in France? The curriculum was designed on a particularly relevant epistemological foundation; numerous researches have widely accompanied its establishment and its evolution for 30 years, and, at the end, we observe the disaffection of pupils and the lack of learning. The answer to such a question is not black and white; however, it is very significant in terms of the evolution of the French educational system during the same period. Many surveys (such as OECD's PISA) show that the gap between the educational attainment of young people is increasing and these differences add to the social inequality. A very old elitist tradition, reinforced by the strong logic of academic subjects, contradicts the democratic principles of free education for every child, whatever his/her social origins, for ensuring equal opportunities.

From the early years of primary education, school performances are the baseline of pupils' assessment, and the school subjects' hierarchy accentuates the place of abstraction and encyclopaedism. In this genesis of education in France, the main way of study is general education. When a pupil fails, he is *oriented* in a technological or vocational course, to reach a professional diploma; the diploma level determines the level of the intended job and therefore the level of social integration of the young graduate. This system was efficient to help France to move from a rural society to an industrial economy, but it has resisted neither mass education nor the economic and social evolution; for 40 years, education policies have tried to make the education system come closer to the needs of contemporary society.

The introduction of TE for all, and we can appreciate the magnitude of the ambition of its designers, was an answer to this evolution. Based on the understanding of the environment, it was also a break with the established hierarchy of school subjects; technology was no longer where pupils made things but the place where they understood why. By articulation of action and reflection in problem-solving situations, it proposed to develop original educational project-based approaches.

This chapter raises some critical issues that allow understanding of the failures and successes of the implementation of such education. Two major lines of action emerge from this educational policy. The first axis concerns the curricula relating to TE but also to other academic disciplines such as mathematics, science, French, history, arts, etc. Curricula should foster broader articulation of individuation and socialisation and be less prescriptive about how to teach and more open on issues of knowledge that make sense socially and culturally.

The second axis concerns the evolution and the increase in skills of teachers, constantly evolving, oscillating between formal academicism and professionalisation. Training should allow future teachers to acquire knowledge and develop the skills necessary to implement changes in curricula, to design and develop educational situations that give each pupil the opportunity to be constructed (individuation) within the framework of social, economic and cultural society of the twenty-first century (socialisation). This second axis means that teaching is a profession that is learned. Teacher education supposes to give meaning to this job; it is not a sum of encyclopaedic knowledge and skilfulness.

In 2013, a big reform opens up these projects explicitly of recasting schools. It aims at a profound change in curricular structures, including emphasising the interdisciplinary, rethinking the academicism of the learned knowledge and paving the way for educational practices that promote a project-based approach and problem-solving. It inscribes teacher training within a university vocational education at master's level. To support these axes, it anticipates the development and structuring of research in education. The goal is explicit: education seeks to educate citizens who can think for themselves, extrapolating from 'I possess therefore I am' to 'I think therefore I am'. The philosophy of the Enlightenment continues to inspire French educational policy!

Beyond this critique of the French curriculum, this chapter brings a contribution to a more general approach of technology education as a tool for critiquing other curricula. The dialectic individuation-socialisation is another way to think of the place of TE as essential part of modern education for all. It highlights the importance of better understanding the teaching-learning process, including through teacher-pupil and pupil-pupil interactions and the effectiveness of organisations implemented. As we have seen, the place and the role of TE in our modern education systems involve many different approaches to be considered and implemented. The construction of this theoretical framework involves references to philosophy, anthropology, sociology, psychology and ergonomics but also the engineering sciences and the sciences in general. Thus posed, this framework is revealed to be a great tool to analyse the actual activity of pupils and teachers and so to query the real curricula. This perspective opens some opportunities for international investigations.

Bibliography

Akrich, M. (1987). Comment décrire les objets techniques ? *Techniques et culture, 9*, 49–63.
Althusser, L. (1986). *Pour Marx*. Paris: Éditions la Découverte.
Althusser, L. (1994). *L'avenir dure longtemps* (Nouvelle édition ed.). Paris: Le livre de poche.
Amigues, R., & Ginestié, J. (1991). Représentations et stratégies des élèves dans l'apprentissage d'un langage de commande. *Travail Humain, 54*(1), 1–19.
Andreucci, C., & Ginestié, J. (2002). Un premier aperçu sur l'extension du concept d'objet technique chez les collégiens. *Didaskalia, 20*, 41–65.
Béguin, P. (2007). The role of innovation and sociocognitive context in worker-designer interactions: A developmental approach. *Travail Humain, 70*(4), 369–390.
Bessot, A., & Vérillon, P. (1992). *Espaces graphiques et graphismes d'espaces*. Grenoble: la Pensée Sauvage.
Bonnardel, N. (2009). Design activities and creativity: From the analysis of cognitive factors to creative design support. *Travail Humain, 72*(1), 5–22.
Borgmann, A. (2001). Opaque and articulate design. *International Journal of Technology and Design Education, 11*(1), 5–11.
Castoriadis, C. (1975). *L'Institution imaginaire de la société*. Paris: Éditions du Seuil.
Castoriadis, C. (1999). *Figures du pensable*. Paris: Ed. du Seuil.
Cazenobe, J. (1987). Esquisse d'une conception opératoire de l'objet technique. *Techniques et culture, 10*, 61–80.
Cheneval-Armand, H., & Ginestié, J. (2009). Des pratiques sociales aux savoirs experts. Une analyse de la prévention des risques professionnels dans les métiers du génie énergétique. *Didaskalia, 35*, 11–36.
Chevalier, A., Anceaux, F., & Tijus, C. (2009). Design activities: Creativity, cooperation, support. *Travail Humain, 72*(1), 1–4.
Christiaans, H., & Venselaar, K. (2005). Creativity in design engineering and the role of knowledge: Modelling the expert. *International Journal of Technology and Design Education, 15*(3), 217–236. doi:10.1007/s10798-004-1904-4.
Clouzot, H.-G. (Writer). (1956). *Le mystère Picasso [DVD]*. In H.-G. Clouzot (Producer). France: Gaumont.
Coles, R., & Norman, E. (2005). An exploration of the role values plays in design decision-making. *International Journal of Technology and Design Education, 15*(2), 155–171. doi:10.1007/s10798-005-8274-4.
Darses, F. (2009). Collaborative design problem-solving. *Travail Humain, 72*(1), 43–59.
De Rosnay, J. (1975). *Le Macroscope: vers une vision globale*. Paris: Éditions du Seuil.
Deforge, Y. (1993). *De l'éducation technologique à la culture technique pour une maîtrise sociale de la technique*. Paris: ESF éditeur.
Dewey, J. (1916). *Democracy and education: An introduction to the philosophy of education (French traduction; (1990); Deledalle, G.; Démocratie et Éducation: introduction à la philosophie de l'éducation)*. Paris: Armand Colin.
Dorst, K., & Cross, N. (2001). Creativity in the design process: Co-evolution of problem–solution. Design Studies, 22(5), 425–437. doi:http://dx.doi.org/doi:10.1016/S0142-694X(01)00009-6.
Ellul, J. (2004). *Le système technicien*. Paris: Le Cherche midi.
Engeström, Y. (2000). Activity theory as a framework for analyzing and redesigning work. *Ergonomics, 43*(7), 960–974.
Fortus, D., Krajcik, J., Dershimer, R. C., Marx, R. W., & Mamlok-Naaman, R. (2005). Design-based science and real-world problem-solving. *International Journal of Science Education, 27*(7), 855–879. doi:10.1080/09500690500038165.
Ginestié, J. (2002). The industrial project method in French industry and in French schools. *International Journal of Technology and Design Education, 12*(2), 99–122. doi:10.1023/A:1015213511549.

Ginestié, J. (2005). La production attendue n'est peut-être pas celle escomptée. In P. Vérillon, J. Ginestié, B. Hostein, J. Lebeaume, & P. Leroux (Eds.), *Produire en technologie à l'école et au collège* (pp. 333–346). Lyon: INRP.

Ginestié, J. (2008a). *The cultural transmission of artefacts, skills and knowledge: Eleven studies in technology education* (R. Watson, Trans.). Rotterdam: Sense Publishers.

Ginestié, J. (2008b). From task to activity, a re-distribution of the roles between the teacher and the pupils. In J. Ginestié (Ed.), *The cultural transmission of artefacts, skills and knowledge: Eleven studies in technology education* (pp. 225–256). Rotterdam: Sense Publishers.

Ginestié, J. (2008c). Konzepte einer Technischen Bildung in Frankreich (C. Vitale, Trans.). In E. Hartmann & W. Theuerkauf (Eds.), *Allgemeine Technologie und Technische Bildung* (pp. 107–125). Frankfurt am Main: Peter Lang.

Haudricourt, A.-G. (1988). *La Technologie science humaine: recherches d'histoire et d'ethnologie des techniques*. Paris: Éditions de la Maison des sciences de l'Homme.

Howes, E. V. (2008). Educative experiences and early childhood science education: A Deweyan perspective on learning to observe. *Teaching and Teacher Education, 24*(3), 536–549.

Johsua, S., & Dupin, J.-J. (2003). *Introduction à la Didactique des Sciences et des Mathématiques* (3e ed.). Paris: Presses Universitaires de France.

Kroes, P. (2002). Design methodology and the nature of technical artefacts. Design Studies, 23(3), 287–302. doi:S0142-694X(01)00039-4.

Latour, B. (1991). *Nous n'avons jamais été modernes essai d'anthropologie symétrique*. Paris: Éd. la Découverte.

Latour, B. (1992). *Aramis ou l'amour des techniques*. Paris: Éd. La Découverte.

Le Moigne, J.-L. (1984). *La théorie du système général* (2ème ed.). Paris: Presses Universitaires de France.

Le Moigne, J.-L., & Morin, E. (2001). *L'Intelligence de la Complexité*. Paris: Éd. l'Harmattan.

Lebahar, J.-C. (1994). *Le design industriel: sémiologie de la séduction et code de la matière*. Marseille: Éditions Parenthèses.

Lebahar, J.-C. (2007). *La conception en design industriel et en architecture. Désir, pertinence, coopération et cognition*. Paris: Hermès-Lavoisier.

Lecourt, D. (1997). *L'avenir du progrès*. Paris: Editions Textuel.

Lecourt, D. (2009). *L'âge de la peur: Science, éthique et société*. Paris: Bayard.

Leroi-Gourhan, A. (1973). *Évolution et techniques: milieu et techniques*. Paris: Albin Michel.

Leroi-Gourhan, A. (1989). *Le geste et la parole*. Paris: Albin Michel.

Leroi-Gourhan, A. (1992). *L'homme et la matière* (1943th ed.). Paris: Albin Michel.

Liu, Y.-T. (1998). Personal versus cultural cognitive models of design creativity. *International Journal of Technology and Design Education, 8*(2), 185–195.

Mathews, M., & Swainston, D. (1992). Business studies within design and technology: Continuity and progression. *Economics, 28*(117), 30–33.

Mauss, M. (1936). Les techniques du corps. *Journal de Psychologie, 32*(176), 279–327.

Mauss, M. (1948). Les techniques et la technologie. *Journal de Psychologie, n° spécial: Le travail et les techniques (dirigé par I. Meyerson et L. Febvre).*

Morin, E. (1991). *La Méthode, Les Idées. Leur habitat, leur vie, leurs moeurs, leur organisation*. Paris: Editions du Seuil.

Morin, E. (1992). *La méthode, La connaissance de la connaissance: anthropologie de la connaissance*. Paris: Éd. du Seuil.

Morin, E. (1994). *La complexité humaine*. Paris: Flammarion.

Morin, E. (1995). *La Méthode: la nature de la nature*. Paris: Seuil.

Perrin, J. (1988). *Comment naissent les techniques: la production sociale des techniques*. Paris: Publisud.

Rabardel, P. (1995). *Les hommes et les technologies, approche cognitive des instruments contemporains*. Paris: Colin.

Rabardel, P., & Béguin, P. (2005). Instrument mediated activity: From subject development to anthropocentric design. *Theoretical issues in ergonomics, 6*(5), 429–461.

Rabardel, P., & Weill-Fassina, A. (1987). *Le Dessin Technique*. Paris: Hermès.

Rashed, R. (1997). *Histoire des sciences arabes, Technologie, alchimie et sciences de la vie*. Paris: Éd. du Seuil.
Roux, J.-P. (2003a). Analyse interlocutoire, dynamiques interactives et étude des mécanismes des progrès cognitifs en situation asymétrique de résolution de problèmes. *L'orientation scolaire et professionnelle, 3*(3), 475–501.
Roux, J.-P. (2003b). Episodes interdiscursifs maître-élève(s) et construction de connaissances dans un dispositif d'enseignement-apprentissage de type socio-constructiviste en CM1. *Bulletin de Psychologie, 56*(4), 531–543.
Roux, J.-P. (2003c). *The interlocutory logic analysis as a methodological approach in studying semiotic mediations: interest, difficulties, limits*. Milan: Paper presented at the XIth European Conference on Developmental Psychology.
Schwartz, Y., & Durrive, L. (2009). *Entretiens sur l'activité humaine: l'activité en dialogues*. Toulouse: Octarès.
Séris, J.-P. (1994). *La technique*. Paris: Presses Universitaires de France.
Sigault, F. (1985). More (and enough) on technology! *History and Technology, 2*(2), 115–132.
Sigault, F. (1990). Folie, réel et technologie. *Technique et culture, 15*, 167–179.
Simondon, G. (1964). *L'individu et sa genèse physico-biologique: l'individuation à la lumière des notions de forme et d'information*. Paris: Presses universitaires de France.
Simondon, G. (1989a). *Du mode d'existence des objets techniques* (Réédition ed.). Paris: Aubier.
Simondon, G. (1989b). *L'individuation psychique et collective à la lumière des notions de forme, information, potentiel et métastabilité*. Paris: Aubier.
Simondon, G. (2004). *Deux leçons sur l'animal et l'homme*. Paris: Ellipses.
Simondon, G. (2005). *L'individuation à la lumière des notions de forme et d'information*. Grenoble: J. Millon.
Simondon, G. (2014). *Sur la technique*. Paris: PUF.
Spengler, O. (1969). *L'homme et la technique* (Réédition du texte en allemand de 1936th ed.). Paris: Éditions Gallimard.
Tricot, N., Rajaonah, B., Popieul, J. C., & Millot, P. (2006). Design and evaluation of an advanced driver assistance system: The case of auto-adaptive cruise control. *Travail Humain, 69*(2), 129–152.
Trognon, A., Ball, M., Schwarz, B., Petrel-Clerraont, A.-N., & Marro, P. (2006). Logique interlocutoire de la résolution en dyade d'un problème d'arithmétique. *Psychologie Française, 51*(2), 171–187.
Verillon, P., & Rabardel, P. (1995). Cognition and artifacts: A contribution to the study of though in relation to instrumented activity (English). *European Journal of Psychology of Education, 10*(1), 77–101.
Vérillon, P., & Andreucci, C. (2006). Artefacts and cognitive development: How do psychogenetic theories of intelligence help in understanding the influence of technical environments on the development of thought? In M. De Vries & E. Mottier (Eds.), *International handbook of technology education: The state of the Art* (pp. 399–416). Rotterdam: Sense Publishers.
Wallon, H. (1979). *De l'acte à la pensée*. Paris: Flammarion.
Weill-Barais, A. (1997). De la recherche sur la modélisation à la formation des professeurs de physique: Comment s'opère la transition ? *Skholê, 7*, 141–155.
Weill-Fassina, A., & Rabardel, P. (1985). Technical drawing, a graphic instrument of professional thinking and communication. *Travail Humain, 48*(4), 301–305.
Wittgenstein, L. (1961). *Investigations philosophiques*. Paris: Gallimard.

Part III
The Application of Critique

Disruptive Technologies

David Barlex

Abstract The question driving this chapter is *How can we use the idea of disruption to enable young people at school to critique technology?* This chapter describes how pupils might develop the ability to critique through a consideration of disruptive technologies in their design and technology lessons. It lists the features of disruption as developed by McKinsey Global Institute and exemplifies how these can play out in society in unpredictable ways. It considers the Gartner Hype Cycle as a way of describing the life cycle of a technology and scenario building as a means of thinking about what might happen when particular technologies are deployed. It identifies and discusses possible approaches by which pupils might critique so-called disruptive technologies. It considers in some depth how these approaches might be applied in critiquing the disruptive technology robotics. The chapter finishes with a commentary briefly discussing how a curriculum and teachers might respond to engaging pupils with critique.

Keywords Disruption • Disruptive technologies • Technology life cycles • Scenario building • Robotics

1 Introduction

The thrust of this book is that teaching young people to critique is an indispensable part of design and technology education. But this critique has to be present for more than its own sake if it is not to be reduced to an academic exercise. This chapter has identified a set of technologies that are likely to appeal to young people and significantly affect their lives, so-called disruptive technologies. Hence, the question driving this chapter is *How can we use the idea of disruption to enable young people at school to critique technology?* The chapter begins with a preamble that discusses the idea of disruption identifying so-called features of disruption and suggesting nine technologies that meet these features and are appropriate for

D. Barlex (✉)
Exeter University Associate, Exeter, UK
e-mail: david.barlex@btinternet.com

P.J. Williams, K. Stables (eds.), *Critique in Design and Technology Education*, Contemporary Issues in Technology Education, DOI 10.1007/978-981-10-3106-9_12

consideration by school pupils aged 11–16 years. Then the chapter considers the 'life cycle' of a technology justifying the use of the Gartner Hype Cycle as a useful and appropriate way for pupils to think about this. Next the chapter considers the building of scenarios which indicate how technologies might play out in society and how young people at school might engage with scenario building as a precursor to critique. Then the chapter discusses possible approaches to critique that pupils might use. Next the chapter considers the applications of these approaches to one specific disruptive technology that is particularly relevant to design and technology education: robotics. The chapter finishes with a commentary which returns to the driving question and discusses briefly how a curriculum and teachers might respond to engaging pupils with critique.

2 Preamble

The introduction of any technology will open up new possibilities for human activity and interaction. The impact of many technologies is limited, perhaps to a subset of the population or to relatively trivial changes in behaviour. For example, the invention of the ballpoint pen made writing in ink an easier, less-messy and, critically, cheaper activity. But, though it probably led to more people writing more easily, the ballpoint didn't significantly change society.

Some technologies, however, have a huge impact on society, changing in significant ways the activity and interactions of people – though these changes may not be immediately apparent. Such technologies can be described as 'disruptive'. A classic example of a disruptive technology is the development of the manufacturing system that enabled the mass manufacture of the motorcar and put the purchase of a car within the means of the average working 'man'. Once the motorcar 'took off' by means of this system, a whole train of events were set in motion, although it took some time for their impacts to be recognised, for example:

- Hugely increased mobility for ordinary people leading to changed work and family patterns.
- Major impacts on the environment, including contributing to climate change.
- Suburban sprawl as the requirement for ordinary people to live close to their workplaces and shops has reduced.
- The creation of whole new industries (such as repair and maintenance garages) whilst essentially killing off others (farriers, blacksmiths).

New technologies can also produce disruption in apparently unrelated social spheres. Johnson (2014) describes how Gutenberg's invention of the printing press in the middle of the fifteenth century led to cheap printed books and a rise in literacy levels—which revealed that a large number of people were myopic, thus leading to a demand for spectacles, a technology that hitherto had really only been used by monks. Within a 100 years of the invention of the printing press, there were thousands of spectacle makers across Europe, and the ready availability of cheap lenses meant that people were able to explore the manipulation of light with them, and soon the telescope and other optical devices were invented. And of course the whole system of book sellers and distribution channels also emerged as a result of the new printing technology.

Clearly, therefore, it is tricky to tell in advance which new technologies might be disruptive, but the ability to look back at technologies we now recognise as having been disruptive helps, and there is a whole industry of futurists working to try and discern the impact of newly emerging technologies. The McKinsey Global Institute (Manyika et al. 2013) has suggested some features that mark out a technology as having the potential to be disruptive. Each feature is exemplified through the impact of the popularisation of photography at the beginning of the twentieth century.

- They upset the status quo, for example, overturning existing hierarchies and offering the possibilities of both more and less democratic hierarchies.

The advent of an easy-to-use camera accompanied by an inexpensive service to develop the negatives and produce black and white prints by George Eastman (Kodak) gave the general populace access to photography which had hitherto only been available to rich people with specialist knowledge.

- They alter the way people live and work, for example, increasing or decreasing employment opportunities, changing the knowledge and skills required for certain kinds of employment, shifting the expectations of education systems and altering relationships.

Enabling ordinary people to take photographs altered the way people worked in providing employment for darkroom technicians who processed the film and the way people lived in providing a popular hobby.

- They reorganise financial and social structures, for example, by redistributing financial rewards towards those who are deploying these technologies.

The Eastman Company became financially very successful in a market that had not previously existed and enabled the employment of photographers in many different industries.

- They lead to entirely new products and services.

Cameras for the domestic market became more sophisticated as people learned more about photography and wanted to take better photographs leading to the development of the single-lens reflex camera, the use of light metres, light metres becoming integrated into the cameras and the availability of coloured film.

Barlex, Givens and Steeg (2015) have identified nine technologies for consideration within design and technology education:

- Additive manufacturing (AM)
- Artificial intelligence (AI)
- Augmented reality (AR)
- Big data

- Intelligent matter
- Internet of things (IoT)
- Neurotechnology
- Robotics
- Synthetic biology

They have justified this selection on two grounds. Firstly, they meet the McKinsey Global Institute features of disruption, and secondly they think they have aspects that are relevant and potentially engaging to young people in schools. A brief description of each technology is included in Table 1.

Table 1 Disruptive technologies briefly described

The technology	The description
Additive manufacturing (AM)	AM involves fabricating physical objects in successive thin horizontal layers, according to digital models derived from CAD designs, 3D scans or video games. Such printing can take place at different scales from nanostructures to complete buildings and may involve a wide range of materials: human tissue, electronics and food as well as traditional industrial product materials such as polymers, metals and ceramics
Artificial intelligence (AI)	AI can be categorised at three different levels. First is 'narrow' AI that specialises in one area, e.g. the AI that plays chess better than humans. The second and third levels are concerned with more general ability. 'General' AI can perform as well as a human across the board, i.e. it is AI that can perform any intellectual task that a human can. Such AI is yet to be developed. Third is 'super intelligent' AI, i.e. an AI that performs better than human brains in practically every field
Augmented reality (AR)	Augmented reality (AR) is a live, direct or indirect view of a physical real-world environment whose elements are augmented (or supplemented) by computer-generated sensory input such as sound, video, graphics or GPS data
Big data	Big data is data that exceeds the processing capacity of conventional database systems. The data is too big, moves too fast or doesn't fit the strictures of standard database architectures. It is collected by large corporations and governments (and, increasingly, open data from 'citizen scientists'), and when interpreted using big data analytics it can be used to give insights into the behaviour of potential consumers and citizens
Intelligent matter	Intelligent matter, sometimes called programmable matter, is matter which has the ability to change its physical properties (shape, density, elasticity, conductivity, optical properties, etc.) in a programmable fashion, based upon user input or autonomous sensing
Internet of things (IoT)	The Internet of Things (IoT) is the networking of physical objects, i.e. things that have embedded within them electronics, software and sensors which are connected to one another over the Internet and can exchange data. This allows extensive communication between the physical and digital worlds, enables remote control of devices across the Internet and produces vast amounts of big data

(continued)

Table 1 (continued)

The technology	The description
Neurotechnology	Neurotechnology is concerned with technologies that inform about and influence the behaviour of the brain and various aspects of consciousness. Current neurotechnologies include various means to image brain activity, stimulation of the brain by magnetism and electricity, measuring the electrical and magnetic brainwave activity, implant technology to monitor or regulate brain activity, pharmaceuticals to normalise erratic brain function and stem cell therapy to repair damaged brain tissue. Recently measurements of brain activity have been used to control real-world artefacts
Robotics	A robot may be defined as 'a machine that carries out a physical task autonomously using a combination of embedded software and data provided by sensors'. This definition embraces relatively simple robots such as the Roomba vacuum cleaner to extremely complex robots such as the Google self-driving car
Synthetic biology	Synthetic biology is the process of designing and creating artificial genes and implanting them in cells. In some cases all existing genes have been removed; in others the new genetic sequences are introduced into the DNA of existing cells. It is far more than simply borrowing existing genes from nature. Synthetic biology is the process by which completely new life forms, i.e. life forms that have never previously existed, are created

3 Describing the 'Life Cycle' of a Technology

Design and technology teachers are well versed in helping young people consider the so-called life cycle of products and have used such teaching to engage students in the environmental impact of not only the manufacture of products but also their use and disposal as a critique of consumerism and the need to move from a linear to a circular economy (MacArthur 2015). Works such as 'The Story of Stuff' (Leonard 2010) have become standard items in teacher education reading lists and are discussed in more detail in chapter "Politicizing the Discourse of Consumerism: Reflections on the Story of Stuff" of this book. Critiquing the emergence of a technology, its adoption and impact on society is less familiar territory but particularly relevant to our concern with disruptive technologies. The Gartner 'Hype' Cycle (Gartner 2015) is an attempt to chart the life of a technology. It provides a graphic representation of the maturity and adoption of technologies and applications and how they are potentially relevant to solving real business problems and exploiting new opportunities. In its general form, it is shown in Fig. 1.

The key features of the cycle labelled in Fig. 1 and taken from Gartner 2015 are as follows:

Technology Trigger A potential technology breakthrough kicks things off. Early proof-of-concept stories and media interest trigger significant publicity. Often no usable products exist and commercial viability is unproven.

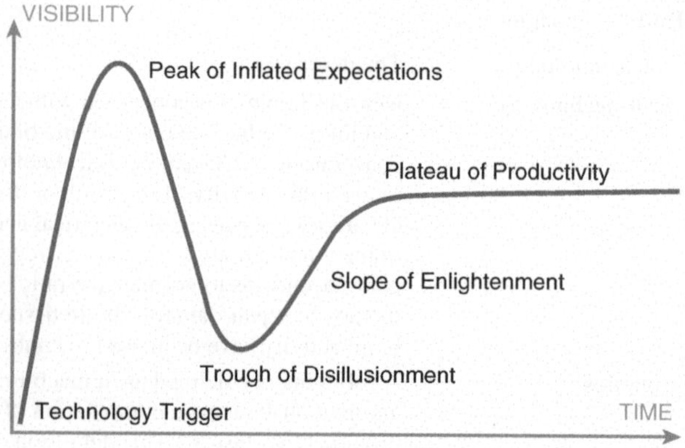

Fig. 1 Gartner 'Hype' cycle in its general form

Peak of Inflated Expectations Early publicity produces a number of success stories—often accompanied by scores of failures. Some companies take action; many do not.

Trough of Disillusionment Interest wanes as experiments and implementations fail to deliver. Producers of the technology shake out or fail. Investments continue only if the surviving providers improve their products to the satisfaction of early adopters.

Slope of Enlightenment More instances of how the technology can benefit the enterprise start to crystallise and become more widely understood. Second- and third-generation products appear from technology providers. More enterprises fund pilots; conservative companies remain cautious.

Plateau of Productivity Mainstream adoption starts to take off. Criteria for assessing provider viability are more clearly defined. The technology's broad market applicability and relevance are clearly paying off.

To those who have followed the success and failure of various technologies, this has a certain intuitive appeal and as such may be a useful device for young people as they try to understand what might be happening around them. It must be acknowledged that the Gartner Hype Cycle was not devised with such educational aims in mind; it is a business tool and is not without its critics (Jorge 2006). One can easily see that there should be a 'tail off' section to the graph after the plateau of productivity indicating when a technology goes into decline.

Despite the 'lack of objectivity' criticism of the Gartner approach, I think it will provide a useful lens through which young people at school can critique in depth the technologies around them.

4 Building Scenarios

Scenario building is difficult and to some extent a dark art (Swartz 1998, Wade 2012). There are several ways to engage young people with future scenarios. They can be presented with scenarios developed by others. These are not difficult to acquire. For example, at TED2014 speakers and attendees were asked to use what they had heard at the conference to suggest what might radically change with regard to society, life and technology in the next 30 years. The result is available at http://ideas.ted.com/2014/03/24/26-ideas-from-the-future/, and many of the comments would provide the basis for interesting discussions. For example:

> Over the next 30 years, humans will widely integrate technology into our bodies for recreational and informational purposes. A teenager in 2044 will marvel at how tech-free our bodies were in 2014, and wonder how we ever managed.

And note that as I write, the BBC Technology Correspondent has just reported on his experience of having a chip implanted under his skin which enabled him to control the opening and closing of doors and a photocopier (http://www.bbc.co.uk/news/technology-31042477)!

If appropriate ready-made future scenarios are not available, teachers can themselves develop scenarios for use with their classes. To do this they will need access to information to inform the scenario they are building. Swartz (1998) recommends the use of filters—i.e. a range of information sources that are likely to provide useful information and prevent one from getting swamped with a plethora of information from too wide a range of sources. Hence it would be useful to have the following items in the design and technology department library to be scanned and relevant content discussed regularly at curriculum update meetings:

The Economist (http://www.economist.com/)
New Scientist (see http://www.newscientist.com/)
MIT Technology Review (see http://www.technologyreview.com/)
Wired (see http://www.wired.co.uk/)
Granta (see www.granta.com/)

The availability of trend data is important, and the above publications might provide links to useful trend data.

Ideally one would want pupils to build scenarios for themselves, but they will not find this easy, and any techniques will require specific teaching in terms of the technique itself and the understanding of specific concepts on which using the technique relies. A general approach often used to build scenarios is to identify two sets of so-called critical or significant uncertainties and to use these as axes to create four quadrants such that located in each quadrant there is a particular scenario (see Fig. 2). Each of these can be fleshed out into a human story which can be explored from various critique perspectives.

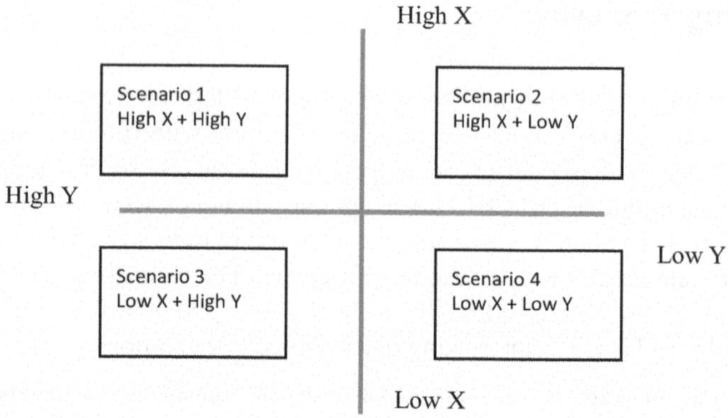

Fig. 2 A general representation of the critical uncertainties X and Y to create four scenarios

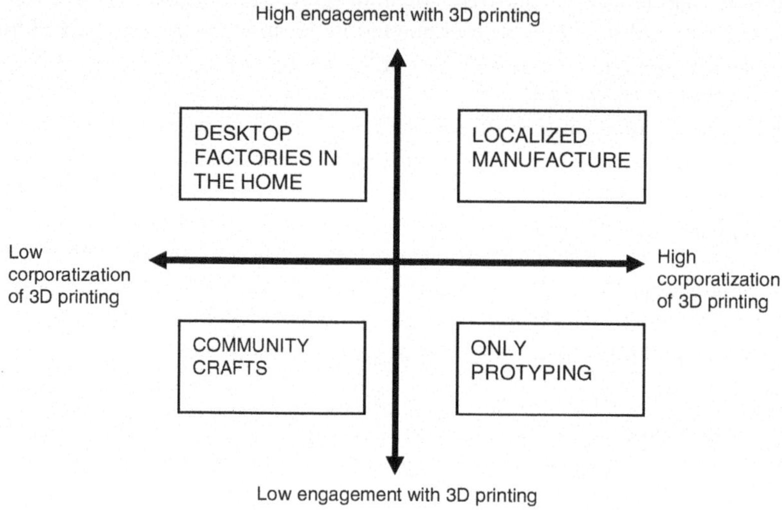

Fig. 3 Four scenarios from axes of uncertainty concerning engagement with 3D printing and corporatisation of 3D printing

Burthchell and Urry (CeMoRe & Centre for Transport & Society 2012) used this approach in their consideration of the impact of additive manufacturing on transport. They created four scenarios based on the set of axes shown in Fig. 3 and provided succinct descriptions of each scenario plus some extended vignettes. The description for 'desktop factories in the home' reads as follows:

Scenario 1 Desktop factories at home

The technical possibilities of fabrication in the home using desktop 3D printers significantly disrupted global systems of production, distribution and retailing. Many factories in the global south closed or downsized. Yet many supply chain and distribution networks remained intact and have even been consolidated due to rapid growth in demand for powders and other feedstocks for 'printers'.

This extract from the vignette about desktop factories considers the use of 3D printing for homework—an aspect of its impact that might appeal to pupils if not their parents.

> My name is Ben and I was born in 2020. I'm trying to finish my homework but my sister, Lucy, is using the printer again for the new bracelet she's been designing all weekend. Everyone at school has a 3D printer at home now and the teachers regularly give us assignments to design and print out all sorts of things to bring to class. Today we worked on a history project to imagine how medieval villagers built their towns. My job in the group was to print out what I thought the village smithy might look like. I got into trouble because I printed out my little smithy nearly 20 times to see what it would look like as I made each modification. My dad says the printer cartridges are really expensive and that it's wasteful and bad for the environment to use so much material on things I don't need.
>
> Page 3

Pupils developing scenarios from scratch for themselves is likely to be too demanding a task, but it is relatively easy to imagine lessons in which pupils are given a set of such axes along with written descriptions of the scenarios within the quadrants plus visualisations of these scenarios as a stimulus for discussion and critique. Pupils could be asked to imagine what it would be like for them if they were a particular person in those scenarios. They might collaborate with one another in writing personal vignettes and use them to critique the technologies in operation. They might turn their personal vignettes into short comic books. At the moment, such activities are not typical for design and technology lessons so collaboration with colleagues who teach English or history where such teaching is commonplace will be useful. In this way it should be possible to develop some very rich homework activities. Once 'inside' a scenario, pupils will be able to critique the impact of the technology on the lives of those living in that scenario.

5 Critiquing Disruptive Technologies: Possible Approaches

An important first step for pupils in acquiring a critical perspective with regard to disruptive technologies is that they become familiar with the features of disruption. To reiterate these are:

- They upset the status quo.
- They alter the way people live and work.
- They reorganise financial and social structures.
- They lead to entirely new products and services.

Hence it will be important for pupils to look at such technologies and convince themselves that they are in fact disruptive. They will need to answer questions such as 'How do they upset the status quo?', 'How have they/are they altering the way people work?', etc. It is important that this isn't done in an onerous and overly time consuming way but with a lightness of touch that is intriguing and gives rise to the desire to know more.

Once they are familiar with examples of current and possible future disruptiveness for a particular technology, it might be useful to situate that technology on the Gartner Hype Cycle. Pupils could be given the general form visualisation as shown in Fig. 1 and asked to position the technology or particular manifestations of that technology on the curve. Is it at the peak of inflated expectations or in the trough of disillusionment? Or maybe it's climbing up the slope of enlightenment? Perhaps some manifestations are on the plateau of productivity, whilst others are still at the very early stages just above the technology trigger. Note that this is in no sense a trivial task. Pupils will need to access information about the technologies under consideration: case studies of their utilisation and commentaries concerning their reception for example. Of course saying where it is in the cycle does not in itself lead to critique, but it does position the technology with regard to potential impact and the necessity to critique.

At this stage it might be useful to return to the features of disruption and consider them more deeply. If the deployment of this technology is altering the way certain people live and work by eliminating their employment opportunities in a particular sector, is this good or bad? If some people are being put 'out of work' and losing out why is this being done? Who is benefiting?

The work of the DEEPEN Project is relevant here. This was a 3-year research project (2006–2009), funded by the European Commission as part of its Sixth Framework Programme, aiming at deepening ethical engagement and participation in emerging nanotechnologies. The core aim of the project was to develop a set of integrated understandings of the ethical challenges posed by emerging nanotechnologies in real-world circumstances and their implications for civil society, for governance and for scientific practice. A distinctive element in the research was a narrative analysis devoted to unravelling the complexity of public responses, and how these were resourced culturally in the form of narratives that became fables and/or posed dilemmas. Macnaghten, Davies and Kearnes (2010) identified five narratives as key resources for negotiating the complex and far-reaching dilemmas that nanotechnologies are likely to pose for social life. These were:

1. Be careful what you wish for—the narrative of Desire
2. Being kept in the dark—the narrative of Alienation
3. Messing with nature—the narrative of the Sacred
4. Pandora's box—the narrative of Evil and Hope
5. The rich get richer—the narrative of Exploitation

Given that these narratives were developed through an exploration of 'lay response' to a complex, new and emerging technology that is probably somewhere near the peak of inflated expectations in the Gartner Hype Cycle, it is reasonable to expect that they will have both appeal and utility for young people at school. Hence in addition to using the lens of 'features of disruption' to critique technology pupils might be encouraged to reveal their initial feelings about a technology and see how they relate to these five narratives. Such narratives are often given voice through the popular press, e.g. Franken Foods (see http://www.medicinenet.com/script/main/art.asp?articlekey=24845 and http://www.newscientist.com/article/

Fig. 4 Using a 'Tetrahedron' for critique

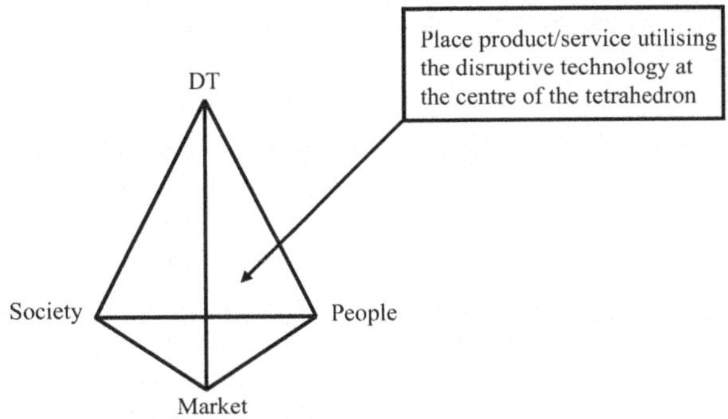

dn9921-instant-expert-gm-organisms.html#.VNN0dijCGQI), and AI will be our doom (see http://gizmodo.com/how-ai-could-ruin-humanity-according-to-smart-humans-1679025876 for a variety of expert views) and will probably have some resonance with the pupils and their families. So the opportunity to use them to scrutinise disruptive technologies is to be welcomed.

Barlex (2003) has reported the following critique tool. The product under consideration is placed at the centre of a tetrahedron as shown in Fig. 4. The vertices of the tetrahedron are labelled: DT (disruptive technology), People, Market and Society. Each of these features will need to be unpacked to some extent, and there will need to be a particular product/service at the centre:

- For the disruptive technology, pupils will need some knowledge of what it does and typical applications.
- For 'People', they will need to consider the following: who will use the product/service, why would they need or want this product/service, how many people are likely to want it, and what sorts of people are likely to want it?
- For 'Market', they will need to consider the following: is there already a market for the product/service—in which case where will your product/service fit in and what will distinguish it from other items in that market, how has this market grown recently, how will potential customers learn about your product/service, and who do you see as your customers?
- For 'Society', they will need to consider the following: what are the prevailing values, hopes and fears currently and in the near future, what are the trends here, are the values moving in a particular direction, and are hopes more significant than fears or vice versa?

These considerations will build into a critique of the product/service and the disruptive technology. This is important as the manifestations of a disruptive technology through different products/services operating perhaps in different contexts may well lead to different critiques.

An interesting adaptation of this approach might be to use the tetrahedron as the basis for asking young people the question 'What do you think needs disrupting?' They could be asked to justify their choices perhaps in terms of helping people in difficulties, making society more equitable or improving the quality of life. Or if this proved too open asking 'What if you wanted to . . . ? They could consider how the applications of particular technologies might give rise to the disruptions they thought should take place. For some young people, this might lead them to take action for themselves. Note that this approach also considers the near as opposed to more distant futures and to this extent may make their speculations more realistic.

In chapter "Hyper Design Thinking: Critique, Praxis and Reflection" Belinda von Mengersen notes that John Wood has used the tetrahedron as a device to reconfigure writing away from conventional linear narrative to one in which the juxtaposition of different elements can be viewed simultaneously.

In the following section, I will explore how the approaches discussed in this section might be applied with regard to one particular disruptive technology, robotics. This technology has been chosen because it is a common feature in the school technology curricula of many different countries.

6 Critiquing Robotics

Hawes (2014) provides a very basic definition of a robot as 'a machine that automates a physical task'. This makes no reference to the form that such a machine might take and indicates that the machine will do something in the real as opposed to a virtual world. Nourbakhsh (2013a) has noted that robots can operate in the real world and at the same time can be fully connected to the digital world. Implied in both definitions is the idea that robots are programmed to do particular tasks. In highly controlled environments with little human presence, they have proved exceptionally successful, but they are already operating in a wider range of arenas, including manufacturing, military operations, surgery, social care, education, transport and domestic services. Manufacturing is probably the most long established field and such robots operate autonomously. DARPA (United States Defense Advanced Research Projects Agency) has pioneered the use of drones in military operations, currently under human control and has developed robots such as Big Dog (http://www.darpa.mil/Our_Work/TTO/Programs/Legged_Squad_Support_System_(LS3).aspx) to support soldiers in the field. The use of drones for a variety of civilian purposes is increasing rapidly. The use of robots in certain forms of surgery is now well established. The use in education is currently quite rare. Considerable effort is taking place in Japan to develop robots that can provide social care especially for the elderly. Currently self-driving cars are available in several states in the USA and are being piloted in cities in England (Baraniuk 2016). The use in domestic service is currently limited to cleaners such as the Roomba (irobot 2013) which have only limited 'intelligence', but, given the recent announcement that James Dyson is investing heavily in robotics research (Wall 2014), it seems likely that intelligent domestic service robots with greater intelligence and capabilities will be developed.

6.1 Will the Use of Robots Meet the Mckinsey Criteria for Disruption?

With regard to *disrupting the status quo*, a key question here is the extent to which robots should be able to make decisions that were formerly made by humans, i.e. could robots be relied upon to make judgements that were driven by the same value systems as would be applied by humans in that situation? There is no doubt that robots already make all sorts of decisions according to their programming. At a very basic level, the Roomba makes decision that allow it to circumnavigate furniture. But as robots become more sophisticated and move into arenas where human and humane decision making is required, there is the distinct possibility that they will be asked to make decisions once made by humans, and this will almost certainly disrupt the status quo.

With regard to *altering the way people live and work*, some argue that robots will replace human workers (Rotman 2013), whilst others (Knight 2012) suggest that robots will become co-workers releasing humans for tasks more suited to human as opposed robot knowledge and skill. In the past the emergence of a technology has usually created more jobs than it has destroyed, but this is being questioned with regard to robotics. Will the domestic robots developed by Dyson increase or decrease opportunities for human cleaner employment? Will being a cleaner involve being in charge of robot workers and being able to maintain, repair and reprogramme them? Will this lead to cleaning being a hi-tech job for those with significant STEM qualifications? If so, where does that leave those who are cleaners today?

Concerning the *rearrangement of value pools*, the McKinsey Global Institute report (Manyika et al. 2013) suggested that by 2025, applications in robotics will have a total direct economic impact of $1.7 trillion–$4.5 trillion. But the report warns that public resistance to job losses and lack of workers educated in mathematics, science and technology are possible barriers to such development. Those companies that successfully enter this new and emerging market are likely to make significant profits, but they will need to make large initial investments to do so. Companies with large financial resources such as Amazon (Rotman 2013) and Google (IEEE Spectrum 2013) are acquiring robotic expertise and capacity.

And finally will there be *new robotic products and services*? The answer is almost certainly yes as with both increased intelligence and manipulative abilities robots will be able to operate outside the highly controlled environments in which they began their development.

6.2 Considering Robotics in the Gartner Hype Cycle

It is clear that the disruption caused by developments in robotics will have diverse effects. Different sorts of robots will occupy different positions on the Gartner Hype Cycle, and one can get some sense of how far into the future it might be for particular robot application to reach the plateau of productivity. Robots used to manufacture cars have been on that plateau for some time now, 20 years at least. Surgical robots may be seen as on the slope of enlightenment, whereas self-driving cars may be at the peak of inflated expectations. How many 'accidents' especially ones involving death or serious injury involving self-driving cars will have to take place before this application finds itself in the trough of disillusionment? The deployment of robots that could undertake mundane tasks, collecting and delivering laundry in hotels or hospitals, for example, might face resistance from the Unions that represent those who currently carry out such work. Embedded in the developments of robotics of course is the development of artificial intelligence, and it has been mooted that the strategic decision making currently carried out by humans will be taken over by AI systems (Nourbakhsh 2013b). Hence it is not just relatively unskilled jobs that could be at risk although these seem to be the most vulnerable in the short term. Automated checkouts in supermarkets and ticket dispensers in railway stations are finding themselves on the plateau of productivity and are perhaps indicative of such facilities moving into other service environments.

6.3 Considering Robotics in Terms of Dilemmatic and Fabalistic Narratives

One way to reveal possible public response to robots is to consider robotics in terms of one or more of the dilemmatic and fabalistic narratives revealed by the DEEPEN Project. It is important to remember that these narratives were derived from concerns with nanotechnology. Such technology is by its very nature is 'hidden' operating at or just above the molecular level in the 'plentiful room at the bottom' identified by Richard Feynman (1959). Hence we must exercise some caution in applying them to robotics where, in many cases, their manifestation will be perfectly visible so concerns derived from 'invisibility' might not apply. We must also exercise caution in making assumptions about young peoples' view of what constitutes a robot. A recent small case study with pupils aged 11/12 in London (Barlex & Steeg 2014) shows a very limited and limiting view of what a robot might be; in drawing of robots by the pupils, the vast majority depict the robot as a 'metal man' with only a few drawings showing female humanoid robots. This is not particularly surprising as popular media reinforces this stereotype.

In the case of 'Be careful what you wish for' (the narrative of Desire), desire for a state in which a technology solves pressing problems can lead to 'throwing caution to the winds', with enthusiastic advocates of the technology giving little consideration of consequences outside intended benefit and an unwillingness to consider possible risks. It is noteworthy that pupils involved in the case study mentioned above did not fall into this trap. They noted the initial benefits of robots completing domestic tasks that many humans find unrewarding, i.e. the removal of drudgery and tasks that require repetition. However, they all expressed reservations with regard to this practice becoming widespread. One pupil wondered how humans would cope if the robots broke down. Other pupils took this further arguing that as robots took over more and more human tasks, humans would lose the ability to do things for themselves, become inactive and unhealthy and lose their sense of being responsible for themselves. One pupil saw this as ultimately leading to a decline in the human population. Only one pupil commented on humans losing employment, the use of flying robots leading to human pilots losing their jobs. Only one pupil considered the possibility of robot revolution—in this case a cleaning robot using hot wax as a weapon against its human owners.

In the case of Pandora's box (the narrative of Evil and Hope), it is in hope for better things that the box is opened, but there is no guarantee that what comes out will lead to a better state of affairs. Application of the precautionary principle would seem to mitigate against technological disaster with the burden of proof that an action, in this case the deployment of robots, would not be harmful falls on those taking the action. But the problem here is that once a technology is deployed, it is difficult if not impossible to put it 'back in the box', and if the precautionary principle has proved false, society is stuck with the consequences. Turney's view (2014) can be seen as a Pandora box dilemma. He has noted that if, in a not too distant future, robots take care of agricultural and industrial production, then theories of value fall apart and work is no longer needed:

> If we abolish work, we abolish exploitation, but also the reward of exercising skill and ingenuity to contribute to the human community. ... can we dispense with one without sacrificing the other? (p. 74)

The film Alien (Scott 1979) provides an intriguing robot example of both being kept in the dark (the narrative of Alienation) and the rich get richer (the narrative of Exploitation). This filmic example has not been chosen lightly. It was inducted into the National Film Registry of the Library of Congress in 2002 for historic preservation as a film which is 'culturally, historically or aesthetically significant'. In a memorable scene, the science officer, Ash, attacks the heroine Ripley who is saved by the intervention of another crewmember, Parker. In the struggle Parker decapitates Ash and realises, to quote, 'Ash! He's a goddamn fucking robot!' So we have a situation in which the rest of the crew don't know that Ash is a robot, appearing amongst them as a human; they have been *kept in the dark* about this and his 'motives'; he had been assigned to the mission to convince the crew to land and capture the creature (the Alien of the title) in order to return it to their employers for analysis, even at the expense of the human personnel. This must surely be an extreme example of *the rich getting richer* at the expense of the poor. This and the subsequent films Aliens, Alien 3 and Alien Resurrection all have the narrative of exploitation enacted by a large corporation pursuing profit from the possibility of utilising the Alien's biochemistry for new technologies whatever the consequences.

Frankenstein (Shelley 1818) can be seen as a warning tale with regard to messing with nature—the narrative of the sacred—with Frankenstein invading territory seen as the province of God in creating a new human, the monster, albeit from the parts of deceased humans. McLeish (2014) has commented that Frankenstein's 'sin' is not so much that he created the monster but that he did not love or nurture his creation. Interestingly the recent TV series Humans (http://www.channel4.com/programmes/humans/on-demand/56459-001) engages with this issue in exploring the relationship between robots and humans as the lines between humans (natural creations) and machines (artificial creations) become blurred.

6.4 Using the Young Foresight Approach to Critiquing Robotics

In using the Young Foresight Approach, we have to put a particular robot in the centre of the tetrahedron, and the resulting critique will be highly dependent on this choice. If we choose a manufacturing robot, used primarily in the car industry, we would get the following results. The disruptive technology with regard to the manufacturing robot is well proven; this application of robotics is on the 'plateau of productivity' with regard to the Gartner Hype Cycle. With regard to Society, the use of robotics in manufacture is now well established and accepted without reservation. It features as a positive in TV advertisements for cars. There is a desire to own and use cars from a relatively early age. Passing the driving test is a rite of passage for most young people in many societies. Without the use of manufacturing robots, it is unlikely that cars could be manufactured at such affordable prices or with such reliability. Hence from the people perspective, the use of manufacturing robots would not be contested. There is clearly an established and effective market for manufacturing robots. The estimated world supply of industrial robots has grown from 69,000 in 1998 to 178,132 in 2013 (worldrobotics.org 2014). If the robot placed in the centre of the tetrahedron were Baxter (Rethink Robotics 2013), a robot developed to work alongside humans, as opposed to replacing them, the critique would be slightly different. Those people who trusted the technology would be inclined to welcome it, whereas those who were suspicious for various reasons might object. In a similar way, the overall societal reaction could be mixed. It is clearly believed that there is a market for such a product; investors in Rethink Robotics, the company that manufactures Baxter, include several well-established venture capitalists including Bezos Expeditions, the personal investment company of Jeff Bezos, founder and CEO of Amazon.com. A comment from Jason Miller, vision and automation technician, at Rodon (Miller 2013), a company that makes considerable use of Baxter, indicates more than workforce acceptance:

> The general sense is that nobody feels it's threatening their jobs. They think he's really bringing jobs back for us. There are several jobs that might be in China if not for a technology like Baxter.

However, if we place an unmanned flying vehicle (UAV commonly known as a drone) at the centre of the tetrahedron, the situation would be infinitely more complex. It would depend on what type of drone. The military use of drones by the USA is coming under severe criticism (Coll 2014) although there seems little likelihood of this criticism leading to a change in policy. This criticism is compounded by the possibility of such drones becoming more intelligent and being able to make 'kill' decisions autonomously without human approval. Note such a killer drone already exists. It will be deployed on the Great Barrier Reef to recognise and kill crown of thorns starfish which are causing damage that threatens the reef's existence (Platt 2016). The civilian use of drones, however, is becoming acceptable for a wide range of applications with the Federal Aviation Administration developing guidelines for their use (FAA 2013), and most recently China's biggest Internet retailer is using them to deliver tea (Kelion 2015).

6.5 Using Scenario Development to Critique Robotics

And finally it is worth exploring the way relationships between humans and robots might play out by means of some scenario developments. The initial task is to identify the critical uncertainties for the axes needed to create the four scenarios. Whether humans are prepared to accept robots into their life and work (as co-workers, substitute workers or helpers) seems a crucial factor. Some people may feel uneasy about a robot presence at work or home especially if they see the robot as somehow 'messing with nature' in that technology has created a sentient being that is not a human but a machine. Others will welcome a robot presence on the grounds that robots tackle tasks they do not wish to and that they can provide companionship. Hence the vertical axis in the scenario development concerns 'acceptance'. Related to acceptance is the possibility that as robots become more sophisticated, to the point where they become first person conscious and moral agents in their own right, they might be granted rights or not. Such rights would curtail the way in which humans could treat robots preventing them, for example, being seen and treated as disposable once they were deemed no longer fit for purpose. Hence, the granting of rights to robots forms the horizontal axis. The resulting four quadrants are shown in Fig. 5.

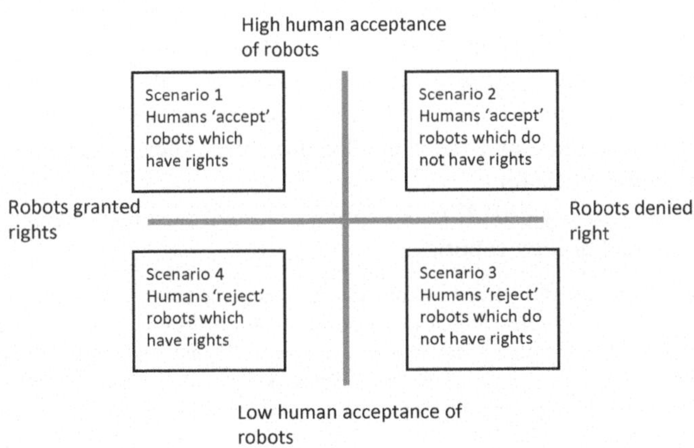

Fig. 5 Four scenarios from axes of uncertainty concerning human acceptance of robots and the granting of rights to robots

It is enlightening to identify recent popular science fiction films which engage with some of the scenarios. In the film Robot and Frank, the character Frank, a grumpy elderly man with dementia, is bought a care robot. Initially Frank is highly suspicious, but as the film unfolds he develops a strong bond with the robot and is aghast when he has to ‘turn off’ the robot, thus destroying all its memories of the relationship with Frank. When Frank says ‘But you’re my friend’ (we are in Scenario 2), Frank has completely accepted the robot as his friend, but the robot does not have rights which prevent his memory of the friendship being destroyed. The film Ex Machina operates in the same scenario but with a very different interpretation of the human-robot relationship. The film explores the way in which a version of the Turing test is applied to a particular robot. The robot, in female form, name Eva, passes the test with flying colours revealing the human abilities to form an empathetic relationship, to lie and be deceitful, to create and pursue her own plans which are very much at the expense of the human protagonists. The exploitation of the Eva and the other robots in her chain of development clearly indicates that the technologist responsible accords them no rights what so ever.

In the Star Wars trilogy (episodes IV, V and VI), the relationship between the humans and the robots C3PO and R2D2 moves us into Scenario 1; the humans and robots work together to accomplish shared goals and show mutual concern for each other’s safety.

The Terminator films move us into an extreme example of Scenario 4. There is outright rejection of the right to exist on both sides. The robots (or machines as they are called in the films) are first person conscious and respond to human existence by a concerted effort to wipe out humanity. Humanity responds by trying to wipe out the machines.

If teachers are going to use science fiction films as resources to support critique in technology education, then they will need to be sensitive to the sorts of films that are appropriate to the ages of the pupils. The film certification will give guidance to some extent, but it is always worth identifying any scenes that might be disturbing whatever the certification to avoid frightening or worrying pupils. Hence for primary the films Wallace and Gromit, ‘The Wrong Trousers’ or WALL-E might be appropriate; for upper primary and lower secondary, the Star Wars series; for upper secondary I Robot, Chappie, AI, Robot and Frank and the later Terminator episodes; and for tertiary and teacher education the Alien Quartet, Prometheus, Blade Runner and Ex Machina.

Lin, Abney and Bekey (2012) make an impassioned plea for a rational approach to the disruptive technology robotics:

> That is why it is important to clear from the field the many incorrect notions about robots—a machine that is so complex that it often becomes unintelligible, even to its designer, but always an artificial product of technology, ontologically and irreparably different from a human being. And that is why it is crucial to tackle not the mythical worries due to ideologies and utopian hopes or dystopian fears, but the real issues facing robotics in the larger society—before it’s too late.

Page 362

7 Commentary

This chapter has considered the following:

- The features of disruption with regard to a technology
- The Gartner Hype Cycle as a way of describing the life cycle of a technology
- Scenario building as a means of thinking about what might happen when particular technologies are deployed
- Possible approaches by which pupils might critique so-called disruptive technologies
- How these approaches might be applied in critiquing the disruptive technology robotics

So where does this leave us with regard to the question *How can we use the idea of disruption to enable young people at school to critique technology?* Echoing the sentiments of Lin, Abney and Bekey (2012), I remain convinced that it is very important for young people at school to engage in critique of technology particularly with regard to those technologies that might have a significant effect on their lives.

This brief excursion into critique via so-called disruptive technologies encourages me to believe that young people will find such activity attractive and worthwhile. But, and it's a big but, they will need to be taught the knowledge and skill that makes it possible to engage with critique with confidence and to deal with the ambiguities and uncertainties that are bound to arise from such work. This will only happen if teachers acknowledge the worth of the endeavour, develop the required expertise and find time to introduce such work into an already crowded curriculum. It will be important for the curriculum to acknowledge the importance of critique; hence I would hope to see critique included as an area to be assessed.

There is the question of the long-term impact of young people learning about disruption and disruptive technologies through critique. Will it lead to a significant increase in awareness of the possible impacts of such technologies? Will this in turn result in the population at large becoming much more active in scrutinising and intervening in the activities of companies engaged in developing and selling disruptive technologies? The prospect of companies having to explain the potential benefits *and* costs of their latest 'products' to a sceptical citizenry is something that those interested in and committed to design

and technology education should welcome. This aspect of design and technology education might, if successful, enable young people to develop a critical frame of mind with which they approach the world and future opportunities.

Acknowledgements The Disruptive Technologies Project is an initiative being taken forward by David Barlex, Nick Givens and Torben Steeg. This chapter is part of that initiative, and although authored by David it could not have been written without our collaborative work. Indeed Nick and Torben acted as critical friends throughout the writing of this chapter.

References

Barlex, D. (2003). Considering the impact of design & technology on society – the experience of the young foresight project. In J. Dakers (Ed.), *The place of design and technology in the curriculum, PATT 13 conference proceedings* (pp. 140–144). Glasgow: University of Glasgow.

Barlex, D., & Steeg, T. (2014). Pupils' understanding of robotics as a disruptive technology. In H. Middleton (Ed.), *Technology education: Learning for life. Proceedings of the 8th Biennial international conference on technology education research* (pp. 1–9). Gold Coast: Griffith Institute for Educational Research: Griffith University.

Barlex, D., Givens, N., & Steeg, T. (2015). Thinking about disruptive technologies. In G. Owen-Jackson (Ed.), *Learning to teach design and technology in the secondary school* (3rd ed., pp. 303–322). Oxford: Routledge.

Baraniuk, C. (2016). *UK driverless car projects get government green light*. Retrieved February 1, 2016, from http://www.bbc.co.uk/news/technology-35461596

CeMoRe & Centre for Transport & Society. (2012). '3D' printing: Four scenarios for 2030. Presented at scenario planning workshop, London

Coll, S. (2014). *The unblinking stare*. Retrieved February 6, 2015, from http://www.newyorker.com/magazine/2014/11/24/unblinking-stare

FAA. (2013). *Integration of civil unmanned aircraft systems (UAS) in the National Airspace System (NAS) Roadmap*. Available at this url: https://www.faa.gov/uas/legislative_programs/uas_roadmap/media/UAS_Roadmap_2013.pdf. Accessed 06 Feb 2015.

Feynman, R. (1959). *Plenty of room at the bottom*. Available at http://www.pa.msu.edu/~yang/RFeynman_plentySpace.pdf. Accessed 03 Feb 2015.

Gartner. (2015). Gartner Hype cycle see this url: http://www.gartner.com/technology/research/methodologies/hype-cycle.jsp. Accessed 29 Jan 2015.

Hawes, N. (2014). Automatic for the people. In S. Westlake (Ed.), *Our work here is done visions of a robot economy* (pp. 45–52). London: NESTA.

IEEE Spectrum. (2013). http://spectrum.ieee.org/automaton/robotics/industrial-robots/google-acquisition-seven-robotics-companies. Accessed 27 Apr 2014.

irobot. (2013). http://investors.irobot.com/phoenix.zhtml?c=193096&p=irol-newsArticle&ID=1897317&highlight=. Accessed 27 Apr 2014.

Johnson, S. (2014). How we got to now; Six innovations that made the modern world. Canada: Particular Books.

Jorge, A. (2006). *Cheap shots at the Gartner Hype curve*, available at https://catenary.wordpress.com/2006/10/22/cheap-shots-at-the-gartner-hype-curve/. Accessed 1 Feb 2015

Kelion, L. (2015). *Alibaba begins drone delivery trials in China*. Retrieved February 1, 2016, from http://www.bbc.co.uk/news/technology-31129804

Knight, W. (2012). *Meet Baxter, a New Kind of Industrial Robot*. MIT Technology Review available at this url: http://www.technologyreview.com/video/429247/meet-baxter-a-new-kind-of-industrial-robot/. Accessed 27 Apr 2014.

Leonard, A. (2010). *The story of Stuff*. London: Constable & Robinson Ltd.

Lin, P., Abney, K., & Bekey, G. A. (2012). *Robot Ethics*. Massachusetts: MIT Press.

Macnaghten, P., Davies, S., & Kearnes, M. (2010). Narrative and public engagement: Some findings from the DEEPEN project. In R. von Schomberg & S. Davies (Eds.), *Understanding public debates on nanotechnologies options for framing public policy*. Luxemberg: European Union.

Manyika, J. et al (2013). *Disruptive technologies: Advances that will transform life, business, and the global economy* McKinsey Global Institute available from http://www.mckinsey.com/insights/business_technology/disruptive_technologies Accessed 1 Feb 2015

MacArthur, E. (2015). *Resources developed by the Ellen MacArthur Foundation* can be found at this url: http://www.ellenmacarthurfoundation.org/schools Accessed 6 Feb 2015.

McLeish, T. (2014). *Faith and wisdom in science*. Oxford: Oxford University Press.

Miller, J. (2013). *Baxter helps the Rodon Group Stay 'Cheaper Than China'*. Retrieved February 1, 2016, from http://cdn-staging.rethinkrobotics.com/wp-content/uploads/2014/08/Rodon_Spotlight_final.9.13.pdf

Nourbakhsh, I. (2013a). *Robot futures*. Cambridge: Massachusetts, MIT Press.

Nourbakhsh, I. (2013b). *It's time to talk about the burgeoning robot middle class Massachusetts MIT review*. Available at this url: http://www.technologyreview.com/view/514861/its-time-to-talk-about-the-burgeoning-robot-middle-class/. Accessed 1 Feb 2015

Platt, John R. (2016). *A starfish-killing, artificially intelligent robot is set to patrol the great barrier reef*. Retrieved February 1, 2016, from http://www.scientificamerican.com/article/a-starfish-killing-artificially-intelligent-robot-is-set-to-patrol-the-great-barrier-reef/

Rethink Robotics. (2013). *'Automation redefined.'* www.rethinkrobotics.com/products/baxter/. Accessed 27 Apr 2014.

Rotman, D. (2013). *How technology is destroying jobs*. MIT Technology Review available at this url: http://www.technologyreview.com/featuredstory/515926/how-technology-is-destroying-jobs/. Accessed 27 Apr 2014.

Schwartz, P. (1998). *The art of the long view* (3rd ed.). Chichester: Wiley.

Scott, R. (1979). *Alien* (film). http://en.wikipedia.org/wiki/Alien_(film). Accessed 05 Feb 2015.

Shelley, M. (1818). *Frankenstein or the Modern Prometheus*. London: Penguin.

Turney, J. (2014) Our work here is done: Robot futures in fiction. In S. Westlace (Ed.), *Our work here is done visions of a robot economy* (pp. 64–74). London: NESTA

Wade, W. (2012). *Scenario planning a field guide to the future*. Chichester: Wiley.

Wall, M. (2014) *Dyson invests £5 m in robotic vision lab with Imperial*. Retrieved February 1, 2016, from http://www.bbc.co.uk/news/technology-26084765

worldrobotics.org. (2014). Industrial robots executive summary available at this url: http://www.worldrobotics.org/index.php?id=downloads. Accessed 02 Feb 2015.

Critiquing Literature: Children's Literature as a Learning Tool for Critical Awareness

Cecilia Axell

Abstract A starting point for this chapter is that children's literature can be a source of reflection and can encourage children to think critically about technology. In contrast to many textbooks and non-fiction books, fictional stories reflect the complexities and contradictions inherent in technology and at the same time reveal its less obvious and concealed aspects and messages. By using books with a strong story line, which are of interest to children, technology can be presented as part of the world around them. Children's literature can thus be seen as a mediator of values and attitudes, which makes it an interesting subject matter for Design and Technology education. This chapter involves an exploration of critical aspects of technology found within a selection of children's books. The stories originate from different historical and cultural contexts, and the basis for the selection is that it represents a variety of critiques and aspects of technology found in children's literature. The conclusion of the analysis is that children's literature can contribute to making technology and the nature of technology more comprehensible and visible to pupils. The ambiguous messages in the books reveal the multifaceted and complex nature of technology and make it possible to problematise it in ways textbooks seldom can. As the stories form the basis for critical discussion about the nature of technology, they could also help to broaden perspectives, thereby acting as a pedagogic tool in fulfilling the aims of Design and Technology education.

Keywords Technology education • Technology • Critique • Children's literature

> Children's literature is one of the means by which children are socialized and are acquainted with important aspects and features of their civilization. It is, therefore, only natural that gadgets and machines should loom prominently in children's books in an era and a society that is, in fact, based on technological development (Schwarcz 1967:82).

C. Axell (✉)
Department of Social and Welfare Studies, Linköping University, S-601 74, Norrköping, Sweden
e-mail: cecilia.axell@liu.se

PJ. Williams, K. Stables (eds.), *Critique in Design and Technology Education*,
Contemporary Issues in Technology Education, DOI 10.1007/978-981-10-3106-9_13

1 Introduction

In the wake of rapid technological change, society increasingly expects more of its citizens, and this is reflected in school policy documents. Becoming an active and responsible citizen with the ability to make well-informed decisions is dependent on basic technological literacy and critical abilities for evaluating technology in everyday life. In the Swedish curriculum for the compulsory school, the aim of teaching technology is to give pupils the "opportunity to develop their understanding of the importance of technology and its impact on people, society and the environment" (Swedish National Agency for Education 2011:254).

Similar writings can be found in curricula in other countries. The importance of critical thinking is highlighted in the English curriculum, where the purpose of design and technology teaching in KS1–3 in schools is to ensure that "[p]upils learn how to take risks, becoming resourceful, innovative, enterprising and capable citizens. Through the evaluation of past and present design and technology, they develop a critical understanding of its impact on daily life and the wider world" (Department for Education 2013).

The importance of history and context in teaching technology is also highlighted in the South African curriculum for Grades 4–6. One of the specific aims of this subject in schools is that "[l]earners should understand the practical uses of Natural Sciences and Technology in society and the environment and have values that make them caring and creative citizens" (Department of Basic Education 2011:11). Moreover, through teaching, learners should be given the opportunity to develop their understanding of "the history of scientific discoveries and technological solutions, and their relationship to indigenous knowledge and different world views [. . .]" (Ibid.).

Since technological knowledge is often a matter of decisions and preferences, it also involves the values in society. Technology, in terms of making and using artefacts, is largely a practical activity. But due to the inherent complexity and practical efficacy of modern technology, we also need to reflect on it and think about it more (Mitcham 1994). Discussion and reflection should therefore be regarded as an important part of a technology syllabus (Dakers 2006; de Vries 2006). Failing to place technology in a broader context ignores the connections between artefacts and human intention, as well as the social implications of how artefacts are used (Axell 2015; Axell and Boström 2015; Klasander 2010; Mawson 2010; Siu and Lam 2005; Turja et al. 2009).

A further important aspect is that technology is global, and knowledge about it should therefore include technology from different cultural contexts. It should not simply focus on modern technologies used in a limited number of parts of the world (Edgerton 2006; Gumbo 2015). However, in order to understand and reflect critically on the impact of technology on people's lives, on society, on nature and the environment, technology needs to be made visible and understandable. In this chapter, knowledge about technology is linked to children's fiction. A starting point is that fiction can be a source of reflection and encourage children to think critically about technology. In this way, it can become an integral part of technology education.

1.1 Technology Landscapes in Children's Literature

Fiction can be considered an arena where the conversation about technology and its impact on humankind takes place in a specific context. Identifying with the characters in a story not only contributes to an understanding of other people, but it also helps develop knowledge and understanding of the world we live in. As Don Norman notes, people are innately disposed to look for causes of events, to form explanations and stories. Stories not only resonate with our experiences, they also provide examples of new instances (Norman 2013). When we engage with a story, we are free to explore our own perceptions of it and follow a variety of suggested prospective realities. These provide an opportunity for alternative possibilities, for visions of the future; an opportunity to reflect on what *is* and what *could be*. The stories can also invite us into a dialogue about the effects of technology on the individual, society and nature. Moreover, in contrast to many textbooks and works of non-fiction, fictional stories highlight the inherent complexities and contradictions of technology, revealing its less obvious and concealed aspects and messages. Children's literature can thus be seen as a mediator of values and attitudes, which makes it a useful subject matter for Design and Technology education (Axell 2015).

Technology landscape is a concept I use to examine the books detailed in this chapter. Technology landscapes vary in different cultural contexts, and also change character over time. In other words, they are a kind of empirical "reality" which includes values and attitudes about technology (Hagberg 2008, 2009; Lindqvist 2011). This chapter can therefore be described as an exploration of critical aspects of technology found within the landscapes in children's literature. The stories originate from different historical and cultural contexts, and they have been selected for the extent to which they represent a variety of critiques and aspects of technology found in children's literature: *mechanising and homogenising aspects of technology*, *older versus newer technology*, *colonising aspects of technology*, *technology versus nature and enduring technology*.

2 Critique of Technology in Children's Literature

2.1 The Mechanising and Homogenising Aspect of Technology

> The people in this country are far ahead of us in everything. All of our remarkable inventions like the telegraph and the telephone and the phonograph and the flying machines and the cinemas, they have had for several hundred years in Kringelkrokien. Skilled as they are at inventing, almost everything in the country is done by machines. (Beskow 1919/1996:55)

The technology landscape in the fairy tale *Doctor Klokamundus' Invention* (1919/1996), written by the Swedish author Elsa Beskow, involves a fantasy land, Kringelkrokien, which is technologically advanced. Inventions such as the telephone, the phonograph, the aeroplane and the cinema have existed there for hundreds of years. Most tasks are done by machines. In Kringelkrokien, the inventor Doctor Klokamundus is the king's closest adviser. The description of Kringelkrokien is reminiscent of Edward Bellamy's (1850–1898) utopian society. Bellamy's utopia is based on rational industrial production, and the capacity of the individual has little value (Frängsmyr 1990). Society in Kringelkrokien is also based on rationality and technological solutions. However, dependence on technology and science has gone too far, and they are used as tools to mechanise people's lives. When the children (boys) start to misbehave, the adults believe that the problem can be solved by technology in the form of a high-tech fostering machine.

Doctor Klokamundus initially constructs a prototype on a small scale. Five baby rabbits are placed in the machine and released after a month. When they are put into the machine, each rabbit has its own individual colour and personality, but after a month in the machine, they are all identically speckled and perform exactly the same tricks. The king is delighted with the small, perfect, well-behaved rabbits and recognises the machine's potential.

In the fostering machine, everything is managed by sophisticated and automated technology, replacing the need for human coexistence. The boys are stuck inside the big, boring machine which is completely lacking in aesthetics and cosiness. Each room is equipped with a clock, and every hour a loud voice announces what needs to be done. If the boys do not obey, they are given an ice-cold shower. In the morning, an alarm clock rings, and as soon as it stops ringing, all the beds are turned upside down and folded into the wall. At 7 o'clock, school starts. The pupils sit at their desks, and a gramophone repeats the lessons over and over again. Twice a week the boys watch a film which teaches them how to behave, such as how to hold a knife and fork in the correct way. Children are therefore brought up with no human contact. If the boys feel they need someone to talk to, they can do so at a certain hour in the afternoon via a receiver on the wall. The answer comes from a gramophone. In contrast to Bellamy's utopia, the machines in the fairy tale take over *all* tasks, not just some of them. A vacuum device sucks up all the dust from the rooms, and the boys use a "clean clothes machine" at a set time every Saturday. They put their dirty clothes in one tray and clean ones appear in another one. There is even a hair-cutting machine which resembles a dentist's chair and which cuts hair "as easily as if it were peeling a potato". However, the children do not have the slightest intention of staying in the fostering machine. They have no desire to be corrected and disciplined in the way the adults wish. "Maybe because they knew that all the baby rabbits had been speckled when they came out of the machine [...], what is certain though is that the children felt a real horror of that machine" (Beskow 1919/1996:64).

As in most fairy tales, the story has a happy ending. The boys escape from the fostering machine and live a "Robinson Crusoe life" in the ruins of an abandoned castle. By using their creativity, they develop new skills and learn how to survive in the wild. The story illustrates the consequences of placing too much faith in technology and shows how it is used as a tool to homogenise people. The implicit message is that there is a risk in a technological world that people's innermost needs and desires will be forgotten. The boys' experiences, however, illustrate that there is an innate social need for human connection and belonging which cannot be fulfilled by technology and that adults need to meet children's emotional needs. The critique in the story can therefore be interpreted as a reflection of our ambivalent view of technology and its consequences. This ambivalence, according to Brian Arthur, comes not from our relationship with technology but from our relationship with nature: "[w]e are caught between two huge and unconscious forces: Our deepest hope as humans lies in technology; but our deepest trust lies in nature" (Arthur 2011:11). Our hope is that technology will solve our problems, make our lives better and help develop the future we want. At the same time, as human beings, we are attuned to nature, a dependence which comes from millions of years of feeling at home with it. Jacques Ellul, on the other hand, traces the roots of this ambivalence towards technology to the shift from spiritual to modern societies which are governed by technological means (Ellul 1964, 2010). George M. O'Har refers to Ellul when he suggests that "[w]hen science and technology replaced magic, what was removed was that physical-mechanical part of the magical system that simply could not compete with a new world based on scientific method and technological efficiency [. . .] Machines cannot calm fears, or provide answers to our deepest questions" (O'Har 2000:864).

Moreover, the way technology and technological development are criticised in the story can be interpreted as the consequence of a deterministic view of technology, in a similar way to Ellul (1964, 2010). Technological development and new technologies in Kringelkrokien follow their own principles based on rationality and efficiency, and development takes place outside people's control. As the solutions in Kringelkrokien's technological landscape have social consequences which go far beyond their intended application, the fairy tale can also be interpreted from the perspective of Langdon Winner. He suggests that if we simply see technology as a neutral tool which can be used for either good or evil, we fail to take into consideration any unintended consequences in its design or construction (Winner 1989).

2.2 *Older Versus Newer Technology*

> Mike Mulligan had a steam shovel,
> a beautiful red steam shovel.
> Her name was Mary Anne.
> Mike Mulligan was very proud of Mary Anne.
> He always said that she could dig as much in a day
> as a hundred men could dig in a week,
> but he had never been quite sure
> that this was true. (Burton 1939/2005:3)

Some children's books carry the message that early technology is better than modern technology and that old technology has greater value. One example is *Mike Mulligan and his Steam Shovel* (1939/2005) by the American writer Virginia Lee Burton, which illustrates several of the problems of a rapidly changing society. In the story's technology landscape, an anthropomorphic steam shovel, "Mary Anne", becomes outdated and is replaced by more modern technology, introducing a message about change and obsolescence. Mike and his steam shovel are described as a team or a couple, as Mary Anne is a female. They have been digging together "for years and years", and Mike has taken such good care of his steam shovel that "she never grew old" (Burton 1939/2005:4). The steam shovel is assigned a feminine gender and embodies values as well as attitudes traditionally associated with women (Lee 1992).

Anthropomorphism, the practice of ascribing human traits and attributes, such as feelings, to animals or things, is quite common in folk tales and children's literature. One suggested reason for the use of anthropomorphic technology is that it helps the reader feel at ease with technology in general as a part of the human world. Where a form of technology is so complex that it is hard to grasp, a way of bringing it closer to us is to depict it as being alive. Anthropomorphism also helps to build an emotional bond between human beings and machines (Schwarcz 1967; Waytz 2013, 2014). This is reinforced by showing how humans and machines can be the best of friends.

In the book, the reader is told that Mike and his steam shovel ("among others") have dug the great canals for the boats to sail through, they have dug out the high mountains so trains can pass through and flattened hills and curves in nature to make roads for the cars. Mike and Mary Anne have also packed down the ground and filled in the holes to make landing strips for the aeroplanes and dug deep holes for the cellars of the tall skyscrapers in the modern cities. Mike and his steam shovel are celebrated as heroes, as they are the basis of industrial society. From this perspective, technology is portrayed as a servant to humanity and seen as a powerful tool in helping human beings achieve their dreams and aspirations. These aspirations include mastering nature. The depiction of how humans use their technology to transform nature for their own needs can be tied to an anthropocentric view of nature. This represents a human-centric approach, where nature is seen as something which exists largely for the benefits of humans. It means that nature itself has no intrinsic value. In the history of Western ideas, Francis Bacon (1561–1626) and René Descartes (1596–1650), more than any other philosophers, provided the basis for consolidating an anthropocentric ethic. Bacon regarded nature as something to be disciplined and harnessed as an obedient slave (Merchant 1989; Sörlin 1991). George Henrik von Wright describes Bacon as "the master philosopher of technology", as he prophesied more eloquently than anyone else how technological inventions, representing the benefits of science, would establish human sovereignty over the universe (von Wright 1987/2010). Descartes held a mechanistic view of nature and believed that animals were nothing more than complex machines. According to Descartes, humans are differentiated from animals by the fact that they have a soul, and this is extended to justify human dominion over animals and nature. Through philosophers such as Bacon and Descartes, the anthropocentric view was therefore lent a kind of scientific legitimacy. A more recent definition of the anthropocentric point of view is that it represents an approach in which each impact on nature should be assessed according to the effect it has on humans. From this point of view, living people must act so that the lives and well-being of future generations is not threatened (Sörlin 1991). This means that an anthropocentric view of nature need not necessarily be the same as the ruthless exploitation of nature. On the other hand, it justifies a view that humans have the right to exploit nature for their own purposes. This message is also incorporated into the story about Mike Mulligan and his steam shovel. The main message in the story, however, focuses on how contemporary technology outperforms and replaces older technology:

> Then came along the new gasoline shovels, and the new electric shovels, the new diesel motor shovels and took all the jobs away from the steam shovels. Mike Mulligan and Mary Anne were VERY SAD. (Burton 1939/2005:13–14)

The steam shovels are sold off for junk or left in gravel pits to rust and fall apart. Mike, on the other hand, loves his machine so much that he cannot do that to "her". However, the "good old days are gone" and no one wants or needs them anymore.

As Kevin Kelly indicates, a technique or artefact which may be rare in the modern urban world can be quite common in the rural developing world (Kelly 2010). In Mike Mulligan, for example, the older technology plays an important role in a smaller town in the countryside. Mike reads in the newspaper that a new town hall is going to be built in Popperville, and he and Mary Anne decide to go there and offer the residents of the town their services. They are given the job of digging the cellar. Mike promises to do all the work in only 1 day, and if he and Mary Anne do not manage it, the town will not have to pay for their work. They succeed in carrying out their task in 1 day, but there is a problem: they have forgotten to leave a way out from the bottom of the pit. A little boy finds a solution, however, when he suggests an alternative use for the steam shovel as a heating plant for the town hall. Mike and his old shovel are required again when they are given the important task of controlling and warming up meetings from the boiler room of the new town hall. As Kelly notes, this kind of anachronistic technology is not at all unusual. For example, as recently as 1962, in what was then called the atomic age, many small businesses in Boston ran machines using steam power delivered to them by overhead drive shafts (Kelly 2010).

From a gender perspective, the fable not only presents an egalitarian relationship between males and females, the female machine in the story paves the way for more democratic social development and, in the end, satisfies a basic human need for warmth when she is installed in the basement of the town hall (Lee 1992). The message about technology in the story also illustrates David Edgerton's argument is that 'new technology' is often a result of transferring existing knowledge and technology to a new use (Edgerton 2006). In the story, the steam shovel is transformed from an instrument for digging to a modern boiler, and what happens in the small town of Popperville can be interpreted as a testament to old-fashioned hard work and ingenuity.

2.3 The Colonising Aspect of Technology

> It was the largest and most famous in the whole world! It was WONKA'S FACTORY, owned by a man called Mr Willy Wonka, the greatest inventor and maker of chocolates that there has ever been. And what a tremendous, marvellous place it was! It had huge iron gates leading into it, and a high wall surrounding it, and smoke belching from its chimneys, and strange whizzing sounds coming from deep inside it. And outside the walls, for half a mile around it in every direction, the air was scented with the heavy rich smell of melting chocolate! (Dahl 1964/1985:18)

Charlie and the Chocolate Factory (1964/1985) by the English author Roald Dahl depicts not only the mechanising function of technology but also its ability to colonise. Charlie's family is poor. Some days Charlie has nothing to eat, and his clothes are dirty and torn. His father, Mr. Bucket, is the only person in the family with a job, and he works in a toothpaste factory, "where he sat all day long at a bench and screwed the little caps on to the tops of the tubes of toothpaste after the tubes had been filled" (Dahl 1964/1985:16). Wonka's Golden Ticket contest, however, brings a sharp rise in sweet consumption, along with a rise in cavities and a subsequent increase in toothpaste sales. To keep up with demand, the toothpaste factory which employs Mr. Bucket mechanises the plant and fires the slower and more expensive human workers. Eventually, the toothpaste factory goes bankrupt. Charlie's father is left jobless, and money for the Bucket family becomes even tighter than before.

The story can be interpreted as a critique of capitalism and consumerism. The mechanisation of people's labour leads to stress, illness, economic inequality and the creation of power structures. The problem is rooted in the fact that human beings no longer develop technology to solve problems but as a basis for greed and a lust for power. Humankind has been reduced to a "cog" in a mechanistic system, and people in industrialised society are identified and valued for what they do and not for who they are (Ewerman 1997). Luckily for the Bucket family, everything changes when Charlie finds his Golden Ticket. On the other hand, their entire future lies in Wonka's powerful hands.

The world inside Willy Wonka's factory, however, differs greatly from other industrial settings. Instead of being dominated by machines in dull colours, Wonka's factory consists of colourful rooms, reminiscent of the countryside:

> They were looking down upon a lovely valley. There were green meadows on either side of the valley, and along the bottom of it there flowed a great brown river. What is more, there was a tremendous waterfall halfway along the river [. . .] Graceful trees and bushes were growing along the riverbanks – weeping willows and alders and tall clumps of rhododendrons with their pink and red and mauve blossoms. (Dahl 1964/1985:73–74)

Willy Wonka explains that he hates "ugliness" and that everything in the landscape is edible and "made of something different and delicious" (Dahl 1964/1985:75). The description of how the factory was transformed into a beautiful and colourful landscape with waterfalls of chocolate and large fields with edible sugar is reminiscent of William Morris' (1834–1896) utopia. As in the technology landscape in Wonka's factory, Morris' ideal society, the factory is a pleasant place with gardens and parks (Ambjörnsson 2004; Frängsmyr 1990; Hård and Jamison 2005). However, it is not a wilderness that is described as beautiful in the story but a form of nature which has been "tamed" and created by humans with the help of technology. A wilderness is described as something dangerous, the place Wonka's labour force, the Oompa Loompas, come from.

The description of the Oompa Loompas supports a view of technology as a colonising force. They are described as "imported directly from Loompaland". Before they moved to Wonka's factory, they spent every moment of their day climbing through the treetops looking for things to mash up with caterpillars to make them taste better. The food they loved more than any other was the cacao bean, but this was hard to find. In Wonka's factory, the Oompa Loompas can have all the cacao beans they want. The Oompa Loompas' place of origin and the description of them as a happy slave workforce make them interchangeable with the cogs in the various machines throughout the factory. This can be traced to Lewis Mumford's idea that early technology was in some ways more democratic, since it was used in a context where human beings were more closely involved in constructing it. In contrast to crafts, industrial technology is autocratic (authoritarian) and therefore a potentially destructive power (Mumford 1964). Based on Mumford's classification, Willy Wonka and his factory can be seen as representing authoritarian technology. However, the message is ambiguous. On the one hand, Wonka and his factory create hierarchies and are sometimes the cause of unemployment. On the other hand, Wonka and his Golden Ticket allow Charlie and his family to rise out of their poverty. Technological development thus creates both winners and losers.

Another ambiguous message can be identified where certain technology is portrayed as revolutionary and amazing, while other technology, like television, is portrayed as harmful. However, Wonka creates a new invention out of television technology, which is depicted as good:

> But first of all, do you know how ordinary television works? It is very simple. At one end, where the picture is being taken, you have a large ciné camera and you start photographing something. The photographs are then split up into millions of tiny little pieces which are so small that you can't see them, and these little pieces go whizzing around all over the place until suddenly they hit the antenna on the roof of somebody's house. They then go flashing down the wire that leads right into the back of the television set, and in there they get jiggled and joggled around until at last every single one of those millions of tiny pieces is fitted back into its right place (just like a jigsaw puzzle), and presto! – the photograph appears on the screen .... (Dahl 1964/1985:136)

Even if Wonka's description of how television technology works is not realistic, this is the way in which his own "television-chocolate" works. On the one hand, the message can be explained by the fact that Wonka does not know how the real technology works, but on the other hand, the message is consistent with Arthur (2011) and Edgerton (2006): that existing technology and technological knowledge are often used to create something new.

As in the story *Doctor Klokamundus' Invention*, technology is used as a tool to homogenise people. Children with Golden Tickets who are nasty and greedy first have to learn a lesson, and then they all become victims of Wonka's technology. The winner in the end is Charlie, who is a kind and considerate boy, and Willy Wonka decides that as soon as he is old enough to run it, the entire factory will become his.

2.4 Technology Versus Nature

I started as a tree. I was a giant Wawa tree in the forest of the hinterland [...] For years, we shared the peace and quiet of our forest with animals and man, who only chopped down small trees for his use. (Asare 1982/1990:1)

The Canoe's Story (1982/1990) written by the Ghanaian writer Meshack Asare is, like *Mike Mulligan and his Steam Shovel*, a story about the tension between old and new technologies. However, it also involves a critique of the fact that humans do not show due respect when nature is used to create technology. The story is anthropomorphic, as it is told from the tree's point of view. The tree tells its life story, from its beginning in the Asante forest to its role in the traditional Ga fishing industry. The story includes descriptions of traditional rituals, like Asante tree-cutting ceremonies and the Ga canoe-naming ceremony.

At the beginning of the story, the Wawa tree is happy to use its branches and leaves to give shade to the humans. But one day, something dramatic happens – people with big, modern machines arrive in the forest and start cutting down the trees. The new machines are destroying everything in their path. The tree can hear the other trees' desperate wailings, and soon the whole forest is "hurt". The animals, which have lived since ancient times among the trees, escape the "iron monsters". As the trees are unable to escape, one by one, they fall victim to the humans' "horrific machines". The Wawa tree realises that soon it will be its own turn. But the tree's fate will differ from that of its friends. One early morning, a couple of men turn up. They have brought with them a piece of fabric, a bird, a bottle of gin and some eggs. These are all votive offerings to the spirit of the tree, which has been with it during the hundreds of years it took to grow into a stately giant tree.

Although the tree is cut down, it is content. It was not cut down by "vicious" machines with jagged chains. Instead, through their gifts, the humans showed respect. By using their tools, axes and chisels, the men give the tree a new guise: "My new shape was not like anything else in the forest and I was very pleased with myself" (Asare 1982/1990:6). By depicting the development from Wawa tree to traditional canoe, and finally to motorised sailing vessel, both the fear of change and the benefits of technological development are illustrated. For example, at the end of the story, the Wawa tree says: "The engine works hard to move me through the water; and for me as fast as a leaf in the wind. So now I'm not afraid of machines anymore" (Asare 1982/1990:20). Technology is the result of evolution, a gradual development, and novel technology (a motorised fishing boat) is descended from earlier forms (Arthur 2011). It is an illustration of how the technological world consists of both modern and indigenous forms of technology (Gumbo 2015). It also portrays the strong tie between humans, technology and nature.

Moreover, the story's critique of exploitation introduces a theme of sustainable development. The tale shows how much the fishermen value the trees from which they indirectly derive their livelihoods. They cannot build their canoes with technological skills alone and know they need the trees. At the same time, the story describes the ongoing mass depletion of the forest, with little consideration for replacing this valuable resource. An implicit message is that we should start asking ourselves what would happen to humanity if this depletion did not stop and all the trees in the rainforest were cut down. The tree's understanding and generosity may cause the reader to rethink the relationship between humans and nature, as well as how humans relate to objects made from nature. There is a spiritual life force in natural objects and materials that must be respected. However, although the story is told from a tree's perspective, there is a message that humans have the right to make use of natural resources if they do so with respect and consideration. This can be described as a *weak anthropocentric view*; a view premised on the centrality of human beings' needs but more than a purely instrumental view of nature (Dobson 2000). Technological progress should be implemented in harmony with nature and with consideration for future generations. Humans are part of nature, and their very survival depends on the preservation and protection of non-human species. By destroying trees, we destroy ourselves.

In the technology landscape of the story, contexts where people have a close relationship with the design process are seen as a beneficial tool which can help human beings solve problems and steer society in a positive direction. In contrast to the design process where fishermen build their own canoes, the modern "iron monsters", or powerful deforestation machines, can be interpreted as representing what Mumford describes as authoritarian technology (Mumford 1964). The monster metaphor is used as a criticism of technological development built on ruthless exploitation of nature. The deforestation machine is related to a techno-centric context, in which human beings use technology as a means of acquiring power or economic resources with no regard for nature or future generations (Axell 2015). The canoe's final thoughts about whether anyone has ever liked him as much as they do now, when he is a useful human tool, is an illustration of the tension between conservation and change. The story presents a careful balance between the values of nature, modern and indigenous technology and technological change.

2.5 The Enduring Aspect of Technology

> Old man Pettson and his cat Findus lived on a little farm deep in the country. They had a few hens in the henhouse and plenty of wood in the woodshed and everything else they needed was in the tool shed. They did not often get visits and that was just as well, thought Pettson (Nordqvist 2000:1)

The technology landscape in the books about Pettson and Findus, written by the Swedish author Sven Nordqvist, involves an idyllic countryside setting with close ties to nature. Rather than highlighting negative images or the consequences of technological development, the books about Pettson and Findus offer alternatives. Their inventions are created from reused artefacts and do not deplete natural resources or require much energy to function. Pettson mends his clothes instead of buying new ones, grows his own vegetables and is self-sufficient in eggs. Everything about Pettson's life on the farm is on a small scale.

What makes the stories unique compared to many other children's books is that they portray the enduring aspect of technology; it is "born" but does not "die", although it is in a constant state of change. During the summer, Pettson travels either on foot or by bike. In winter, he uses his sledge, and Findus uses skis. Sledges, skis and kick sleds are all examples of artefacts which have a long history and which are still in use today. In this sense, they are not linked to a particular time and, as artefacts, have hardly changed through time. By underlining the enduring dimension of technology, it becomes easier to appreciate technologies which have been an important part of people's lives for many generations and continue to serve a purpose today. New technologies emerge and others disappear, but some continue to be used over time. This dimension is also discussed by Kelly when he notes that very few technologies disappear once they have been established. His conclusion is partly based on a review of nearly 600 pages of the Montgomery Ward Catalog for the years 1894–1895. He concludes that a majority of items sold in the catalogue are still available for purchase today. The actual design or style may have changed, but the underlying technological function remains the same (Kelly 2010).

Edgerton notes a similar aspect to Kelly but from a different angle. He challenges the common practice of categorising technological development in terms of a historical timeline, where each invention is denoted by the year it was invented. This presents each technological development as "new", even if "new" often involves building on existing knowledge and applying it in new ways (Edgerton 2006). Kelly's description is similar to how Pettson creates his technology. In the stories, technology is the result of a creative design process in which Pettson is a *bricoleur*. The term *bricoleur* was first introduced by anthropologist Claude Lévi-Strauss (1962), who considered it to be in some ways the opposite of an engineer. He/she is skilled in a variety of tasks but, unlike the engineer or scientist, uses only what is available, both in terms of tools and material. A bricoleur is a "Jack of all trades", and Pettson solves his own technological problems. A great deal of technological evolution occurs, as Arthur notes, when components are improved and used in other applications (Arthur 2011). For example, Pettson invents a device to scare away foxes, made from a balloon, pepper, a roll of steel wire and firecrackers. Another of Pettson's inventions is a "fishing bow":

> Pettson had invented a fishing bow. Down by the lake he explained to Findus how it worked. The hook and the float were attached to an arrow. The arrow was attached to the fishing line. The rest of the line was wound onto a reel, which was attached to a bow. With it he could shoot the arrow with the hook far out into the water, much further than he could reach with the fishing rod. It worked quite superbly. (Nordqvist 2003:11)

Pettson and his cat also use their artefacts differently from how they were originally intended to be used: an electric jigsaw becomes a breadknife, a brace and bit become a cup holder, a hat is transformed into a lampshade and a small plane is used as a cheese slicer. Don Ihde (2006) uses the term "the designer fallacy" to explore whether a designer can really include the purpose of an artefact in the design itself. Technology is embedded in different cultural contexts, writes Ihde, which affect both the design and its applications. The same kind of technology can apply to different but specific contexts, and artefacts can incorporate a variety of technological uses and trajectories of development (Ihde 2006). This is an aspect noted by Kay Stables (chapter "Critiquing Design: Perspectives and World Views on Design and Design and Technology Education, for the Common Good") when she discusses how we expect an object to be designed with a particular purpose in mind, such as meeting a human need or solving a problem. Sometimes, its purpose may be to protest, and one person's purpose or need may be completely the opposite to that of another person. From this perspective, Pettson's technological solutions can be construed as a protest against the mainstream expectation that the design of an artefact must be linked to a specific technical function. Daniel Dennett sees this in a similar way to Ihde and Stables and gives the example that old irons are frequently used today as bookends and doorstops. He concludes that "[. . .] the inventor is not the final arbiter of what an artefact is, or is for; the *users* decide that" (Dennett 1990:186). Norman uses the terms "affordance" and "signifier" to describe the relationship between an object which has been designed and the agent with which it interacts. Affordances are, according to Norman, the possible interactions between people and the environment, while signifiers signal what actions are possible and how they should be carried out. Affordances determine possible actions, while signifiers communicate where these actions should take place (Norman 2013). The books about Pettson show that there are many different affordances in Pettson's technology. The signifiers are to be found in the illustrations, for example, how a clamp on the kitchen table holds a loaf of bread and how a jigsaw beside the bread slices indicates how the tool can be used.

Building on the discussion above, the books about Pettson and Findus can also be said to represent a view of technology consistent with Joseph C. Pitt's approach that technology should be considered a neutral tool which can be used in a "good" or "evil" way (Pitt 2014). As Pettson is using and developing his technology with good intentions, his technology can also be considered "good".

A further dimension of technology in the books about Pettson is that it does not have to solve problems, be "useful" or lead to the performance of a task in a simpler or more efficient way. Pettson's Santa Claus machine (Nordqvist 1994) is an example of an artefact that neither rationalises nor improves the efficiency of human activity.

Some of Bruce Archer's, Ken Baynes' and Phil Roberts' characteristics of design can be identified in the way Pettson makes his Santa Claus machine. Pettson "envisages" what he is going to create, and his invention is also going to meet a particular need: to make Findus happy on Christmas Eve. Pettson's design is *intentional*, *integrative* and *inventive*, but in the end, it turns out not to be *useful* or *expedient*, and the machine can hardly be described as *productive* (Archer et al. 1992). The Santa Claus machine, as well as some other of Pettson's and Findus' inventions, can therefore be described as "Rube Goldberg machines" which perform tasks in a more complicated way than is necessary (Acharya and Sirinterlikci 2010). Many of their bricolages are only good for doing things in an alternative way or for "solving" what could hardly be regarded as a technological problem. Examples of these are pots on wheels and tea cups with "teaspoon holders".

In the stories about Pettson and Findus, crafts and early technology are given a higher value than industrial and more technically advanced inventions. As in Morris' utopia, there is a belief in technology in Pettson's world, but only in the technology that relates to crafts, and is handmade. Morris distanced himself from both the pessimistic and the uncritically positive settings of technology and industrial society. Instead, he wanted to combine the old and the new, the innovative and the traditional and the functional and the aesthetic (Frängsmyr 1990; Hård and Jamison 2005).

There is no explicit critique of the impact of technology on society in the books, but they could be interpreted as an implicit critique of industrialisation, harking back to the past in a nostalgic way with an inherent message that "it was better before". They could also be read as depictions of a technology landscape which represents an alternative and more sustainable way of living and which stands in contrast to a modern society built on consumption (Axell 2015; Axell et al. 2014). What places the Pettson and Findus books in a special category in their relationship with technology is that the aim of their creations is not to produce something perfect, but something unique.

2.6 Concluding Analysis

An examination of the technology landscapes in the books included in this chapter shows that a critique of technology is very much present in many children's books and involves discussions about the advantages and disadvantages of technology. The focus is on the creative process, rather than making a technological process or artefact more effective or profitable. From a design perspective, the stories illustrate the critical point that it is the user of an artefact, not the designer, who ultimately decides how it is defined and can be used (Dennett 1990; Ihde 2006).

Furthermore, anthropomorphism is common in children's fiction. Ascribing human traits to technology and aspects of nature helps built an emotional attachment to machines and/or the natural environment. Caring relationships contribute to bridging barriers between human beings, technology and nature (Schwarcz 1967; Waytz 2013, 2014). A caring relationship with machines also prevents technology from becoming obsolete; there are always new applications for old technologies (Edgerton 2006; Kelly 2010).

The stories convey a belief in technology but mainly in technologies which are related to crafts: handmade and indigenous technologies which have a long history. Well-designed objects can bring a sense of pride and enjoyment and a feeling of being in control (Norman 2013). This can be related to the idea that early technologies are in some ways more democratic, since they are used in contexts where human beings share a closer relationship with the artefact and the process of making it (Mumford 1964). The books therefore involve a nostalgic, but not dystopic, view of technology. Some of the stories can be interpreted as calling for a shift in how we see technological development, as well as in the design process: from the new to the old, from the big to the small and from the spectacular to the mundane (Edgerton 2006). What is missing, however, is a context that shifts the focus away from a technology dominated by men to one which also includes women.

All in all, the critique of technology and technological progress in the stories is often ambiguous. It advocates that a life in harmony with nature is a path to the future. On the other hand, there is a message that as long as humans aim to create technology that can satisfy their needs and desires, they are morally justified in using and transforming nature as they see fit. However, this must be done with consideration and respect towards nature. The stories also seem to have a built-in duality, describing how technology can establish individual freedom but also weaken human bonds. Like the Roman God Janus, technology is two-faced, with a positive and negative side, one constructive and the other destructive.

So finally, how can children's fiction be used as a learning tool for developing critical thinking?

2.7 Children's Literature as a Learning Tool for Critical Thinking

Thinking critically about technology is about drawing conclusions, evaluating and seeing things from different angles, i.e. being open-minded and considering alternative ways of looking at something. Questioning is therefore an important part of learning. By asking questions that challenge children's conceptions and ideas, teachers can help children to continue developing their critical thinking abilities. Examples of activities where teachers can use fictional children's books as springboards for critical discussions about technology include:

- *Identifying.* Identifying how technology is represented in the story. "What is technology?" "What kind of technological artefacts, activities, processes and systems can be found?"
- *Making comparisons and contrasting.* Identifying similarities and differences related to different technologies. "What are the similarities and what are the differences?" Comparing designs of the same type of artefact. "What are the recurring characteristics?" "For example, how do we know from the design that an artefact is a chair?" This helps children to analyse and categorise.
- *Exploring technology from a cultural and temporal perspective.* Cultural and historical perspectives can help children discover the nature of human desires,

needs and aspirations throughout history and in different cultural settings, and how these needs have been fulfilled using different types of technology. "How do people solve this problem in other parts of the world?" "Are there any similarities and differences?" "How did people solve this problem a hundred years ago/a thousand years ago, etc.?" "How could this problem be solved in the future?" "What are the positive and negative aspects of the different solutions?"

- *Asking questions which have no direct answers.* This helps children infer and draw conclusions based on their own understanding of technology. Questions starting with "Why do you think . . . ?" encourage them to think more freely.
- *Relating the technological content in the story to children's own lives and outside events.* This can encourage children to use their own knowledge about technology in new ways and apply it to different ideas and contexts.
- *Not telling the whole story.* Asking the children to finish a story encourages them to use critical thinking skills. When they are not given an ending, they have to develop the story in a creative way, drawing conclusions, coming up with their own solutions and finding counter-arguments. "What do you think happened next?" "How can X solve this problem? Are there different ways?"
- *Providing cooperative learning opportunities.* Reading with peers promotes the development of critical thinking skills in children. When children read together, they share ideas and learn from one another. Encouraging them to identify technology in the stories, examine its role and find alternative solutions can form a basis for creative discussion.

To summarise, an examination of children's fiction shows that by reading children's literature through a "Design and Technology education lens", we discover that technology plays an important role in many children's books. By using books that have a strong story line and are of interest to children, technology can be presented as part of the world that surrounds them. The stories can lead to critical discussions about the nature of technology but can also contribute to an understanding of how human beings have related to technology in other cultures and in earlier generations. The ambiguous messages in the books reveal the multifaceted and complex nature of technology, which makes it possible to problematise it in ways that textbooks seldom can. The technology landscapes in children's literature could thus contribute to making technology and the nature of technology more comprehensible and visible to pupils. These landscapes could also help broaden their perspective and therefore act as a pedagogic tool in fulfilling the aims of Design and Technology education.

References

Acharya, S., & Sirinterlikci, A. (2010). Introducing engineering design through an intelligent rube Goldberg implementation. *Journal of Technology Studies, 36*(2), 63–72.

Ambjörnsson, R. (2004). *Fantasin till makten! Utopiska idéer i västerlandet under fem hundra år.* Stockholm: Ordfront.

Arthur, W. B. (2011). *The nature of technology: what it is and how it evolves*. London: Allen Lane.
Archer, B., Baynes, K., & Roberts, P. (1992). *The nature of research into design and technology education*. Loughborough: Loughborough University.
Asare, M. (1982/1990). *Kanoten [the Canoe's story]*. Stockholm: Carlsen/if.
Axell, C. (2015). *Technology landscapes in children's literature. A didactic journey from Nils Holgersson to Pettson and Findus*. Dissertation. Linköping: Linköping University.
Axell, C., & Boström, J. (2015, April 6). Facts for youngsters – Contextualised technology or fragmented artefacts? A study on portrayals of technology in picture books from a gender perspective. In M. Chatoney (Ed.), *PATT 29 plurality and complementarity of approaches in design and technology education* (pp. 42–48). Marseille: Aix Marseille University.
Axell, C., Hallström, J., & Hagberg, J.-E. (2014). Images of technology and sustainable development in Swedish children's literature. *Australasian Journal of Technology Education, 1*(1), 1–9.
Beskow, E. (1919/1996). Doktor Klokamundus' uppfinning [Doctor Klokamundus' invention]. In *Muntergök*. Stockholm: Bonnier Carlsen.
Burton, V. L. (1939/2005). *Mike Mulligan and his steam shovel*. London: Frances Lincoln.
Dakers, J. R. (2006). Toward a philosophy for technology education. In J. R. Dakers (Ed.), *Defining technological literacy: towards an epistemological framework* (pp. 145–158). New York: Palgrave Macmillan.
Dahl, R. (1964/1985). *Charlie and the chocolate factory*. New York: Puffin Books.
Dennett, D. C. (1990). The interpretation of texts, people and other artifacts. *Philosophy and Phenomenological Research, 50*(Supplement), 177–194.
Department of Basic Education. (2011). *Curriculum and Assessment Policy Statement (CAPS): Natural sciences and technology. Intermediate phase grades 4–6*. Pretoria: Government Printing Works.
Department for Education. (2013). *National curriculum in England. Design and technology programmes of study (Statutory guidance)*. Available from: https://www.gov.uk/government/publications/
Dobson, A. (2000). *Green political thought* (3rd ed.). London: Routledge.
Edgerton, D. (2006). *The shock of the old: Technology and global history since 1900*. London: Profile Books.
Ellul, J. (1964). *The technological society*. New York: Vintage Books.
Ellul, J. (2010). The autonomy of technology. In C. Hanks (Ed.), *Technology and values: Essential readings* (pp. 67–75). Malden: Wiley-Blackwell.
Ewerman, A. (1997). Den informationssociala revolutionen. Historiens cykler och vår tids epokskifte. In G. von Wright (Ed.), *Handla! Om förändring, välfärd, arbete, lärande, konsumtion, arkitektur, design, kultur, framtid* (pp. 87–102). Nerenius/Santérus: Stockholm.
Frängsmyr, T. (1990). *Framsteg eller förfall: Framtidsbilder och utopier i västerländsk tanketradition*. Stockholm: Allmänna förlaget.
Gumbo, M. T. (2015). Indigenous technology in technology education curricula and teaching. In P. J. Williams, A. Jones, & C. Buntting (Eds.), *The future of technology education* (pp. 57–75). Singapore: Springer.
Hagberg, J.-E. (2008). *Livet genom tekniklandskapet: Livslopp, åldrande och vardagsteknikens förändring. The National Institute for the Study of Ageing and Later Life (NISAL)*. Linköping: Linköping University.
Hagberg, J.-E. (2009). Att lära i teknikens rum och landskap. En metadidaktisk betraktelse. In P. Gyberg & J. Hallström (Eds.), *Världens gång – teknikens utveckling: Om samspelet mellan teknik, människa och samhälle* (pp. 41–60). Lund: Studentlitteratur.
Hård, M., & Jamison, A. (2005). *Hubris and hybrids: A cultural history of technology and science*. New York: Routledge.
Ihde, D. (2006). The designer fallacy and technological imagination. In J. R. Dakers (Ed.), *Defining technological literacy: Towards an epistemological framework* (pp. 55–59). New York: Palgrave Macmillan.
Kelly, K. (2010). *What technology wants*. New York: Viking.

Klasander, C. (2010). *Talet om tekniska system: Förväntningar, traditioner och skolverkligheter. Dissertation*. Linköping: Linköpings universitet.
Lee, J. Y. (1992). The feminization of technology: Mechanical characters in picture books. In M. L. Greenberg & L. Schachterle (Eds.), *Literature and Technology* (pp. 206–224). Bethlehem: Lehigh University Press.
Lévi-Strauss, C. (1962). *The savage mind*. London.
Lindqvist, S. (2011). *Changes in the technological landscape: Essays in the history of science and technology*. Sagamore Beach: Science History Publications.
Mawson, B. (2010). Children's developing understanding of technology. *International Journal of Technology and Design Education, 20*(1), 1–13.
Merchant, C. (1989). *The death of nature. Women, ecology and the scientific revolution*. San Fransisco: Harper & Row.
Mitcham, C. (1994). *Thinking through technology: The path between engineering and philosophy*. Chicago: University of Chicago Press.
Mumford, L. (1964). Authoritarian and democratic technics. *Technology and Culture, 5*(1), 1–8.
Nordqvist, S. (1994). *Tomtemaskinen [The Santa Claus machine]*. Stockholm: Opal.
Nordqvist, S. (2000). *The fox hunt*. Bromma: Opal.
Nordqvist, S. (2003). *Pettson goes camping*. Bromma: Opal.
Norman, D. A. (2013). The design of everyday things. [Revised and expanded edition]. New York: Basic Books.
O'Har, G. M. (2000). Magic in the machine age. *Technology and Culture, 41*(4), 862–864.
Pitt, J. C. (2014). 'Guns don't kill, people kill'; values in and/or around technologies. In P. Kroes & P.-P. Verbeek (Eds.), *The moral status of technical artefacts* (pp. 89–101). Dortrecht: Springer.
Schwarcz, H. J. (1967). Machine animism in modern children's literature. *The Library Quarterly, 37*(1), 78–95.
Siu, K. W. M., & Lam, M. S. (2005). Early childhood technology education: A sociocultural perspective. *Early Childhood Education Journal, 32*(6), 353–358.
Swedish National Agency for Education. (2011). *Curriculum for the compulsory school system, the pre-school class and the leisure-time centre*. Stockholm: Fritzes.
Sörlin, S. (1991). *Naturkontraktet: Om naturumgängets idéhistoria*. Stockholm: Carlsson.
Turja, L., Endepohls-Ulpe, M., & Chatoney, M. (2009). A conceptual framework for developing the curriculum and delivery of technology education in early childhood. *International Journal of Design and Technology Education, 19*(4), 353–336.
Vries, d. M. (2006). Technological knowledge and artifacts: An analytical view. In J. R. Dakers (Ed.), *Defining technological literacy: Towards an epistemological framework* (pp. 7–30). New York: Palgrave Macmillan.
Waytz, A. (2013). Making meaning by seeing human. In K. D. Markman, T. Proulx, & M. Lindberg (Eds.), *The psychology of meaning* (pp. 135–146). Washington, DC: American Psychological Association.
Waytz, A. (2014). The mind in the machine: anthropomorphism increases trust in an autonomous vehicle. *Journal of Experimental Social Psychology, 52*(3), 113–117.
Winner, L. (1989). *The whale and the reactor: A search of limits in an age of high technology* (5th ed.). Chicago: The University of Chicago Press.
Wright, v. G. H. (1987/2010). *Vetenskapen och förnuftet: Ett försök till orientering*. Stockholm: Bonnier.

Modelling as a Form of Critique

Niall Seery

Abstract This chapter is concerned with the cognitive and related physical manifestations utilised to further insight and refine cognitive processes. Modelling in all its forms is considered as a support for critique. Therefore modelling and models are seen as a critical aspect of the external and internal dialectic that supports new and better capacities to create and synthesise knowledge and meaning.

Regardless of trying to understand the world *as it is* or *as it could be*, navigating the unknown is variable. Modelling is a generative process that functions as a means of making explicit or externalising the variability in thinking. Directly associated with this capacity to make thinking visible (inside or outside the head) is the opportunity to critique and reason.

The relationship between modelling and behaviour is discussed, and as a result some of the key issues associated with the cognitive processes that support modelling as a form of critique are highlighted. Seeing in the *mind's eye* is a natural human capacity that describes a broad cognitive skill that in general includes imagination, memory and visualisation. The capacity to utilise this skill as the basis for meaningful learning in Design and Technology is considered.

This chapter considers modelling in all its forms and discusses the speculative and enquiring nature of modelling as a critical feature of critique.

The chapter concludes by considering some of the implications for practice and highlights the need to consider the role of modelling within contemporary understandings of teaching and learning.

Keywords Modelling • Critique • Design education • Natural enquiry

1 Introduction

It's not apparent if it was *hard coded* by evolutionary processes or learnt behaviour, but without prescription, target or compulsion, we played as children. We naturally created a world of make-believe, creating contexts, personas and scenarios,

N. Seery (✉)
Department of Design and Manufacturing Technology, University of Limerick, Limerick, Ireland
e-mail: niall.seery@ul.ie

PJ. Williams, K. Stables (eds.), *Critique in Design and Technology Education*,
Contemporary Issues in Technology Education, DOI 10.1007/978-981-10-3106-9_14

sometimes conjured from caricatures of the known or more often simply unique imaginative creations. We designed and created accessories, costumes and environments to help represent our world and utilised accessible resources to produce props to reinforce that reality. In doing so, we framed, reframed and solved problems, often creating new problems, but largely unaware of the existence of a problem. At times actions were intentional and at other times intuitive (separated constructs in the context of experimental processing), but always purposeful. This was spontaneous, unbounded, voluntary, developmental and normal!

As a child natural enquiry was driven by need, speculation or sheer whimsy, where imagining new possibilities, conceptions or realities served as enrichment, at least from the perspective of the creator. But regardless of intention or motivation, the need to acquire additional height to access the Sweet Jar, wondering if the iPhone would float in the bath or assuming the persona of Spiderman, these all have the same thing in common, speculation, that can lead to the furthering of insight. This externalisation of imagination creates rich evidence of instinctive and vicarious learning that permeates early development. Curiously, this limitless preoperational development (Piagetian) is unbounded by logic.

However, this development does not happen in isolation; there is a situational, contextual and environmental influence. The external mediation of the creative endeavours helps to refine the initial model. For example, parental intervention will address the structural integrity of acquiring the additional height, possibly with a cautionary note of *don't fall*. The iPhone experimentation results in a determination of the parameters of *good* and *bad* ideas. The humorous reaction to a miniature Spiderman can facilitate more elaborate mimicry of factitious capacities. This interplay forms a type of metadiscourse between modelled and critiqued actions, mediated by what is socially acceptable.

This hyperbole of childhood antics serves as a useful metaphor for the nature of the activity that predicates effective modelling in Design and Technology. This chapter explores the nature of modelling with respect to inhibitors and influencers and highlights the need to consider the relationship between modelling and critique.

The chapter focuses on the generative and critical processes of modelling in its various forms and considers implications for practice. In particular this chapter is concerned with the cognitive and related physical manifestations utilised to further insight and refine cognitive process. Modelling is not considered a panacea, but a critical aspect of the external and internal dialectic that supports new and better capacities to create and synthesise knowledge and meaning.

2 Models

Models are created in multiple forms, with variable functions, and accompanied by associated conceptions, motivations or agendas. Archer (1992b, p.7) describes a model as "... anything which represents anything else for informational, experimental, evaluative or communicative purposes". A useful classification is to consider models as either a physical or mental representation of thinking at a given time, during any given process. A model is a result of a purposeful or speculative act that generally serves one of two functions: to support further insight or to externally communicate. Physical models are a critical part of early childhood play and development. The fortress built behind the sofa and the depiction of the pet cat sitting on the moon (possibly only decipherable by the creator) are examples of models. These props act as an intermediary between existing realities and imaginative possibilities. The manipulation and refining of physical models facilitate further insight, exploration and communication.

Cognitive models are a more complicated idea. The capturing, interpretation and synthesis of sensory data enable memory and imagination to model an individual's external reality and further conceptual realities. As a result of this processing in the *mind's eye*, there is an unbounded myriad of perspectives and possible realities. Cognitive models are the product of creating personal realities from experiences, consciousness and cognitive capacity. In terms of purpose, they are similar to physical models; they function as the mechanism to support natural enquiry, refine thinking and effectively communicate (usually with the self). Archer and Roberts (1992) consider cognitive models as being language and symbol system independent. However, imagining or seeing in the *mind's eye* is not an entirely independent act. Baynes (1992b, p.12) argues "... that the models available to us are determined on the one hand by deep structures in the mind and on the other by the content of our culture". Acknowledging influence, it is also interesting to consider our predisposition to develop and translate cognitive models through language and notation. Our capacity to decode and encode imaginative thinking is a key designerly/creative capacity.

The term *model* is used to represent something. For example, the codification of ideas into a comprehensible scientific or mathematical language is an important capacity. The use of certain codified models furthers our ability to coherently enquire within specific degrees of abstraction, within relevant disciplines. For example, the use of mathematical models can be used to study systems, explain performance or make predictions. Baynes (1992b, p.12) argues that "All models are an abstraction from the chaos of the real world – this is their value – we can isolate the relevant variables for specific treatment". These mathematical, scientific or conceptual models augment thinking and are "... powerful because they isolate an aspect of reality and allow us to represent, interpret, manipulate or control it" (Baynes 1992b, p.12). However, the functions of a model are not exclusive, regardless of being a stimulus for internal dialogue or an externalisation of thinking; models also form the basis of critique.

Models are used to describe and manipulate different aspects of reality. It is the manipulation and separation of these aspects that make models and modelling such a powerful tool. With this in mind, models should be considered as just that – a model – not the best model or the only model but a focus of enquiry and critique, constrained by internal (cognitions, beliefs, values, etc.) and external (interpretations, cultural, economics, ethics, etc.) factors.

3 Modelling

Regardless of trying to understand the world *as it is* or *as it could be*, navigating the unknown is variable. Modelling is a generative process that functions as a means of making explicit or externalising the variability in thinking. In broad terms *modelling* is the process of developing, expanding and encoding a theory or idea. The product of the modelling process is an imperfect model; there are always better models, "... the optimum obtained from this model does not pretend to be the absolute best, but only the best relative to the society to which it applies" (Koen 1985, p.15). The process of generating a model can be described as a search for new meaning, a means of better understanding the world. Although cultural and social implications are critical aspects of any discussion on modelling, initially it is important to frame modelling as an intrapersonal and interpersonal communicative activity.

The dependent variables that govern the modelling process are predicated on the intentionality of the produced model, for example, the act of modelling is influenced by motivation, need, agenda and context (ideas that will be explored later in the chapter). To model a representation of a conceptual design, the task of modelling is structured and constrained by time and resources and the fitness of purpose to effectively communicate the *idea* or *agenda* of the design. The modelled reality forms a target for critique, where scale, function, performance, aesthetics, etc. can be questioned, clarified and refined. Interestingly, throughout the discussion, the focus can shift from intrapersonal emphasis to interpersonal dialogue with every iterative evolution of the externalised artefact. Producing physical models helps to create a discursive medium. Interacting with physical representations of reality is a natural (or at least socially conditioned, e.g. playing with dolls or toy trucks as imaginative stimulus) behaviour that stimulates our imaginative capacities. The objective of the modelled thinking is to share a view of the world that is difficult to communicate through the use of natural language or scientific notation. The architect's scaled model of the new shopping complex, the fashion designer's croquis and the engineer's prototype control board are good examples of physical models. Although these examples are constrained, they were constructed through critique. These examples are governed or influenced by the parameters of the design brief; the design of the shopping centre must consider location, environment, population, etc., with these considerations forcing the direction of the critique. The designer's croquis captures essential design elements and features within the context of materials, function and aesthetic, so as to support further dialogue around the initial concept. The engineer's control board is equally influenced by context of the model, with the design critique not concentrating on the best prototype, but the model that can best support considering the implications for cost, application, reliability, manufacturability, etc.

Modelling as a form of critique must play a significant part in learning challenges and tasks. When considering modelling as a function of learning, it is important to be clear about our assumptions and expectations. Do we expect learners to get the correct answer? Is the correct answer that valuable (from either a pedagogical or learning perspective)? How do we expect the learner to behave (assuming that there is an intention to learn)? Do we expect the same cognitive reaction from each learner? Do we want the same approach to learning? Often practices are predicated on what Brousseau (1980, p.127) calls the *didactic contract*; this defines "the set (of specific [knowledge taught]) behaviours of the teacher expected by the pupil and the set of pupil behaviours expected by the teacher". Questions are asked in anticipation of getting the correct answer, and learners respond only if they know the answer. This would appear to subscribe to a transmission model where the room for plurality and alternatives in meaning is not supported. An alternative expectation is that when confronted with a learning challenge, the learner is expected to model a conceptual understanding of the problem and then a proposed exploration or examination of the parameter, before modelling a possible solution. For example, solving an abstract linkage problem could stimulate a number of fluid discussions and modelling behaviours:

- It is similar to how the wipers work on a car, so I imagined it with that reference and could see it work.
- I had to see it, so I made the linkages from lollipop sticks and simulated the motion, and then it was easy to plot its loci.
- The schematic sketch allowed me to determine the parameters of the movement and the conditions that controlled it.
- After discussing its application, it was easy to imagine its movement.
- I used the simulated CAD model to investigate the key elements of the linkage.

Johnston-Wilder and Mason (2005) describe these types of actions as important in helping to get a *sense of the problem*, and although this is not an exhaustive list of potential human reactions and behaviours, it helps in demonstrating the variability, purposefulness and utility of modelling. What are the considerations that need to be addressed to support this epistemic approach to learning? What are the characteristics that need to be supported to facilitate better and more effective modelling?

Modelling is variable but in general begins with speculative generation of auxiliary information, experiences and insights through a personal dialectic. This process, regardless of being intentional or instinctive, is always purposeful and creates the necessary insights to make critical decisions and form rational thought. Modelling can be described as the unconscious interplay between top-down (evidence and data) and bottom-up (theories and underlying physical properties) enquiry (Scholl and Phelan 2004). Archer (1992b) describes modelling as being placed between natural language and tacit knowledge, and although it does not meet the technical criteria of a language, it is a relevant and effective way of communicating imaginative and designerly thinking. Modelling as an alternative language or a type of metalanguage is a useful way of framing the concept.

4 Modelling as a Metalanguage

Richard (cited in Archer (1992a, p.9)) gives a useful definition of communication: "Communication takes place when one mind so acts upon its environment that another mind is influenced in a relevant way". This provides an important foundation for considering the nature of communication in Design and Technology. Modelling as a metalanguage describes a clear paradigm: there is something that I want to communicate (either with self or other) and the product of my intent should be comprehensible.

It is well argued that Design and Technology education requires a distinct type of activity that calls on a distinct type of cognitive capacity (Baynes 1992a; Norström 2013), where learning challenges are largely ill-defined and the resolutions and responses are often difficult to determine in terms of the *proper* answer. Kimbell (2007) argues that pupils in Design and Technology can be excellent in distinctly different ways. Understanding the nature of Design and Technology activities requires us to consider the language of the discipline. Technical communication of terms, processes and materials is a critical part of any discipline understanding; however, recognising the analogous basis of modelling as a means of communication in a comprehensible way can be considered a distinct language. The culmination of sensory information captured in models and prototypes and symbolic representations developed through ideation sketches, when consolidated through language and notations, function as a type of metalanguage. Although this does not satisfy the technical characteristics (lexicon, morphology, syntax, etc.) of a language and although this culmination is often unorthodox, it has proven to be essential in supporting designerly thinking and activities.

There are a number of models in education (e.g. Argyris and Schon 1974; Kolb 1984) that are particularly useful in Design and Technology education as they advocate for the bidirectional relationship between doing and thinking. Kelly et al. (1987) highlight the importance of this interaction with the dialectic model, formed on the bases of critique and enquiry. The deeper the thoughts, the more purposeful your actions, and the more purposeful your actions, the more significantly the actions can result in deeper thinking. This conversation with the self is mediated on many levels. The conversation with the self is a direct causal discussion. Actions are purposeful and are generally either aimed at manipulating the known based on a thesis or idea or a less formal speculative *what if*. Craft and physical modelling actions are a normal part of Design and Technology education, and as a result we can often assume that there is an undercurrent of conversation. This is not always the case and should not be an assumed part of practice. Critical review of actions is the foundation of the conversation. The learner must engage with the evidence of the actions in a way that invokes critical review. This is not necessarily to require the learner to reflect, as reflection is often an affirming process, sometimes speaking to erroneous criteria and not elaborating on the cognitive act of critique. McGarr and McCormack (2014) highlight this when discussing the nature of reflecting to conform. Their research suggests that conversations with self when prescribed can be relatively ineffective. The learning process is less epistemic and more criterion referenced when the learner is engaging in their perceptions of social norms or assumed expectation surrounding the evidence of effective learning.

The process of modelling (in all its forms) is a values-laden activity and is influenced by underpinned beliefs and principles. Furthermore, considering modelling as a form of critique requires the development of skills of appraisal and abilities to critique. The variable nature of modelling and the contextual nature of critique make their relationship a complicated amalgam of evidence, theories, beliefs and speculation. Learning and developing attitudes, skills and knowledge to succeed in Design and Technology is the true value of the discipline. It is therefore the focus of this chapter to explore modelling from the perspective of learning and teaching, based on a number of assumptions:

- Modelling is framed in a metacognitive paradigm, when the learner is trying to understand better and when the relationship between construction and critique is intentional.
- Modelling is an unadulterated act of enquiry and insight.
- Modelling represents plurality of meaning and fluidity in alternatives.

5 Modelling as Learning

All modelling is purposeful and goal orientated, although the motivation for modelling is variable. This section explores modelling within the context of learning, highlighting the specific context of where modelling is an appropriate act, the associated faculties and the resultant insights.

Do we change the nature of human enquiry when we formalise modelling acts for curricular learning purposes? The following section explores and unpacks some of the key behaviours and processes that impact on the nature and quality of modelling when considered within the context of teaching and learning. Modelling as a process of influence can be considered from two key perspectives:

1. The lived experience – embedded cultural and social cues that are subtle messages from influencers
2. The cognitive generative process of modelling resulting from an experienced disequilibrium

Therefore, it is important to look at modelling in the broadest sense.

6 Modelling as a Social, Vicarious Act

Modelling is a ubiquitous process that can be considered from the perspective of our lived experience. Bandura's early work on modelling (Bandura et al. 1961, 1963) provides a basis for us to consider modelling as a form of critique. He suggests that behaviours are learnt through a self-regulated and considered response that is influenced by *reciprocal determinism*; this is the relationship between the environment, the behaviour and the person (values, beliefs and cognition). Bandura (1977) recognises that there are internal factors that influence our behaviour, and it is not always necessary to reinforce or punish behaviours to achieve a modelled response. This recognition is a distinct departure from behaviourist theory and demonstrated the inclusion of critique as a critical feature of *vicarious* learning. The modelled behaviour forms a catalyst for consideration and supports the learner in forming critical judgement in order to develop or curb certain behaviours. Regardless of whether you are developing practical craft skills or refining your imitation of Spiderman, vicarious learning is particularly important. However, there are a number of critical considerations that are of interest. Observations of a modelled behaviour are subject to interpretation and may not direct the critique appropriately. For example, deploying a guard on a machine tool, although good practice, can be interpreted as the initial act in the safe use of that machine, when in actual fact it should be considered the last resource from a developed safety culture perspective. Furthermore, the credibility of the modeller is also a key influencing factor in relation to the success of the learning. Two distinct characteristics determine the appropriateness of the modeller, the esteem of the observed, e.g. the credible teacher, and the relevance of the modeller to the observer, i.e. can the observer relate to them is a way that is commensurate. In addition, adopting and sustaining modelled behaviour is also determined by the perceived reinforcement that the behaviour invoked. Judgement as to social acceptability of the behaviour is determined, and consideration is given as to whether in fact this reinforcement is desirable. This is of particular importance within Design and Technology education as the norms and practices that establish the ways of thinking and working can be defined and reinforced through practice. What is interesting in relation to the habitual nature of modelled behaviours is its reach and influence. The idea of habits can also be extended to *habits of the mind*; it is this idea that Claxton (2008) highlights through his metaphor of apprenticeship, arguing that an epistemic apprenticeship would support the young learner (through osmosis) learn relevant skills and the values and beliefs of their culture as enacted and modelled by those around them.

Within the context of Design and Technology education, observational learning is a powerful means of influencing norms and practices. Supplemented by direct instruction and demonstration, the lived experience (in the practical classroom) is shaped by what Bandura calls the *reciprocal causation model*: the continuous interaction between behaviours, personal factors and the environment. Critically, the development of physical craft and manipulative skills is framed by *models of psychomotor development* (Dave 1970; Dawson 1998) that highlight the importance of observation and mimicry as the underpinning to mastery and naturalisation.

7 Modelling and Behaviour

Modelling is a natural behaviour designed to support human enquiry into the unfamiliar. Navigating the unknown can be full of potential pitfalls, uncertainty, errors, misconceptions and dead ends, none of which are problematic outside of compulsory education. Speculation by its nature accepts plurality of outcomes with logic of *best fit* helping the speculator frame their understanding of the world. This is comfortable, flexible and low stakes as there is no absolute or correct examinable views of their world. This is not always the case in school; there is a way to see the world and there are particular allocations of marks to see it in that way. The creation of unique and personally authentic mental model as a synthesis of experience, reference, comparison and knowledge is very different for the interpretation of preconstructed, optimal, *off-the-shelf* model.

There is little doubt that there is a significant relationship between mental models and how they guide behaviour. Kempton's (1986) thermostat example illustrates how two varying mental models can influence human behaviour. The operation of the thermostat can be modelled in two distinct ways. One model is to analogise it with the accelerator of the car; if you want the room to heat up, you turn up the thermostat beyond the required temperature, similar to pressing your foot further on the accelerator to make the car go faster. The second common mental model was built on the conception of a switch. This model conceives the operation of the thermostat correctly; once the temperature drops below a predefined value, the heating system will turn on and regulate the temperature of the room. Both models were constructed by assimilating the new information into pre-existing schema. The selection of the schema as an appropriate analogous model is where the divergence emerged, despite the capacity to build a mental model. What is interesting about this example is the resultant behaviour. People who analogised it to the accelerator tended to frequently adjust and readjust the setting; when the room was too cold, it was adjusted, and when too hot as a result of the adjustment, it was then readjusted. People that supported the switch model set the controls infrequently. It is clear from this example that past experiences act as schema.

The incorrect analogous reference for the thermostat was based on the effectiveness of the critique surrounding the appropriateness of the analogy. The lack of knowledge associated with the function of the thermostat coloured by the model, manifested in suboptimal use. There is much evidence to suggest that the capacity to effectively critique is dependent on knowledge (Christodoulou 2014; von Aufschnaiter et al. 2008). Although the nature of critique is categorised by higher-order thinking, it is subordinate to lower-level skills such as remembering and understanding. This is interesting in relation to our educational expectation surrounding discipline specific knowledge – when is modelling utilising knowledge, constructing knowledge, or is modelling a utility that functions as the catalyst for acquiring new knowledge?

Design education is immersed in the resolving of ill-defined problems. The behaviours of learners are not well known in terms of efficiency and effectiveness. Possibly this is due to the erroneous constraints that are put on design practice within formal schooling. Largely what we are concerned with in Design and Technology education in the unknown world is what the world could be like based on the *what if* enquiry model. Critiquing our understanding of the natural world tends to be a defined act that follows a logical and reasoned pattern. Modelling the unbounded is a much more difficult construct to manage. Gigerenzer, Todd and the ABC Research Group's (1999) work on *bounded rationality* considers the nature of decision-making in an uncertain world, when optimisation is out of reach. They ask a number of key questions:

- What is in the adaptive tool box of a person or culture?
- What normative world will heuristics work or fail?
- How does using these insights help to make better (ecologically rational) decisions?

It is the response to this space that makes us consider the relationship between modelling and behaviour and the role modelling plays in relation to critique.

Interestingly, at a point everything we learn was once to us unknown. Therefore, the idea of certainty is difficult to comprehend when subscribing to an enquiry-based model of education. This is also emphasised by the work of Henderson et al. (2015) in Science Education, who argue for the need to engage students in the idea of *plural* alternatives. It is accepted that there is a passive consumption of theories and principles that form the learner's understanding of the natural world, and there is a clear need to build understanding from first principles and support the construction of appropriate analogous models. The expectation is that the learner would engage with critique and counter-critique to develop comprehensive insights into the natural world. This shift is difficult, as the natural human default tends to be *confirm* rather than *construct*. Bearing in mind that the focus and outcome of Science Education and Design and Technology education are not commensurate, there is common ground in relation to the need to develop the capacity to conceive, create and critique possible models of reality. Understanding the behaviours of people when confronted with this challenge further develops the discussion in relation to modelling as a form of critique.

Especially within the context of Design and Technology education, modelling and its relationship with heuristic techniques are particularly interesting. Heuristics are often described as a process of achieving cognitive shortcuts through a practical methodology that is neither perfect nor optimal, but rather the application of simple and efficient rules. There is a misconception that heuristics are only relevant when we reach our cognitive capacity (Gigerenzer 2008). Heuristics are mental shortcuts that allow us to reduce cognitive load in making decisions. Considering heuristics in the context of modelling is useful as they form the *satisficing* (Simon 1956) conditions for building models that are neither perfect nor optimal; instead, they are sufficient for the immediate goal. Although all modelled insights are not guaranteed to contribute directly to a solution, their importance for learning, particularly as a potential precursor to critical decision-making, is significant. Although heuristics are shortcuts and largely avoid critique, their relationship with modelling is interesting. There are a number of well-known heuristics, for example, the *availability heuristic* describes how people make judgements based on how easy something comes to mind, the *anchoring and adjustment heuristic* describes that tendency to make decisions relevant to a given or known piece of information (the anchor) and *familiarity heuristic* is where it is assumed that the conditions that underpinned a previous behaviour apply to the present and future situations. These behaviours can be viewed in two ways, one they are insufficient as they lack the necessary critique to advance thinking and behaviour and two they are useful in speculating elements of a more complex (or in the case of design education an ill-defined) problem. What is significant about considering heuristics, specifically *availability*, *anchoring* and *familiarity*, is their relationship with the environment and the conditioned norms and practices that the learner is exposed to. This would suggest that it is possible to influence the effectiveness of heuristic techniques to better serve the speculative design agenda. This would strengthen the capacity to engage in more sophisticated concepts and principles as the shortcuts become a more accurate representation of reality.

This suggests that design education to some degree provides an alternative perspective on modelling and its relationship with knowledge. Archer and Roberts (1992, p.3) argue that design activity is concerned more with the attainment of a result than with the acquisition of knowledge. This idea represents an approach that goes beyond plurality of meaning and supports a designerly approach that promotes fluidity in alternatives.

Kimbell (2011) gives a good example of this by illustrating the work of a pupil designing a kitchen utensil. The pupil, who was not bounded by not knowing the technical name for a given material, resolved their design by describing the nonslip material as *grippy stuff*. Although it is sensible to argue that the correct identification of materials is an important part of discipline competency, it would be difficult to argue that it is the most important. The pupils' capacity to select an appropriate material in relation to function and analogies is based on existing experience, and resulting schema demonstrates more sophisticated qualities of critique than skills of identification and recall. It is interesting to think of limited knowledge as cognitively liberating, from a speculative, design perspective. Therefore, modelling as a responsive, reflexive, speculative tool to generate better insight needs to be considered with respect to the context, situation and environment, with specific focus on how the learner frames the educational transaction.

8 Cognitive Functions of Modelling

Modelling as a process can be considered as the connection between the perceptual and processing domains. Predominantly, focusing on a *processing continuum*, modelling as a form of critique, is positioned within the experiential learning framework. This enables us to focus on the bidirectional relationship between abstract conceptualising and active experimentation while taking cues from concrete experience, observation and reflection (Kolb 1984). Based on earlier assumptions, effective and purposeful modelling is predicated by intent or need and results in a generative process that supports the creation of new insights. Learning something new, although subjective, can often be a comfortable assimilation into existing schema. This is where new information will easily fit with information that is already understood. However, the process of *adaptation (*Piagetian theory), triggered by the need for new information that does not fit into existing schema, requires a more uncomfortable accommodation or even the building of entirely new schema. How we facilitate this adaptation within formal education is of interest.

Having discussed models and modelling and the influence and impact that they have on our behaviour, it is important to consider the role that modelling plays in refining our cognitive processes. The capacity of models to control, separate or speculate about elements of the natural world is helpful in creating opportunities for *transitivity*. We use models to represent something, and we then reason the relationships between the model and its relevance to new meanings and insights. This analogous process utilises symbol systems, sketches, mental representations and physical models to go beyond given information and evolve new or existing schema.

Modelling is underpinned by one's capacity to build effective representations of the world. The capacity, nature, accuracy and efficacy of these representations are variable. Personal experiences, interpretation and processing of sensory data coupled with intuitive theories can all influence the creation of mental representations. It is important to consider representations as the fundamental building blocks of mental models, and it is this representation that controls and frames the known, so as to be used as an analogous means of transfer. Novick and Bassok's (2005) summary of work on analogous problem-solving is of particular relevance, as it demonstrates the need to build robust representations of the problem to ensure effective transfer. This work highlighted that poor or underdeveloped representations of a problem result in a surface engagement with its critical features and therefore lessen the learners' capacity to solve similar problems. Where representations were more comprehensive, the student understood the structural features of a problem and as a result had better transfer capacities.

Although mental models are inherently flexible, they are also a type of information filter that is prone to *selective perception* and as a result error. The relationship between representation and knowledge is well documented (Alibali et al. 2009) and can limit the value of the model in relation to meaningful enquiry and critique. From both the perspective of constructing effective models and critical reasoning, it is apparent that knowledge is a critical variable. As highlighted in the previous section, knowledge can influence not only the efficacy of the model but also the resultant behaviour (e.g. the Thermostat experiment). What is questionable is the process by which one constructs a representation and model.

The work of Öllinger et al. (2013) on insight problems highlights another dimension that helps us consider behaviours and problem representation as a means of furthering the discussion on modelling as a form or critique. An insight problem is where there is "a high probability of triggering an initial representation which has a low probability of activating the knowledge needed to solve the problem" (Ohlsson 2011, p.10). Either inappropriate or inaccurate representation quickly results in an impasse and prevents the learner developing an associated strategy to solve the given problem. It is possible to argue multiple reasons for the construction of the initial representation, for example, conditioned response, heuristic, misinterpretation of the problem, memory system error or lack of knowledge. What is important is the capacity to reframe and restructure the original model to overcome insight problems. The determination that the constructed model is no longer sufficient creates the need for critical review. Where the model is determined to be no longer sufficient, the resultant restructuring can often lead to a sudden *aha* moment (Kounios and Beeman 2009).

It is not sensible to think that overcoming *insight* and *impasse* is as simple as a restructuring of a mental model. Often this alteration is a difficult conceptual change (Duit and Treagust 2003) and may involve fundamental changes to one's assumptions or knowledge. Posner et al. (1982) argued that there are generally two factors that spark conceptual change, anomalies and fundamental assumptions. Akin to the concept of *disequilibrium*, anomalies provide a cognitive conflict that requires the learner to rethink or remodel their initial conception, controlling some element of the model so as to enable a more *satisficing* representation. Where fundamental assumptions are built on rational theories, the need to reframe, reconceive or restructure is more difficult. However, often learners will have no rational basis for theories and as a result will subscribe to irrational ones, for example, because they were told it was true.

Cognitive adaptability is the aim of effective education. The ability to engage with critique, counter-critique and associated reasoning to build both representative and speculative models is an important skill. Maladaptive and narrowly conditioned behaviours are the antithesis to effective modelling and performance. The work of Delahunty (2014) demonstrates the varying levels of adaptability of learners when solving convergent STEM problems. His work utilised a triangulated method that included EEG monitoring to explore the cognitive functions that were exercised during problem-solving episodes. This work highlighted a number of significant findings. Participants that were successful demonstrated a capacity to alter their approach to solving the problem, once they reached an impasse. Unsuccessful participants repeated strategies that did not lead to a successful result, with no evidence of a capacity to restructure or remodel their strategy. What were of particular interest in this study were the cognitive functions that emerged during the problem-solving episodes. Participants that were successful tended to utilise visuospatial cognition (by definition of location, synchronisation and desynchronisation of cognitive faculties). The capacity to build mental imagery and as a result rely on appropriate semantic (rather than episodic) memories suggested a greater level of adaptive behaviours. In contrast, unsuccessful participants tended to rely on memory and reasoning with no mental model, resulting in a maladaptive approach. For example, a particular participant in the study performed poorly in a set of tasks designed to be optimally solved using a mathematical conception. One of the key reasons for this poor performance was the inappropriateness of the initial problem conception which was based on an episodic process (by definition of the EEG data) that featured a reliance on a past process experienced during their schooling. The problem triggered a familiarity with the student that bypassed the necessity to model the task in real time and resulted in a reliance on an approach that was ill-matched to the task.

Regardless of considering an observed model as with vicarious learning, or the creation of models to better understand the natural world, or developing new models of how the world could be, they all converge on the capacity to *see in the mind's eye*.

9 How Fuzzy Is the Picture?

Seeing in the *mind's eye* describes a broad cognitive skill that in general includes imagination, memory and visualisation. Depending on discipline or perspective, this human capacity can be described in many ways. Archer's (1992a, p.5) explanation of *imagining* is a useful starting point as he describes it as the "... part of cognitive modelling which construes sense data and constructs representations spatially and presentationally, rather than discursively and sequentially". The idea of *spatially* is considered significant, as it suggests a working area that is open and unbounded, with room to create, manipulate and refine any addition to that space. The ability to effectively function in this cognitive space is an area that relies on numerous cognitive processes from sensory data to imagination to perception and can facilitate much of the logical and reasoning capacities that are not exclusively spatial. Since its conception as a primary cognitive ability, the definition of spatial ability has been a contentious issue. However, despite the range of descriptors, there are dominant definitions offered. Lohman (1979, p.126) defines spatial ability as "the ability to generate, retain, and manipulate abstract visual images", and Gaughran (2002) defines it as "the ability to visualise, manipulate and interrelate real or imaginary configurations in space". These definitions offer explicit insights in relation to the cognitive functions involved in spatial reasoning.

The ability to perform mental rotations is considered an important skill and is specifically described as the *spatial visualisation* factor (Uttal et al. 2013). The malleable characteristics of this factor are important from an educational perspective, as good spatial skills have been proven to predict achievement and success in STEM education (Kozhevnikov et al. 2007; Shea et al. 2001; Smith 1964; Sorby et al. 2014; Wai et al. 2009). Elevated spatial capacities have also been linked to creativity and technical innovation (Jones and Burnett 2008; Kell et al. 2013). The work of Lane et al. (2009) has gone some way further in developing this relationship and looked at how the development of sketching skills advances relevant cognitive activities that resulted in evidence of creative and designerly thinking. Also, Fish and Scrivener (1990) identify a continuum that ranges from *imagination* to *memory* to *observation*, and it was this relationship that allowed Lane et al. (2010) to exploit elevated observational skills so as to help pupils build robust *graphical libraries*. This is important from a design perspective; being able to call on an extensive resource of mental imagery enables both the recall of images and their synthesis, so as to form new conceptual creations of value (Lane 2011). Borst and Kosslyn (2008) highlight the importance of accessing *graphical libraries* through visual mental imagery to support cognitive function. Again the elaboration of the act of sketching is not necessary to present information to others, but a communication with the self in an attempt to refine thinking. The work of Kosslyn et al. (2001) and Pearson (2007) describe the ability to generate images across sensory modalities as *creative visualisation*. This process specifically refers to the ability of a person to generate and process visual mental imagery. Pinker and Kosslyn (1978) argue that these mental images are representations of our experiences of the world.

Visualising images in the mind, constructing objects for manipulation or rotation and synthesising imagery to produce new conceptions of reality are all related to elevated performance in STEM education. Congruently, the capacity to utilise mental models requires an ability to formulate representation of a problem or situation and be flexible enough to tailor or reconceive the models in response to new insights or stimuli. It is the ability to generate vivid imagery that supports these processes. The *imagery* factor which refers to the ability to mentally produce vivid images has been suggested as an important cognitive factor for solving geometric problems (Schneider and McGrew 2012). It is the image that is the critical part of modelling. Archer and Roberts (1992, p.4) argue that it is the image that we are externalising through the various forms of modelling (e.g. sketches, diagrams, mock-ups, prototypes and language and notation). They claim that it is the image that is "embodied in the construction or enactment of the emerging responses". The power of the image is its capacity to adapt and refine before, during and after its externalisations, and it is this externalisation that makes this image comprehendible.

10 Implications for Practice

When considering the influencers and inhibitors of effective modelling, they tend to converge on a specific set of conditions and attributes:

- The learner's reaction to the learning task that employs modelling above other reasoning and logic strategies
- The cognitive faculties that are then activated in response to the modelling act
- The actions and behaviours that are enacted as a result of conceiving and refining a cognitive model and the iterative nature of modelling building towards *equilibrium*
- The capacity to effectively utilise the modelling process to highlight anomalies, intuitive theories and reason relevance and appropriateness

For the most part, influencers and inhibitors of effective modelling are two sides of the same coin. Technology education frames its contribution as a design-driven education that explores the world *as it could be* not *as it is*. *As it could be* requires a cultural acceptance that modelling as an exploratory generative process results in trialling, repeating, affirming, refining, visualising and framing, not the execution of a predetermined model. This iterative process requires pedagogical fluidity and the recognition that learners' behaviours are driven by the delineation between need (problem) and capacity (existing schema). Design and Technology is about realising ideas, creating artefacts and solving problems; it is about modelling thinking around what is effective and efficient for a particular purpose (Norström 2013, p. 378).

It is the freedom in the definition of Design and Technology practice that necessitates a fluid, adaptive and flexible means of representing, configuring and manipulating the world. Dependent on a predisposition, to see the world *as it could be* represents opportunities or ambiguities. Modelling is a natural response to ambiguity. It is personal, subjective and generative and is acknowledged as being incomplete, where the expectations of the self are centred on the meaning of the model and not the model. Its analogous premise is polymorphous and responsive to interpretations from multiple perspectives. In practice, the process is messy, but a key utility is in its ability to help formulate an insight that may not be possible, efficient or complete by reasoning.

This variability in definition and ambiguity in practice is to the advantage of modelling. The ability to use models to relax constraints, control variables or even invent conditions is a rich capacity in an uncertain world. Yes, modelling is imperfect and prone to errors and can result in suboptimal thinking and behaviour, but in the context of constructivism and critique, is this not the underpinning of learning? Arguments for critique and counter-critiques that frame the agenda that supports plurality are laudable. The extension of this approach recognises the nature of Design and Technology learning as fluid.

11 Summary

The nature of models and modelling considers a broader conception of human capacity. This chapter discusses some of the cultural, social and cognitive implications of recognising effective learning in support of modelling. How often in practice are pupils asked to close their eyes and build a mental image of their conception of a problem, product, etc.? Have we expectations of pupils gesturing to simulate motion in an attempt to refine their thinking? Is sketching seen as an intrapersonal communication tool, with outputs often only decipherable by the creator? Is there explicit recognition in Design and Technology education that recognises modelling as the cornerstone of effective leaning?

Baynes (2014, p.96) separates the design education curriculum from the sciences and humanities based on the need for imaginative modelling. This framing highlights design as the third culture that has a distinct way of thinking, acting and learning, separate from scientific and linguistic competencies. Designerly thinking liberates a unique reasoning capacity that relies on spatial and presentational qualities. The interaction between hand and mind forms a critical part of an enquiring process that is driven by imagination and bounded by the *content of culture*.

Throughout this chapter I have drawn heavily from the expertise, thinking and models captured in the *Orange Series*. Again with reference to the work of Roberts (1992), there is a critical message for the practicum. He highlights that a call for a "'definitive design vocabulary' – meaning definitions without ambiguity - rather than for a useful meta-language, is misinformed" (p.40). He develops this idea by acknowledging the importance of meaning making as to be truly involved in learning and that this active enquiry has a *polysemous quality*. Modelling as a form of critique is messy! Actions are difficult to predict, behaviours are chaotic, outcomes are often unexpected and standardisation is impossible.

Yet, "... to say that this should not be the case appears as a challenging and curious proposition" (Roberts 1992, p.40).

References

Alibali, M., Phillips, K., & Fischer, A. (2009). Learning new problem-solving strategies leads to changes in problem representation. *Cognitive Development, 24*(2), 89–101.

Archer, B. (1992a). A definition of cognitive modelling in relation to design activity. In P. Roberts, B. Archer, & K. Baynes (Eds.), *Modelling: The language of design, Design: Occasional paper No. 1* (pp. 5–6). Loughborough: Loughborough University.

Archer, B. (1992b). As Complex as ABC. In P. Roberts, B. Archer, & K. Baynes (Eds.), *Modelling: The language of design, Design: Occasional paper No. 1* (pp. 7–11). Loughborough: Loughborough University.

Archer, B., & Roberts, P. (1992). Design and Technological awareness in education. In P. Roberts, B. Archer, & K. Baynes (Eds.), *Modelling: The language of design, Design: Occasional paper No. 1* (pp. 3–4). Loughborough: Loughborough University.

Argyris, C., & Schon, D. (1974). *Theory in practice: Increasing professional effectiveness*. California: Jossey-Bass.

Bandura, A. (1977). *Social learning theory*. Jersey: Prentice-Hall.

Bandura, A., Ross, D., & Ross, S. (1961). Transmission of aggression through imitation of aggressive models. *Journal of Abnormal and Social Psychology, 63*(3), 575–582.

Bandura, A., Ross, D., & Ross, S. (1963). Imitation of film-mediated aggressive models. *Journal of Abnormal and Social Psychology, 66*(1), 3–11.

Baynes, K. (1992a). Research into primary design and technology. In B. Archer, K. Baynes, & P. Roberts (Eds.), *The nature of research into design and technology education, Design curriculum matters: Occasional paper No. 1* (pp. 15–21). Leicestershire: Loughborough University.

Baynes, K. (1992b). The ethics of representation. In P. Roberts, B. Archer, & K. Baynes (Eds.), *Modelling: The language of design, Design: Occasional paper No. 1* (pp. 12–17). Loughborough: Loughborough University.

Baynes, K. (2014). *Design: Models of change: the impact of designerly thinking on people's lives and the environment*. Loughborough: Loughborough Design Press.

Borst, G., & Kosslyn, S. (2008). Visual mental imagery and visual perception: Structural equivalence revealed by scanning processes. *Memory & Cognition, 36*(4), 849–862.

Brousseau, G. (1980). Les Échecs Électifs dans l'Enseignement des Mathématiques à l'École Élémentaire. *Revue Di Laryngology Otology Rhinology, 101*(3–4), 107–131.

Christodoulou, D. (2014). Minding the knowledge gap: The importance of content in student learning. *American Educator, 38*(1), 27–33.

Claxton, G. (2008). *What's the point of school? rediscovering the heart of education*. London: Onework Publications.

Dave, R. (1970). Psychomotor levels. In R. Armstrong (Ed.), *Developing and writing behavioral objectives* (pp. 33–34). Arizona: Educational Innovators Press.

Dawson, W. R. (1998). *Extensions to bloom's taxonomy of educational objectives*. Sydney: Putney Publishing.

Delahunty, T. (2014). *Investigating conceptualisation and the approach taken to solving convergent problems: Implications for instructional task design*, PhD thesis, University of Limerick.

Duit, R., & Treagust, D. (2003). Conceptual change: A powerful framework for improving science teaching and learning. *International Journal of Science, 25*(6), 671–688.

Fish, J., & Scrivener, S. (1990). Amplifying the mind's eye: Sketching and visual cognition. *Leonardo, 23*(1), 117–126.

Gaughran, W. (2002). Cognitive modelling for engineers. In *2002 American society for engineering education annual conference and exposition*. Montréal, Canada, 15–19 June: American Society for Engineering Education.

Gigerenzer, G. (2008). Why heuristics work. *Perspectives on Psychological Science, 3*(1), 20–29.

Gigerenzer, G., Todd, P., & ABC Research Group (Eds.). (1999). *Simple heuristics that make us smart*. Oxford: Oxford University Press.

Henderson, J. B., MacPherson, A., Osborne, J., & Wild, A. (2015). Beyond construction: Five arguments for the role and value of critique in learning science. *International Journal of Science Education, 37*(10), 1668–1697.

Johnston-Wilder, S., & Mason, J. (2005). *Developing thinking in geometry*. London/California/New Delhi: SAGE Publications.

Jones, S., & Burnett, G. (2008). Spatial ability and learning to program. *Human Technology: An Interdisciplinary Journal on Humans in ICT Environments, 4*(1), 47–61.

Kell, H., Lubinski, D., Benbow, C., & Steiger, J. (2013). Creativity and technical innovation: Spatial ability's unique role. *Psychological Science, 24*(9), 1831–1836.

Kelly, A. V., Kimbell, R., Patterson, V. J., Saxton, J., & Stables, K. (1987). *Design and technology: A framework for assessment*. London: HMSO.

Kempton, W. (1986). Two theories of home heat control*. *Cognitive Science, 10*(1), 75–90.

Kimbell, R. (2007). E-assessment in project e-scape. *Design and Technology Education: An International Journal, 12*(2), 66–76.

Kimbell, R. (2011). Wrong... But Right Enough. *Design and Technology Education: An International Journal, 16*(2), 6–7.

Koen, B. V. (1985). *Defintion of the engineering method*. Washington, D.C.: American Society for Engineering Education.

Kolb, D. (1984). *Experiential learning: Experience as the source of learning and development*. Jersey: Prentice-Hall.

Kosslyn, S., Ganis, G., & Thompson, W. (2001). Neural foundations of imagery. *Nature Reviews Neuroscience, 2*(9), 635–642.

Kounios, J., & Beeman, M. (2009). The aha! moment: The cognitive neuroscience of insight. *Current Directions in Psychological Science, 18*(4), 210–216.

Kozhevnikov, M., Motes, M., & Hegarty, M. (2007). Spatial visualization in physics problem solving. *Cognitive Science, 31*(4), 549–579.

Lane, D. (2011). *Developing sketching expertise within technology education*, PhD thesis, University of Limerick.

Lane, D., Seery, N., & Gordon, S. (2009). The understated value of freehand sketching in technology education. *Engineering Design Graphics Journal, 73*(3), 13–22.

Lane, D., Seery, N., & Gordon, S. (2010). A paradigm for promoting visual synthesis through freehand sketching. *Design and Technology Education: An International Journal, 15*(3), 68–90.

Lohman, D. (1979). *Spatial ability: A review and reanalysis of the correlational literature*. Stanford: California.

McGarr, O., & McCormack, O. (2014). Reflecting to conform? Exploring Irish student teachers' discourses in reflective practice. *The Journal of Educational Research, 107*(4), 267–280.

Norström, P. (2013). Engineers' non-scientific models in technology education. *International Journal of Technology and Design Education, 23*(2), 377–390.

Novick, L., & Bassok, M. (2005). Problem Solving. In *The cambridge handbook of thinking and reasoning* (pp. 321–369). New York: Cambridge University Press.

Ohlsson, S. (2011). *Deep learning: How the mind overrides experience*. New York: Cambridge University Press.

Öllinger, M., Jones, G., Faber, A., & Knoblich, G. (2013). Cognitive mechanisms of insight: The role of heuristics and representational change in solving the eight-coin problem. *Journal of Experimental Psychology: Learning, Memory, and Cognition, 39*(3), 931–939.

Pearson, D. (2007). Mental imagery and creative thought. *Proceedings of the British Academy, 147*(1), 187–212.

Pinker, S., & Kosslyn, S. (1978). The representation and manipulation of three-dimensional space in mental images. *Journal of Mental Imagery, 2*(1), 69–84.

Posner, G., Strike, K., Hewson, P., & Gertzog, W. (1982). Accommodation of a scientific conception: Toward a theory of conceptual change. *Scientific Education, 66*(2), 211–227.

Roberts, P. (1992). Of models, modelling, and design: An applied philosophical enquiry. In P. Roberts, B. Archer, & K. Baynes (Eds.), *Modelling: The language of design, Design: Occasional paper No. 1* (pp. 32–42). Loughborough: Loughborough University.

Schneider, J., & McGrew, K. (2012). The Cattell-Horn-Carroll model of intelligence. In D. Flanagan & P. Harrison (Eds.), *Contemporary intellectual assessment: Theories, tests, and issues* (3rd ed., pp. 99–144). New York: Guilford Press.

Scholl, H., & Phelan, S. (2004). Using integrated top-down and bottom-up dynamic modeling for triangulation and interdisciplinary theory integration: The case of long-term firm performance and survival. In M. Kennedy, G. Winch, R. Langer, J. Rowe, & J. Yanni (Eds.), *Proceedings of the 22nd international conference of the System Dynamics Society* (pp. 1–20). Oxford: System Dynamics Society.

Shea, D., Lubinski, D., & Benbow, C. (2001). Importance of assessing spatial ability in intellectully talented young adolescents: A 20-year longitudinal study. *Journal of Educational Psychology, 93*(3), 604–614.

Simon, H. (1956). Rational choice and the structure of the environment. *Psychological Review, 63*(2), 129–138.

Smith, I. M. (1964). *Spatial ability: Its educational and social significance*. Michigan: R.R. Knapp.

Sorby, S., Nevin, E., Mageean, E., Sheridan, S., & Behan, A. (2014). Initial investigation into spatial skills as predictors of success in first-year STEM programmes. In *SEFI 2014* 42nd annual conference european society for engineering education. Birmingham: SEFI.

Uttal, D., Meadow, N., Tipton, E., Hand, L., Alden, A., Warren, C., & Newcombe, N. (2013). The malleability of spatial skills: A meta-analysis of training studies. *Psychological Bulletin, 139*(2), 352–402.

von Aufschnaiter, C., Erduran, S., Osborne, J., & Simon, S. (2008). Arguing to learn and learning to argue: Case studies of how students' argumentation relates to their scientific knowledge. *Journal of Research in Science Teaching, 45*(1), 101–131.

Wai, J., Lubinski, D., & Benbow, C. (2009). Spatial ability for STEM domains: Aligning over 50 years of cumulative psychological knowledge solidifies its importance. *Journal of Educational Psychology, 101*(4), 817–835.

Politicizing the Discourse of Consumerism: Reflections on *The Story of Stuff*

Terry Wilkinson

Abstract What might a critical pedagogy of consumption mean for design and technology education? In response to fervent calls for politicized forms of consumer, environmental, and science education, I submit that we also need to politicize design and technology education by providing learning experiences that encourage young people to critically analyze and question ecologically unsound processes of a market economy and, in particular, the relationship between technology and consumerism. In this chapter, I consider what a critical approach might offer to teaching for a critical literacy of the built world. First, a small section of the Ontario Elementary School curriculum is analyzed to identify how children consumers are discursively positioned and in whose interests these constructions work. Drawing on key ideas put forth by a number of critical scholars, I next consider the merits of using Annie Leonard's video, *The Story of Stuff*, as a resource for learning about technological design processes—including the motives underpinning increasingly short product life spans and externalized production costs. Presented as a quasi-case study, I suggest the video serves more importantly as a model for critiquing that aims to help young people prepare for and take responsible action on issues relating to their social and ecological well-being. The chapter concludes by proposing that the politicizing of discursive and technological practices in education—while challenging—will be necessary to foster critically literate, empathic, and confident problem-solvers and designers for social good.

Keywords Technology education • Design education • Story of stuff • Critical literacy • Critical pedagogy

1 Introduction

In response to fervent calls for politicized forms of consumer education (Farahmandpur 2010; McGregor 2010; Sandlin 2010), environmental education (Clover 2002; Hodson 1992; Jensen 2004; Kahn 2008a, b), and science and

T. Wilkinson (✉)
York University, Toronto, ON, Canada
e-mail: terry_wilkinson@edu.yorku.ca

PJ. Williams, K. Stables (eds.), *Critique in Design and Technology Education*, Contemporary Issues in Technology Education, DOI 10.1007/978-981-10-3106-9_15

technology education (Hodson 1994, 1999, 2003), I submit that we also need to *politicize* design and technology education by providing learning experiences that encourage young people to critically analyze and question ecologically unsound processes of a market economy and, in particular, the relationship between technology and consumerism (Elshof 2005; Margolin 1998; Petrina 2000a). I locate my work within a critical practice perspective and stand with others who argue that conventional technological practices that narrowly address "needs-wants issues" (Keirl 2007, p. 310) can no longer be ethically justified and are therefore inadequate in terms of providing an education for the future (Elshof 2006, 2009; Huckle 2010; Petrina 2000a). Like Sue McGregor (2010), my utopian vision for the future is an education that empowers "citizens concerned with sustainability, solidarity, justice, peace, and the human condition" (p. 122). By "utopian," I mean the expression of desire—not in the form of unbridled fantasizing but as a concrete utopian imagining of how life could be *otherwise* (Bloch 1986)—that simultaneously anticipates and effects the future (Levitas 1997). This is a critical praxis-oriented project of hope that "reaches forward to a real possible future, and involves not merely wishfull [*sic*] but willfull thinking" (Levitas, p. 67).

In this chapter, I will first consider what Jennifer Sandlin and Peter McLaren's (Sandlin and McLaren 2010) call for a "critical pedagogy of consumption" might offer to teaching and learning in design and technology education. I will draw on the scholarly works of Darlene Clover and Katie Shaw (Clover and Shaw 2010) and John Huckle (2010) to think about the challenges of teaching for a critical literacy of the built world. Next, I will examine a small section of the Ontario Elementary School curriculum for science and technology in an attempt to answer two questions posed by Sandlin and McLaren: (1) "What kind of consumers are being created?" and (2) "In whose interests do those constructions work?" (p. 15). Following this analysis, I draw on key ideas put forth by these and other scholars to consider the merits of using Annie Leonard's (2007a) video animation entitled *The Story of Stuff* as a teaching resource for introducing students to sophisticated understandings of externalized costs and product obsolescence to problematize the discourse of production and consumption and to reorient design thinking for longer-term prospects (Pilloton 2009). My paper concludes by suggesting that the politicizing of discursive practices—while challenging—will be necessary to prepare informed, critical, and empathic problem-solvers and designers for social good (Chochinov 2009).

2 A Critical Pedagogy of Consumption

Drawing on the work of the Brazilian educator Paolo Freire and the radical consumer research of Norman Denzin (2001), Sandlin and McLaren (2010) call on educators to trouble the naturalization of consumption with its acquire-use-dispose logic of products. They imagine school as a place of contestation in which consumer capitalism is questioned and consumer resistance works as a space of learning "where power, ideology, gender, and social class circulate and shape one another" (Denzin 2001, p. 325). Clover and Shaw (2010) also wish to interrupt the dictates of a consumer ideology that are tied to notions of "free and abundant choice of goods" as symbols of "freedom, affluence, and the good life" (p. 204). Others have argued that technology education with a cultural studies perspective can play a key role in exploring how the making of artifacts and consumption-driven lifestyles contribute to the sustainability problem (Elshof 2005, 2006; Petrina 2000a, b).

It has been noted by many critics (for instance, Foster 2002; Hoechsmann 2007; Molnar et al. 2010; Schor 2004) that commercial advertising promoting the consumption of goods and services has saturated our cultural, economic, and social worlds. Clover and Shaw (2010) have gone as far as to claim that learning to consume has been "one of the deepest and most pervasive educative processes at work since the Second World War" (p. 203). They and other scholars (e.g., Kahn 2008a, b, 2010a, b) have rightly argued that the lack of emphasis on political literacy in environmental education today is problematic because it enables, at least in part, large trans- or multinational corporate involvement and responsibility for socio-environmental impacts (e.g., unsafe working conditions, worker exploitation, pollution, natural resource depletion, species extinction) to remain hidden from public scrutiny. With specific reference to Canadian education today, Clover and Shaw have asserted,

> Problematically, much of what passes for public environmental education in this country has been woefully inadequate in responding appropriately to consumerism. In one regard, it ignores the politics of over-consumption and waste, choosing instead to focus on the individual and leaving corporations to carry out their activities unencumbered by critique or challenge from a politicized public. (p. 203)

The shortcomings of environmental education highlighted by Clover and Shaw seem to substantiate Sandlin and McLaren's (2010) critique of how the "misidentification" and "protect[ion of] the individual as the foundation of entrepreneurial capitalism" serves to replace "the well-being of the collectivity" with the "politics of consumption" (p. 14). Whether it is down to unintentional or willful blindness in school curriculum studies, the occlusion of corporate and government culpability works to sustain a capitalist orthodoxy of consumerism and profiteering in which neoliberal notions of free choice "celebrate the singularities of individuals by valorizing the desire to obtain and consume objects of pleasure" (Clover and Shaw, p. 206). Moreover, when the consumer is blamed for making a *bad choice*, companies again evade responsibility for their poor quality or unhealthy products (Jensen 2004) or the harm caused to others or the environment. Consumer blame and guilt were also the subject of an online article in *The Huffington Post*, in which the self-described unapologetic activist Annie Leonard (2012, ¶4) wrote:

> ... companies target consumers by creating desires we didn't know we had and meeting them with cheap shiny gadgets we didn't know we needed. And when the companies get caught trashing the environment or mistreating their workers, everyone blames the customers – that's us – for demanding cheap shiny gadgets... Sometimes it seems everything we buy is tarnished by guilt. Whether it's electronics from unsafe factories, clothes from oppressive sweatshops or coffee from the rainforest, we blame ourselves and our fellow consumers for our complicity in an unjust and unsustainable system.

Returning to Sandlin and McLaren's (2010) question about what kind of consumers are being created, Leonard's article would suggest that we are apolitical and dupable pleasure-seekers who, with the help of psychologists (Leonard and Conrad 2010), can be manipulated by the mediated arts of persuasion into thinking we need the goods and services we are offered. This may be an oversimplification but it behooves us to ask, in whose interests do these constructions of blame and guilt work?

For about 20 years, so-called green marketing strategies have been criticized for exacerbating the environmental dilemma. For example, while "eco-design" packages intended to appease consumer guilt may be lucrative for manufacturers, many such "greenwash" products do little to protect the environment (Lahaye 1995). Calling for better public education, Marie-Christine Lahaye (1995) suggests that only when advice on green consumption is "independent of industry and government" and includes "stakeholders from all sectors of society" will consumers be able to make informed and responsible purchases (p. 61). Twenty years on, a critical understanding of how green consumerism operates as a public pedagogy (Giroux 2005) still seems to elude many of us. The term "public pedagogy" refers to the life-shaping "educational forces" of culture that operate extensively in the sphere of formal schooling and increasingly across a wide variety of public sites of knowledge and meaning production (Giroux, ¶13–19). The mainstream culture of green consumerism is certainly one form of public pedagogy that requires our critical due diligence. As Richard Kahn (2010b) has rightly argued, "our educational relationship with the ecological issues that these products purport to help solve is reduced and cheapened when we accept that buying the new "eco-friendly" formula thereby absolves us of deeper levels of social inquiry and political action" (p. 49). Kahn openly questions how "endless repetitions of spending" on green products "in any way represents real opposition to either a culture defined by hyperconsumption or an economic structure that demands it" (p. 40).

While Leonard and Conrad (2010) do not outrightly reject the practice of "greensumption," they suggest that "an informed and engaged consumer is not a substitute for being an informed and engaged citizen" (p. 175). This philosophy is shared by Darlene Clover (n.d.), who envisions education as a transformative project for change in understandings (http://www.uvic.ca/education/psychology/people/home/faculty/cloverdarlene.php). Both she and Shaw (Clover and Shaw 2010) call for a stronger emphasis on the powerful influence corporations have in political, social, and environmental matters, as well as "on what needs to be done to change things around and return the blame to where it belongs" (p. 206). The goal of transforming understandings about the way things work—with the intent to tackle issues—is strongly resonant in Sandlin and McLaren's (2010) critical pedagogy that encourages learners to "question assumptions and challenge the status of existing structures as natural" (p. 16). The authors argue that locating human experience within a "specific social relations of production" framework will enable students to "see how, through the exercise of power, the dominant structures of class rule protect their practices from being publicly scrutinized as they appropriate resources to serve the interests of the few at the expense of the many" (p. 14).

3 Toward Political Literacy

If we are to educate for political literacy and citizen engagement, there are at least two assumptions currently operating in our hypercapitalist world that need to be challenged: (1) the idea that individuals have the freedom of choice when it comes to choosing goods and services and (2) the idea that the demand for products drives the supply. Leonard (2012) questions the logic of supply and demand by rhetorically asking, "Before single-serving plastic bottles, who wanted to carry around a throwaway container of water that, despite no guarantee of being cleaner or safer, costs thousands of times more than what comes out of the tap?" (¶6). Now some people might push back on this by pointing out that the manufacture of plastic bottles is a designed response to individuals' preferences and willingness to pay for lifestyle convenience and expression. However, when it comes to selecting goods and services, the idea of consumer choice also needs to be reexamined. As Matthew B. Crawford (2009) astutely noted, many commercial products are marketed with promises of greater personal choice, and yet the important design decisions have been remotely controlled—leaving only a "playground-safe field of options" (p. 69) for narcissistic gratification. In other words, aside from a few style elements such as color and shape, there is very little choice at all. Political theorist Benjamin Barber (2007) has also countered the notion that the essence of liberty comes from the "the right to choose from a menu," arguing instead that "the real power, and hence the real freedom, is in the determination of what's on the menu. The powerful are those who set the agenda, not those who choose from the alternatives it offers" (p. 139). If product choices are not consumer-driven but rather profit-driven (as Lahaye (1995), Leonard (2012), Crawford (2009), Barber (2007) and others have suggested), then herein lies an answer to Sandlin and McLaren's (2010) question: "*In whose interests are consumers constructed?*" (p. 15). When the creation of wants, proliferation of pseudo choices, and promotion of unconstrained acts of consumption generate profitable markets, it could be argued that willing consumers (Eastwood 2006) are primarily constructed to serve economic and corporate interests.

4 Constructing Consumer Identity Through Curriculum

Children are socialized into their consumer identities through advertising, marketing, and television shows (Denzin 2001; Foster 2002; Giroux and Pollock 2010; Steinberg 2011). To this I would add (along with Darder (2010), Elshof (2006, 2009), Kahn (2010a), Petrina (2000a), Schor (2004), and Spring (2003) that consumer identity, behavior, and consciousness are also constructed in schools. This is illustrated by taking a small excerpt from Ontario's science and technology curriculum (MoE 2007a) to analyze how middle school children in grade seven (year 7) are discursively positioned as consumers. Motivated by a desire to better understand how teachers (as curriculum mediators) are implicated, my search for moral and ethical grounding comes from a deep concern I share with Steve Keirl (Chapter "Critiquing as Design and Technology Curriculum Journey: History, Theory, Politics, and Potential"), Susan McLaren (Chapter "Critiquing Teaching: Developing Critique Through Critical Reflection and Reflexive Practice"), and David Spendlove (Chapter "The Identification and Locationof Critical Thinking and Critiquing in Design and Technology Education"), who also believe critical questions must be asked to make visible—and possibly challenge—interests, viewpoints, and assumptions underlying educational policies and pedagogical practices. To this discussion I bring a very situated viewpoint, informed by personal classroom experiences and reflections on how particular activities and artefacts support or constrain the enactment of a prescribed curriculum (Edwards 2011). I have chosen to focus on this particular policy text for the following reasons. Firstly, Ontario is my home province where, over the course of my teaching career, I have engaged with different curricular formulations of technology education. Since its revision in 2007, the science and technology (S&T) document has been the *official* curriculum policy I know best. I am aware that my comments might invite the response that this case may not be generalizable beyond the Ontario context and even less so outside North America. This is certainly a reasonable response. However, Ontario's "post-positivist vision for science education" (Pedretti and Nazir 2011, p. 602) draws inspiration from the STS[E] (science, technology, society, and the environment) education movement, and while there is no unanimous agreement on what STSE is (Pedretti and Nazir p. 602), its "*science for all*" philosophy is strongly supported in other countries (e.g., Fensham 1988; Layton 1988; Solomon and Aikenhead 1994; Yager 1996; Ziman 1980). Another compelling reason why this analysis has relevance is that during and since the Decade of Education for Sustainable Development (2005–2014), Ontario— like many other provinces, states, and nations—has made efforts to introduce and/or revise existing environmental education and sustainability initiatives. Recommendations put forth in the Working Group report entitled *Shaping our schools, shaping our futures* (MoE 2007b) were purportedly based on "the successful practices of other jurisdictions in Canada and around the world" (p. 7), including three Canadian provinces, several states in the USA, Australia, Finland, Ireland, Israel, New Zealand, Sweden, and the "United Kingdom" (pp. 20–21). What is most interesting here is not that it is a case of "lesson drawing" (Rose 1991), but that it shows a particular interpretation of sustainability education, at a time when there was (and continues to be) no consensus on what sustainability means (Jickling and Wals 2008; Lee et al. 2015). To illustrate how sustainability has been conceptualized by Ontario's Ministry of Education, I will now examine a section of the S&T curriculum document.

5 Problematizing the Discourse of Technological Design

The fundamental concepts of sustainability and stewardship, embedded within an STSE framework, cover a number of social, economic, and environmental considerations as is shown in the following grade seven objective for the study of "form and function":

By the end of Grade 7, students will:

> 1.1 evaluate the importance for individuals, society, the economy, and the environment of factors that should be considered in designing and building structures and devices to meet specific needs (e.g., *function; efficiency; ease of use; user preferences; aesthetics; cost; intended lifespan; effect on the environment; safety, health, legal requirements*) (MoE 2007a, p. 130; italics in original).

Sample guiding questions are suggested to teachers for scaffolding critical inquiry. However, upon closer reading of these questions, the seemingly well-intentioned STSE approach takes a market-driven turn. For example, the question, "Why is it **important for companies to find out what consumers want now** and what they might want and/or need in the future?" (MoE 2007a, p. 130; bold added) prioritizes the perspective of "companies" while it positions consumers as users of products and services. From a critical practice perspective, I think that other equally important questions could, and should, be: "What are some of the ways companies externalize their costs of production?" and "Why is it necessary for consumers to pay the true cost of a product? Justify your answer from an eco-justice standpoint." Following the earlier question is another related to life cycle analysis:

> What things **might a company need to take into account** when considering the construction of a new structure **that consumers might not consider** (e.g., the environmental impact of using certain resources to make the structure, the eventual disposal of the structure)? (MoE 2007a, p. 130; emphasis added)

This question is clearly posed from a hypothetical "company" or corporate perspective. The suggestion that "consumers might not consider" resource and waste management issues effectively positions commercial business as sole decision-maker, while consumers who lack agency or technological literacy are left out of the process. One might also wonder why industrial designers, engineers, tradespeople, and employees—among other invested citizens—would not be identified. This simplistic, and arguably perhaps, false dualism between business and consumer offers no place in which students can participate as "collective caretakers of the planet" (Darder 2010, p. xv). Granted, given only two choices, it is more likely that 12-year-old children would identify their interests as entitled consumers rather than as corporate executives. Still, what is lacking is any question for reciprocal accountability to offset the company perspective. One suggestion could be, "What courses of action could be taken by consumers, workers, citizens and governments to hold parties accountable for violating environmental protection laws or contravening workers' basic rights to safe and fair labour practices?"

The lack of a strong critical and ethical focus is also evident when, in the basic concepts section of the curriculum, the learning expectations are framed in a technocentric design and manufacturing process. Concerning the suitability of materials for use, the document states,

By the end of Grade 7, students will:

> 3.7 identify the factors *(e.g., properties of the material as they relate to the product, availability, costs of shipping, aesthetic appeal, disposal)* that determine the suitability of materials for use in manufacturing a product *(e.g., a running shoe)*. (MoE 2007a, p.131; italics in original)

Here, the imperative to "identify factors" utilizes a distinct linguistic form commonly associated with school science. Such commands or directives imply that the knowledge students are expected to master is already known by the asker of the question (Olson 1989). No critique is explicitly called for—scientific certainty is the hallmark of well-established matters of fact and unproblematic objects (Latour 1987, 2004). Without question, telling the complex life cycle story of a running shoe made by a multinational corporation would require much study. However, it seems surprising that there is no accounting for fair trade and social justice "factors" anywhere in this entire grade seven strand, considering that Petrina (2000a) and others have reported how:

> [m]ost of the assembly is done through the labour of children and women cutting, gluing, and sewing under sweatshop conditions of high temperatures (100 degrees F) and toxic fumes from solvent-based toluene glues and paint. Their average wage is about 15 cents per hour over their 65 hour work week... (p. 217)

The failure to acknowledge exploitive practices serves to keep the study safe from moral or ethical redress and is one more instance where child-consumer identities are constructed as "future technologists rather than technologically capable critical thinking citizens" (Elshof 2009, p. 138). With the exception of the environmental design "factor" of disposal, decisions are presented as straightforward and value-neutral considerations with little regard for quality-of-life issues for humans and nonhumans. Moreover, I would add that the relatively uncritical and apolitical activity of identifying "factors" runs the unintended risk of sanctioning ignorance (Andreotti 2006) of the role of colonialism in wealth creation for the so-called First World, not to mention the role of the "color-coded international division of labor" (Wright 2012, p. 49) to maintain that wealth.

6 Multimedia Classroom Resources for Critiquing Consumption

Teachers have the difficult task of devising situations, to use Maxine Greene's (2000) words, "in which the young will move from the habitual and the ordinary and consciously undertake a search" (p. 24). Despite our current environmental dilemma (and some would argue we are facing an ecological crisis), I am not aware of very many multimedia classroom resources that both address these complex issues from a critical perspective and in a mode that young middle school learners can comprehend.

Contemporary forms of design activism (Fuad-Luke 2009) can help to raise public awareness of the impact of overconsumption (e.g., culture jamming), as well as afford new and imaginative ways to communicate complicated, statistical data information in ways that are easily understood by non-experts. Alastair Fuad-Luke (2009) has highlighted a small number of successful projects that draw on the strength of visual representation. These include Worldmapper's (n.d.)

distortions of map projections based on population and Giraffe Innovation's (2015) interactive project called Changing Habbits where the environmental impact of personal consumer habits is represented by 3D humanoid renderings. I also recommend for school-age children, in particular, two clever animations and e-learning resources that promote life cycle thinking by disclosing the hidden impacts of manufactured products: Leyla Acaroglu's *The Secret Life of Things* (http://www.thesecretlifeofthings.com/) and Annie Leonard's *The Story of Stuff* (http://storyofstuff.org/). Both of these projects creatively illustrate how symbolic imagery, metaphors, and humor can be utilized as powerful storytelling tools. In the next section, I describe how *The Story of Stuff* provides a springboard for critiquing issues of consumption, injustice, and environmental degradation.

6.1 A Creative and Critical Teaching Resource: The Story of Stuff

As a quasi-case study of the materials economy, *The Story of Stuff* offers a positive and engaging alternative to less effective "expository and didactic" approaches to information sharing (Clover and Shaw 2010, p. 206). Leonard's (2007a) short and fast-paced video animation of simple black and white cartoon figures is an edgy, visually entertaining, and humorous anti-capitalist critique of the problems inherent in the linear production-consumption-disposal mindset of American consumer culture. The sophisticated and creative use of visual metaphors helps to explain social and psychological concepts related to the design and marketing of familiar everyday products. Leonard deconstructs the discourse of consumption by describing a "system in crisis": the exploitation and overconsumption of the world's resources, the use of toxic chemicals in manufacturing, the externalized costs of production for profit, the planned and perceived obsolescence, and the unsustainable cradle-to-grave approach to waste management. While the scope of this paper does not permit a very detailed description or in-depth analysis, a few screenshots selected from the video will illustrate how critical literacy is fostered.

7 A Critique of Hyperconsumerism

Figure 1 shows a person caught in a nonstop "work-watch-spend treadmill" (Leonard 2007b, p. 13). This clever visual metaphor depicts a perpetual cycle of consumption driven by the desire to seek happiness through the accumulation of products. The image ties in remarkably well with Allan Schnaiberg's (1980) concept of the "treadmill of production" along with John Foster's (2002, p. 45) characterization of the system as a "giant squirrel cage" and John Huckle's (2010, p. 136) "capitalist treadmill in crisis." Driven by the desire to accumulate wealth, the

Fig. 1 Work-watch-spend treadmill. (Screenshot taken from the Story of Stuff video). Retrieved July 22, 2013

treadmill "manufactures consumer wants in a way that creates an insatiable appetite for more" (Huckle, p. 137). Children are not immune to the imperatives of capitalist production (Langer 2002, p. 72). Driving their consumptive behaviors are feelings of longing and dissatisfaction which, according to Beryl Langer (2002) and Jeremy Seabrook (1985), are produced and manipulated by corporate advertising.

8 Critique of Design's Role in Product Obsolescence

Many young people are not familiar with the manufacturing and marketing strategies of product obsolescence. Leonard (2007a) is able to problematize the discourse of consumerism with the clever alliteration "designed for the dump" (see Fig. 2). Quoting Lahaye (1995), she explains that by the 1950s, "forced consumerism was extolled by the markets as a must: things had to be consumed, burned, used, replaced and discarded at a constantly accelerating pace" (p. 60). In his book, *Made to Break: Technology and Obsolescence in America*, Giles Slade (2006) traced the history of product design and the underlying profit motive for technological, psychological, and planned obsolescence. Noting that industrial designers like Brooks Stevens were unapologetically open in those years about the underlying profit motive (see Adamson 2003), Slade stated, "Not only did we invent disposable products, ranging from diapers to cameras to contact lenses, but we invented the very concept of disposability itself, as a necessary precursor to our rejection of tradition and our promotion of progress and change" (pp. 3–4). (see Figs. 3 and 4).

I have found that many 12-year-olds' level of understanding of the materials economy goes as far as thinking that *cheap things usually break because they are cheap* and they, as consumers, are the victims of a "rip-off." *The Story of Stuff* informs them that they are not the only "victims." The reason why many products can be sold so cheaply is that the true costs are externalized, which means that other people are "paying" through poor wages, dangerous working conditions (see Fig. 5), and destruction of their local environment. The exploitation of others struck a chord with some of my students who expressed their understanding and empathy in terms

Fig. 2 "Designed for the dump." (Screenshot taken from the Story of Stuff video). Retrieved July 22, 2013

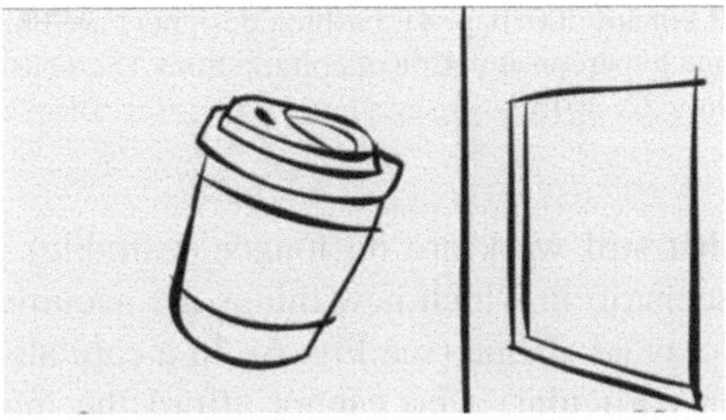

Fig. 3 Products are designed to be useless as quickly as possible (Leonard 2007a, b). (Screenshot taken from the Story of Stuff video). Retrieved July 22, 2013

Fig. 4 The profit motive underlies deliberate obsolescence and the promotion of progress and change. (Leonard 2007a, b). (Screenshot taken from the Story of Stuff video). Retrieved July 22, 2013

Fig. 5 Factory workers of reproductive age are exposed to many toxic chemicals. (Leonard 2007a). (Screenshot taken from the Story of Stuff video). Retrieved July 22, 2013

of "fairness." They thought that people should have the right to live in a safe place and they should be paid fairly for their labor.

Young people are often surprised and perturbed to learn of the possibility that the products they buy are deliberately designed to break. Another revelation for many has been the issue of perceived obsolescence (i.e., the notion that things

Fig. 6 In our capitalist system, "if you don't own or buy a lot of stuff, you don't have value" (Leonard 2007b, p. 4). Fashion designers are implicated in the arousal of desire, mass production, and hyperconsumption of commodities. (Screenshot taken from the Story of Stuff video). Retrieved July 22, 2013

that still work are no longer desirable). The video depicts a familiar peer group scenario in which new things are acquired either to establish one's social status or to avoid shame (see Fig. 6). In a capitalist system, those (poor and minority youth in particular) who cannot afford the money, resources, and leisure time to shop for new things are considered "failed consumers" (Giroux 2015), or as Leonard (2007b, p. 4) explains, "If you don't own or buy a lot of stuff, you don't have value." Research on the influence of peer groups and the mass media on commodity consumption supports my personal observation that middle- and upper middle-class children are quite aware of the rapid turnover of digital technologies and clothing, and many readily admit to their desires to purchase the newest models and stylish brand-name fashions of "coolness" as markers of self-identity and group identity (Hoechsmann 2010; Willis 1991). Very often, material culture is used to demarcate social difference but it can also lead to the social exclusion of others (Martens 2005, p. 355). Conspicuous consumption, as noted by Slade (2006), is in part manipulated by marketers who psychologically target people's anxiety and "desire not to lose face" (p. 51). Corporate advertising is one redoubtable pedagogue (Steinberg 2011). In 2002, it was estimated that the average American watched 21,000 television commercials a year and that in 1 year alone, the $1 trillion marketing expenditures aimed at consumers by American businesses exceeded the total spent on public and private education by about $600 billion (Foster 2002, pp. 46–47).

9 A Call to Action

Pronouncing that even our existing recycling practices are unsustainable, Leonard (2007b) concludes her critique with an open-ended call for collective action and "a new school of thinking" (p. 15) based on principles of sustainability and equity. Clean production, green chemistry, zero waste, closed loop production (see Fig. 7), and renewable energy are some of the initiatives she identifies. In effect, the viewers are socially positioned—not as compliant shoppers but as capable agents of change.

Fig. 7 True recycling. Closed loop production seeks to eliminate natural resource input and waste output (Leonard 2007b, p. 15). (Screenshot taken from the Story of Stuff video)

10 Critiquing the Critique

So far, I have focused solely on the merits of *The Story of Stuff* but would be remiss if I did not also acknowledge that there has been some opposition to its use in schools. In this section, I will first briefly identify key criticisms offered by those in the fields of science, interest group politics/public policy, and education, followed by a carefully considered rejoinder. Extending the critique, I will then reflect on the difficult challenge teachers face when negotiating the tensions between indoctrination and empowerment (Pedretti and Nazir 2011).

10.1 Chemophobic and Anti-capitalist Propaganda

Leonard's detractors dismissed the film as misleading, factually inaccurate, statistically skewed, leftist, and anti-capitalist propaganda (e.g., Baum 2009a; Cooper 2009; Doren 2009; Kaufman 2009). The issue of public risk in the production, use, and disposal of carcinogens was particularly provocative for some members of the American Chemical Society, who expressed consternation over what they considered to be "chemophobic" propaganda that deployed "scare tactics" of misrepresentation, exaggeration, or oversimplification of certain chemical effects (Baum 2009b, p. 3; Frantom 2009, p. 4; Canan 2009, p. 4). Among the most outspoken critics were members of self-described conservative think tank organizations who subscribe to libertarian principles of free enterprise, limited government, environmental skepticism, and a strong national defense: the Heritage Foundation, the National Center for Public Policy Research, and the Competitive Enterprise Institute. Using the logic of market self-regulation, outsourcing is viewed not as exploitation, but as an example of mutually beneficial global cooperation (Doren 2009). The issue of environmental degradation was deflected with the argument that technological innovations had, among other things, increased food production, improved water quality, and reduced the volume of trash going into landfill dumpsites (Doren 2009).

From a critical literacy perspective, such technocentric, salvationist, and paternalistic rebuttals are extremely problematic in the way they celebrate technical fixes, gloss over the complexities of social problems, forget past legacies and complicities, and thereby serve to reproduce—albeit perhaps inadvertently—what Vanessa Andreotti (2012) has described as "problematic historical patterns of thinking and relationships" (p. 2).

10.2 More Pedagogical Considerations

As a cautionary reminder, teachers need to be sensitive to their students' cognitive and emotional readiness to grasp relatively complex concepts of systems thinking (Pitt and Heinemeyer 2015) or handle stark and scary facts (Kaufman 2009) to avoid undue anxiety, fear, anger, guilt, or paralysis. It was reported in *The New York Times*, for instance, that one 9-year-old boy was "worried it might hurt the environment if he bought a new set of Legos" (Kaufman, ¶20). Social class may be another consideration as one teacher warned that "students, particularly affluent ones, might take the critique personally," become offended, and "turn off the learning button" (Kaufman, ¶20). Little research exists on the consumption experiences of children; however Lydia Martens (2005) and Leonard et al. (2003) have argued against treating young consumers as a singular homogeneous (i.e., white, middle class) social entity and call for more exemplary studies (e.g., Chin 2001) to better understand how socioeconomic and cultural background influence consumer-related learning, conduct, and values. And by the same token, one might wonder how differently *The Story of Stuff*'s unmistakably middle-class American-centric message has been received not only by different groups across the USA but in different countries as well.

10.3 Teaching or Indoctrination?

Probably the most common criticism against the use of Leonard's video in schools is that it proselytizes or indoctrinates school children (e.g., http://www.groupsnoop.org/Greenpeace). While there are many different interpretations of the meaning of indoctrination, the main concern is that students are left open to manipulation when only one side of a political or controversial issue is being presented (Dobson 2003, p. 196). At the heart of this debate are larger questions about the very purpose of education and whether the enactment of curricula can be neutral. Space limitations here preclude the detailed treatment such a discussion deserves, but for the record, I am not alone in my view that education is not a neutral process (e.g., Shaull 2000) and that curricula and resources carry (often implicit) values and beliefs (e.g., Apple 1990, 2014; Jenkins 1992; Layton 1988; Vasquez 2014). This is not to deny that there is a need for the continual reexamination of one's own educational practices (Pedretti and Hodson 1995). I fully recognize that classrooms cannot be *value-free* environments, but maintain (along with Loving et al. (2003) and Zeidler et al. (2005)) that teachers should strive to make them *value-fair*. I believe it is important to move classroom discussions beyond simplistic and unhelpful binaries of good/bad, capitalist/anti-capitalist binaries by exposing students to a wide range of ideological perspectives and encouraging them to deconstruct conventional wisdom. There are environmental education critics (e.g., Jickling 1992, 1994, 2005; Jickling and Spork 1998; Jickling and Wals 2008, 2012) who are adamantly opposed to instrumental aims of an "education for" and argue that working toward prescribed ends is counterproductive to broad, pluralist conceptualizations of "critical literacy as a practice of opening to the world" (Nicholson et al. 2012, p. 75).

It is for these reasons that I include *The Story of Stuff* as one of many resources in my teaching toolkit. As a pedagogical model, the video is instructive because it avoids the pitfalls of what Clover and Shaw (2010) identify as the "stifling, limited, and pedantic aspects of so much environmental education" (pp. 206–207). In contrast, Leonard (2007a) maintains the "ordinary" citizen consumer status that she establishes in her introduction (as an iPod owner) and sidesteps the higher status generally attributed to experts. I think Sandlin and McLaren (2010) would applaud the way in which *The Story of Stuff* locates human experience "within specific social relations of production" (p. 14). Leonard (2007b) does this by tracing back through the life cycle of a portable radio from the shelf of a big box store, through the hands of a minimum wage cashier, shelf stocker, transport driver, ocean freight handler, "some 15 year old in a maquiladora [factory] in Mexico," and "the kids in parts of the Congo ... [who] have had to drop out of school to mine coltan" (p. 8). Rather than *The Story* ending, she enthusiastically invites her viewers to join in; to "reclaim and transform this linear system .... Remember that old way didn't just happen by itself. It's not like gravity that we gotta live with. People created it. And we're people too. So let's create something new" (Leonard 2007b, pp. 15–16). Here Leonard uses a "language of possibility" (Giroux 1988) to promote social change through collective action. With strong leanings toward a "praxis-oriented pedagogy" (Farahmandpur 2010, p. 66), the video discursively brings together critical knowledge and social practice. It is now up to the teacher to extend the gesture by creating opportunities for students to participate in reconstructive efforts for a fairer society.

Granted, simply watching this video does not turn viewers into active citizens, or activists, or artists, or designers, but it does introduce educators and students to a new discourse and critical framework for thinking about how they might take "more informed, responsible and ethical action" to help solve the problems of inequality and injustice (Andreotti 2006, p. 48). According to Vanessa Andreotti (2006), reflexivity and dialogue are basic principles for personal change in a critical citizenship education that promotes "engagement with global issues and perspectives and an ethical relationship to difference, addressing complexity and power relations" (p. 48). By identifying problems inherent in a linear production-consumption-disposal paradigm, *The Story of Stuff* does have the potential to inspire young people's future design activities. But at the very least, critical discursive practices could serve to interrupt the ideology of consumption. As Leo Elshof (2009) persuasively argues:

> Although young people are not responsible for designing or creating the technological systems within which they live, they are nonetheless active participants in its evolution. Long before they have become technologically literate, they are active as young citizen consumers. In this sense they do become co-creators of the world and technology education can help them understand why they must begin to share responsibility for its care. (p. 138)

11 Implications for Technology Education

With the ecological health of our planet in jeopardy, our Western "throwaway ethic" (Slade 2006, p. 281; see also Packard 1960) is no longer sustainable. As Elshof (2009) asserts, countries like Canada and the USA—which create the largest ecological and carbon footprints on the planet—have the added responsibility to encourage their young people "to think and act differently in terms of the ways they use, consume and design technologies" (p.134). A critical design education can play a key role in contesting the manufacturing of desires and to "problematize the interrelations between conspicuous consumption and ecological death" (Petrina 2000a, p. 212). While I am encouraged by the latest revisions of the Ontario curricula for science and technology education that boldly introduced an STSE foundation, I am troubled by the prevailing ideological discourse of neoliberalism that continues to prioritize values of individualism and economic competitiveness. More than a decade ago, Petrina (2000a) argued for a "political ecology of design" (p. 218) where ecological values of care, complex life cycles, and interconnectedness work in tandem with "political values such as control, distribution, equity, interests, justice, liberty, and power" (p. 218). Sadly, as the horrific deaths of more than 1100 Bangladeshi garment workers in April of 2013 tragically remind us, criticality is urgently needed to problematize the interrelations between hyperconsumption and what Foster (2002) believes are "issues of economic justice—the exploitation of the poor by the rich" (p. 49).

12 The Challenges of Critical Literacies for Design and Technology Education

Design and technology education, from a critical literacy perspective, has tremendous potential as a site for transformative learning in which young people are encouraged to develop the intellectual tools to *critique* and *act* "to transform the world around them in ways that make a more just and democratic society for everyone" (Saltman 2005, p. 119). By carrying out their own design projects, students can develop their capacities to imagine the future "as something more than a repeat of the present" (Giroux 2005, ¶34). As Derek Hodson (1994) argued, "Politicisation of science [and technology] education can be achieved by the provision of opportunities for confronting issues that have a scientific, technological or environmental dimension" (p. 84) and maintained that young people are more likely to become active citizens if they are encouraged to take "suitable action" (p. 87) in their local schools and communities. As teachers, we need to deepen our own criticality so that we can recognize how we are both part of the problem and how we can be part of the solution (Andreotti 2006). If we do not, then we "run the risk of (indirectly and unintentionally) reproducing the systems of belief and practices that harm those [. . . we] want to support" (Andreotti 2006, p. 49–50). The challenge to D&T teachers then is to create safe spaces where students can critically engage and reflect on how we came to think, feel, and act the way we do and then try out other ways of being and acting in the world. This pedagogical approach encourages but does not impose change on the learners—the decision of whether or how to change or take action is ultimately to be made by the individual student (Andreotti 2006). Otherwise, if "correct" readings of the world are determined by teachers, there is a danger that critical literacy's project of reconstruction could lead to one of indoctrination (Nicholson et al. 2012; Pedretti 2003; Pedretti and Nazir 2011).

Developing a critical technological literacy about how systems currently work begins with knowledge and understanding of how local everyday consumption practices are tightly linked to global processes of capitalist production. From a social responsibility perspective, it is incumbent upon "citizen designers" (Heller & Vienne 2003) to consider objects—not in isolation, but as constituent elements of a dynamic "macrocosm" that includes, to quote Véronique Vienne (2003, p. 244), "all the befores and afters of the manufacturing process." To effect social and environmental change, Vienne is among those who propose that designers must, for example, trace the origins of specific food or other material chains, consider whether or not their projects are part of a renewable energy system, design waste out of products' life cycles, and create artifacts for easy disassembly, remanufacture, or recovery (see McDonough and Braungart 2002; Pitt and Heinemeyer 2015).

Teachers also need to appropriate the powerful communication techniques that marketers and advertisers employ so well in order to create "a counter-ideology" (Freire 1985, p. 18) that will challenge taken-for-granted beliefs that serve only the interests of the socially powerful (Clover and Shaw 2010). I propose that through "purposeful critiquing" (Keirl 2007), we help young people engage in problem-posing (Freire 2000) for positive social and environmental change (Hodson 2003). Instead of creating consumers, we create consumer advocates and cultural critics (Denzin 2001, p. 326). Instead of reproducing a mindset for designing objects as solutions, we reorient design thinking as an ethical solution-building process for "social good" (Chochinov 2009, p. 8; Fuad-Luke 2009; McCoy 2003; Riley 2003; Vienne 2003) that may not even create more products (Keirl 2007; Pilloton 2009). And instead of preparing future technologists, we prepare critical and technologically literate citizens who will question and challenge our "*existing* technologies, systems and worldviews [that] contribute to the global environmental crisis" (Elshof 2009, p. 142; italics in original). Informed by a "politicized ethic of care" (Hodson 1999, p. 789), the rejection of rampant consumerism for "a more environmentally sustainable lifestyle that promotes appropriate technology" is, in itself, a kind of social reconstruction. Hodson (1994, 1999) propounds that education for developing critical scientific and technological literacy entails helping students to develop a deep understanding of socially and personally relevant issues, *as well as* learn how to translate their concern into responsible actions (Hodson 1999, p. 789). Increasing competence to take some form of action to solve student-identified problems is key for countering "action paralysis" (Jensen 2004). With specific reference to environmental education, Bjarne Jensen proposed that teachers begin with the "views, concerns and anxieties of students" in order to "[transform] the sense of powerlessness into the desire and ability to act" (p. 405).

13 Concluding Remarks

I would like to think that my students' attitudes as expressed through classroom talk and design projects have been influenced in part by some of the messages presented in *The Story of Stuff.* As Keirl (2007) points out, it is through (de)constructive critique that students develop their voices "as would-be democratic citizens" (p. 310). At the very least, I am confident that many young people are able to understand and are eager to participate in discussions and debates, as well as take action to address issues related to planned and perceived obsolescence, the ethics of fair trade, the externalization of costs, and the impact of technology on the environment (see Wilkinson and Bencze 2015; Chappell 2015). Also encouraging are teachers' stories about "children who become environmental advocates at home after seeing the video" (Kaufman 2009, ¶21; see also Sperling et al. 2014; Zoras and Bencze 2014). Active engagement in school and community design projects that address issues is central to students' personal transformation (Taylor 2008) as they learn to learn, do, be, live, and work collaboratively, "translate their intentions into actions" (Pavlova 2015, p. 96), and ultimately bring about social change (Hodson 2003). I am also mindful, however, that while there are benefits to be gained by teaching for a critical literacy of the built world (Petrina 2000b), we need to be prepared for the possibility of unintended emotional fallout when learners may experience feelings of "guilt, internal conflict and paralysis, critical disengagement, feeling[s] of helplessness" (Andreotti 2006, p. 48). In the unsettling process of critical self-reflection, great courage and humility are required when we recognize "how we are [all] implicated or complicit in the problems we are trying to address" (Andreotti 2012, p. 2). For this reason, I think there is much to learn from multimedia resources like *The Story of Stuff* in which alliterative, metaphoric, and visual forms of conversational storytelling are cleverly utilized to raise consciousness and stimulate imaginative critique. Perhaps it is the video's creative use of animation and humor—along with Leonard's message that we are all in this mess together—that helps to inspire and motivate people to become part of the solution. My speculations invite further inquiries into the affordances of word play and what Åsa Wettergren (2009) refers to as "fun/humour" to "[open] up the present and [show] that 'another world is possible'" (p. 6). Critical explorations into the pedagogical role of "utopian laughter" (Wettergren, p. 6) that arises from a reflexive stance with the world would certainly add to the existing but limited educational research on the use of humour (such as Boyle and Stack (2014), McGhee (1989), and Zuk and Dalton (1998)).

Imagination and creativity are also resources with great emancipatory potential for developing new methods for "seeing, exploring and challenging the world being created for us" (Clover and Shaw 2010, p. 206; Greene 2000). From an action-oriented perspective, designing, like problem-posing education (Freire 2000), is based on creativity and calls for reflection and action upon reality (p. 84). Critical design thinking aims to break free from what is considered fixed or perhaps escapes our notice entirely. Imagination is what makes empathy and entering others' worlds possible (Greene 2000). It enables us to look at things with the view to make them *otherwise* (Bloch 1986; Greene 2000) and, as such, is hopeful. Also encouraging is Clover and Shaw's (2010) work in arts-based communities of practice which can remind us "[the] arts are far more than mere self-expression; they are tools of emancipation and critical learning that can inflame politicians, force people to see the taken-for-granted differently, and engage the imagination in explorations of consumerism" (p. 211). I am reminded of Friere's (1985) understanding of education as "simultaneously an act of knowing, a political act, and an artistic event" (p. 17). He believed that helping children "shape themselves as beings" (p. 17) constitutes the dynamic, difficult, and aesthetic work of the teacher learner. Stressing the politicity of education, Friere wrote,

> When we try to be neutral, ... we support the dominant ideology. Not being neutral, education must be either liberating or domesticating [or a mixture of both]. ... Thus, we have to recognize ourselves as politicians. It does not mean that we have the right to impose on students our political choice. .... Our task is not to impose our dreams on them, but to challenge them to have their own dreams, to define their choices, not just to uncritically assume them. (pp. 17–18)

While I am not claiming that a 20 min "cartoon" will inspire life-changing world views, I do think *The Story of Stuff* is a creative pedagogical resource for developing critical learning. It acquaints viewers with a language of possibility (Giroux 1988), models purposeful critiquing (Keirl 2007) of our cultural patterns of material consumption, and stimulates ethical design thinking with a care-driven sensibility for ecological and social justice (Noddings 2010; Orr 2004). The level of thoughtfulness and ethical concern my own students express for the welfare of others and the natural world keeps me hopeful about the future for active and critical citizenship—just as the work of critical education scholars (including, but not limited to, those identified in this chapter) strengthens my resolve to advocate for a politicized form of design and technology education.

References

Acaroglu, L. (n.d.). *The secret life of things: Uncovering the hidden environmental impacts of everyday things!* Retrieved from http://www.thesecretlifeofthings.com/

Adamson, G. (2003). *Industrial strength design: How Brooks Stevens changed your world* (p. 2003). Cambridge, MA: MIT Press.

Andreotti, V. (2006). Soft versus critical global citizenship education. *Policy and Practice: A Development Education Review, 3,* 40–51. Retrieved March 2012, from http://www.developmenteducationreview.com/issue3-focus4

Andreotti, V. (2012). Editor's preface: HEADS UP. *Critical literacy: Theories and practices, 6*(1), 1–3.

Apple, M. (1990). *Ideology and curriculum.* [EBook]. New York: Routledge.

Apple, M. W. (2014). *Official knowledge: Democratic education in a conservative age* (3rd ed.). New York: Routledge.

Barber, B. R. (2007). *Con$umed: How markets corrupt children, infantilize adults, and swallow citizens whole.* New York: W.W. Norton & Co..

Baum, R. (2009a, February 23). More on limits. *Chemical and Engineering News, 87*(8), 3. Retrieved from

Baum, R. (2009b, May 18). The story of stuff. [Editor's Page]. *Chemical and Engineering News, 87*(20), 3. Retrieved from http://cen.acs.org/articles/87/i20/Story-Stuff.html

Bloch, E. (1986). *The principle of hope.* (1 ed.). (trans: Plaice N., Plaice S. , & Knight P.). Cambridge, MA: MIT Press.

Boyle, F., & Stack, N. (2014). An explorative study into the possible benefits of using humor in creative tasks with a class of primary five pupils. *Humor, 27*(2), 287–306. doi:10.1515/humor-2014-0029.

Canan, S. S. (2009, August 3). The story of stuff. [Letter]. *Chemical and Engineering News,* 87(31), 4. Retrieved from http://cen.acs.org/articles/87/i31/Story-Stuff.html

Chappell, B. (2015). Sustainability + fun = A change in behavior. In K. Stables & S. Keirl (Eds.), *Environment, ethics and cultures: Design and technology education's contribution to sustainable global futures* (pp. 271–284). Rotterdam: Sense Publishers.

Chin, E. (2001). *Purchasing power: Black kids and American consumer culture.* Minneapolis, MN: University of Minnesota Press.

Chochinov, A. (2009). A good long tradition. [Foreward]. In E. Pilloton (Ed.), *Design revolution: 100 products that empower people* (pp. 6–9). New York: Metropolis Books.

Clover, D. (2002). Traversing the gap: Concientizacíón, educative-activism in environmental adult education. *Environmental Education Research, 8*(3), 315–323. doi:10.1080/1350462022014546 5.

Clover, D. (n.d.). *Darlene Clover.* [Research statement]. Retrieved from http://www.uvic.ca/education/psychology/people/home/faculty/cloverdarlene.php

Clover, D. E., & Shaw, K. (2010). Re-imagining consumption: Political and creative practices of arts-based environmental adult education. In J. Sandlin & P. McLaren (Eds.), *Critical pedagogies of consumption: Living and learning in the shadow of the shopocalypse* (pp. 203–213). London: Routledge.

Cooper, R. (2009, May 11). The story of lies: Greenpeace in your kid's school. *The Daily Signal.* [The Heritage Foundation]. Retrieved from http://dailysignal.com/2009/05/11/the-story-of-lies-greenpeace-in-your-kids-school/

Crawford, M. B. (2009). *Shop class as Soulcraft: An inquiry into the value of work.* New York: Penguin Press.

Darder, A. (2010). Preface. In R. Kahn (Ed.), *Critical pedagogy, ecoliteracy, & planetary crisis: The ecopedagogy movement* (pp. ix–xvii). New York: Peter Lang.

Denzin, N. K. (2001). The seventh moment: Qualitative inquiry and the practices of a more radical consumer research. *Journal of Consumer Research, 28*(2), 324–330.

Dobson, A. (2003). *Citizenship and the environment.* Oxford, England: Oxford University Press.

Doren, L. (2009, May 13). *Story of stuff, the critique part 1 of 4, Lee Doren.* [YouTube]. Retrieved from https://www.youtube.com/watch?v=c5uJgG05xUY

Eastwood, L. (2006). Contesting economic order and media construction of reality. In S. Best & A. J. Nocella II (Eds.), *Igniting a revolution: Voices in defense of the earth* (pp. 114–126). Oakland, CA: AK Press.

Edwards, R. (2011). Translating the prescribed into the enacted curriculum in college and school. *Educational Philosophy and Theory, 43*(S1), 38–54. doi:10.1111/j.1469-5812.2009.00602.x.

Elshof, L. (2005). Teacher's interpretation of sustainable development. *International Journal of Technology and Design Education, 15*(2), 173–186. doi:10.1007/s10798-005-8277-1.

Elshof, L. (2006). Productivism and the product paradigm in technological education. *Journal of Technology Education, 17*(2), 18–32.

Elshof, L. (2009). Toward sustainable practices in technology education. *International Journal of Technology and Design Education, 19*(2), 133–147. doi:10.1007/s10798-008-9074-4.

Farahmandpur, R. (2010). Teaching against consumer capitalism in the age of commercialism and corporatization of public education. In J. Sandlin & P. McLaren (Eds.), *Critical pedagogies of consumption: Living and learning in the shadow of the shopocalypse* (pp. 58–66). London, England: Routledge.

Fensham, P. J. (1988). Approaches to the teaching of STS in science education. *International Journal of Science Education, 10*(4), 346–356.

Foster, J. B. (2002). *Ecology against capitalism*. New York: Monthly Review Press.

Frantom, P. (2009, August 3). The story of stuff. [Letter]. *Chemical and Engineering News, 87*(31), 4. Retrieved from http://cen.acs.org/articles/87/i31/Story-Stuff.html

Freire, P. (1985). Reading the world and reading the word: An interview with Paulo Freire. *Language Arts, 62*(1), 15–21.

Freire, P. (2000). *Pedagogy of the Oppressed* (30th Anniversary ed.). (M. B. Ramos, Trans.). New York: Continuum International.

Fuad-Luke, A. (2009). *Design activism: Beautiful strangeness for a sustainable world*. London: Earthscan.

Giraffe Innovation. (2015). *Changing habbits*. Retrieved from http://www.changinghabbits.co.uk/

Giroux, H. A. (1988). *Teachers as intellectuals: Toward a critical pedagogy of learning*. Granby: Bergin & Garvey.

Giroux, H. A. (2005). Cultural studies in hard times: Public pedagogy and the challenge of neoliberalism. *Fast Capitalism, 1*(2). Retrieved from: http://www.uta.edu/huma/agger/fastcapitalism/1_2/giroux.htm

Giroux, H. (2015, April 17). Disposable youth. *Ideas*. [CBC Radio Broadcast: SPUR lecture and interview in Toronto]. Retrieved from: http://www.cbc.ca/radio/ideas/disposable-youth-1.3036140

Giroux, H. A., & Pollock, G. (2010). *The mouse that roared: Disney and the end of innocence* (Updated & expanded ed.). Lanham: Rowman & Littlefield.

Greene, M. (2000). *Releasing the imagination: Essays on education, the arts, and social change*. San Francisco: Jossey-Bass.

Hodson, D. (1992). Politicizing environmental education. *Crucible, 23*, 14–19.

Hodson, D. (1994). Seeking directions for change: The personalization and politicization of science education. *Curriculum Studies, 2*(1), 71–98. doi:10.1080/0965975940020104.

Hodson, D. (1999). Going beyond cultural pluralism: Science education for sociopolitical action. *Science Education, 83*(6), 775–796. doi: 10.1002/(SICI)1098-237X(199911)83:6<775::AID-SCE8>3.0.CO;2-8

Hodson, D. (2003). Time for action: Science education for an alternative future. *International Journal of Science Education, 25*(6), 645–670. doi:10.1080/09500690305021.

Hoechsmann, M. (2007). Advertising pedagogy: Teaching and learning consumption. In D. Macedo & S. R. Steinberg (Eds.), *Media literacy: A reader* (pp. 653–666). New York: Peter Lang.

Hoechsmann, M. (2010). Rootlessness, reenchantment, and educating desire: A brief history of the pedagogy of consumption. In J. Sandlin & P. McLaren (Eds.), *Critical pedagogies of consumption: Living and learning in the shadow of the shopocalypse* (pp. 23–35). London: Routledge.

Huckle, J. (2010). ESD and the current crisis of capitalism: Teaching beyond green new deals. *Journal of Education for Sustainable Development, 4*(1), 135–142.

Jenkins, E. W. (1992). School science education: Towards a reconstruction. *Journal of Curriculum Studies, 24*(3), 229–246.

Jensen, B. B. (2004). Environmental and health education viewed from an action-oriented perspective: A case from Denmark. *Journal of Curriculum Studies, 36*(4), 405–425.

Jickling, B. (1992). Why I don't want my children educated for sustainable development. *Journal of Environmental Education, 23*(4), 5–8. doi:10.1080/00958964.1992.9942801.

Jickling, B. (1994). Studying sustainable development: Problems and possibilities. *Canadian Journal of Education, 19*(3), 231–420.

Jickling, B. (2005). Sustainable development in a globalizing world: A few cautions. *Policy Futures in Education, 3*(3), 251–259.

Jickling, B., & Spork, H. (1998). Education for the environment: A critique. *Environmental Education Research, 5*(3), 309–28.

Jickling, B., & Wals, A. E. J. (2008). Globalization and environmental education: Looking beyond sustainable development. *Journal of Curriculum Studies, 40*(1), 1–21. doi:10.1080/00220270701684667.

Jickling, B., & Wals, A. E. J. (2012). Debating education for sustainable development 20 Years after Rio: A conversation between Bob Jickling and Arjen Wals. *Journal of Education for Sustainable Development, 6*(1), 49–57. doi:10.1177/0973408211006001111.

Kahn, R. (2008a). From education for sustainable development to ecopedagogy: Sustaining capitalism or sustaining life? *Green Theory & Praxis: The Journal of Ecopedagogy, 4*(1), 1–14. doi:10.3903/gtp.2008.1.2.

Kahn, R. (2008b). Introduction. *Green Theory & Praxis: The Journal of Ecopedagogy, 4*(1), i–iv. doi:10.3903/gtp.2008.1.1.

Kahn, R. (2010a). *Critical pedagogy, ecoliteracy, & planetary crisis: The ecopedagogy movement.* New York: Peter Lang.

Kahn, R. (2010b). Producing crisis: Green consumerism as an ecopedagogical issue. In J. Sandlin & P. McLaren (Eds.), *Critical pedagogies of consumption: Living and learning in the shadow of the shopocalypse* (pp. 47–57). London: Routledge.

Kaufman, L. (2009, May 10). A cautionary video about America's 'stuff'. *New York Times.* Retrieved from http://www.nytimes.com/2009/05/11/education/11stuff.html?_r=2&pagewanted=print

Keirl, S. (2007). Critiquing in a democratics of design and technology education. In J. R. Dakers, W. J. Dow & M. J. de Vries (Eds.), *PATT 18: Pupils attitudes towards technology international conference on design and technology educational research* (pp. 306–312). Glasgow: Faculty of Education, University of Glasgow. Retrieved from http://www.iteea.org/Conference/pattproceedings.htm

Lahaye, M. C. (1995, Spring). The consumer and green products. *Ecodecision, 16*, 60–62.

Langer, B. (2002). Commodified enchantment: Children and consumer capitalism. *Thesis Eleven, 69*, 67–81.

Latour, B. (1987). *Science in action: How to follow scientists and engineers through society.* Cambridge, MA: Harvard University Press.

Latour, B. (2004). Why has critique run out of steam? From matters of fact to matters of concern. *Critical Inquiry, 30*(2), 225–248.

Layton, D. (1988). Revaluing the T in STS. *International Journal of Science Education, 10*(4), 367–378.

Lee, J. C.-K., Williams, M., & Stimpson, P. (2015). Preface by the series editors. In R. Jucker & R. Mathar (Eds.), *Schooling for sustainable development in Europe: Concepts, policies and educational experiences at the end of the UN decade of education for sustainable development* (pp. v–vii). Springer: Cham.

Leonard, A. (2007a). *The story of stuff* [Video]. Retrieved from http://www.storyofstuff.org/movies-all/story-of-stuff/

Leonard, A. (2007b). *Story of Stuff* [Referenced and annotated script]. Retrieved from http://www.storyofstuff.org/2011/03/14/story-of-stuff/

Leonard, A. (2012, March 30). The iPhone and consumer guilt. *The Huffington Post.* Retrieved from http://www.huffingtonpost.com/annie-leonard/the-iphone-and-consumer-g_b_1391324.html?ref=tw

Leonard, A., & Conrad, A. (2010). *The story of stuff: The impact of overconsumption on the planet, our communities, and our health—and how we can make it better*. New York: Free Press.

Leonard, D., Frankham, J., & Bragg, S. (2003). Consuming children: Education-entertainment-advertising—Kenway & Bullen. [Book Review]. *British Journal of Sociology of Education, 24*(4), 515–526.

Levitas, R. (1997). Educated hope: Ernst Bloch on abstract and concrete utopia. In J. O. Daniel & J. Owen (Eds.), *Not yet: Reconsidering Ernst Bloch* (pp. 65–79). New York: Verso.

Loving, C. C., Lowy, S. W., & Martin, C. (2003). Recognizing and solving ethical dilemmas in diverse science classrooms. In D. L. Zeidler (Ed.), *The role of moral reasoning on socioscientific issues and discourse in science education* (pp. 183–194). Dordrecht, NL: Kluwer Academic Publishers.

Margolin, V. (1998). Design for a sustainable world. *Design Issues, 14*(2), 83–92.

Martens, L. (2005). Learning to consume—consuming to learn: Children at the interface between consumption and education. *British Journal of Sociology of Education, 26*(3), 343–357. doi:10.1080/01425690500128882.

McCoy, K. (2003). Good citizenship: Design as a social and political force. In S. Heller & V. Vienne (Eds.), *Citizen designer: Perspectives on design responsibility* (pp. 2–8). New York: Allworth Press.

McDonough, W., & Braungart, M. (2002). *Cradle to cradle: Remaking the way we make things*. New York: Northpoint Press.

McGhee, P. E. (1989). The contribution of humor to children's social development. *Journal of Children in a Contemporary Society, 20*(1–2), 119–134. doi:10.1300/J274v20n01_09.

McGregor, S. L. T. (2010). Politicizing consumer education: Conceptual evolutions. In J. Sandlin & P. McLaren (Eds.), *Critical pedagogies of consumption: Living and learning in the shadow of the shopocalypse* (pp. 122–133). London, Eng: Routledge.

Ministry of Education [MoE]. (2007a). *The Ontario curriculum grades 1–8: Science and technology*. Toronto: Queen's Printer for Ontario. Retrieved from https://www.edu.gov.on.ca/eng/curriculum/elementary/scientec18currb.pdf

Ministry of Education [MoE]. (2007b). *Shaping our schools shaping our future: Environmental education in Ontario schools*. Toronto: Queen's Printer for Ontario.

Molnar, A., Boninger, F., Wilkinson, G., & Fogarty, J. (2010). Schools inundated in a marketing-saturated world. In J. Sandlin & P. McLaren (Eds.), *Critical pedagogies of consumption: Living and learning in the shadow of the shopocalypse* (pp. 83–96). London: Routledge.

Nicholson, M., Last, A., & Widell, N. (2012). Response to 'gaps and silences'. *Critical literacy: Theories and practices, 6*(1), 75–79.

Noddings, N. (2010). Moral education in an age of globalization. *Educational Philosophy and Theory, 42*(4), 390–396. doi:10.1111/j.1469-5812.2008.00487.x.

Olson, D. R. (1989). On the language and authority of textbooks. In S. de Castell, A. Luke, & C. Luke (Eds.), *Language, authority and criticism: Readings on the school textbook* (pp. 233–244). London: Falmer Press.

Orr, D. W. (2004). *Earth in mind: On education, environment, and the human prospect*. Washington, DC: Island Press.

Packard, V. (1960). *The waste makers*. New York: David McKay Co., Inc..

Pavlova, M. (2015). Design and technology education for sustainable futures: In preparation for global citizenship. In K. Stables & S. Keirl (Eds.), *Environment, ethics and cultures: Design and technology education's contribution to sustainable global futures* (pp. 87–99). Rotterdam, NL: Sense Publishers.

Pedretti, E. (2003). Teaching science, technology, society and environment (STSE) education: Preservice teachers' philosophical and pedagogical landscapes. In D. L. Zeidler (Ed.), *The role of moral reasoning on socioscientific issues and discourse in science education* (pp. 219–239). Dordrecht, NL: Kluwer Academic Publishers.

Pedretti, E., & Hodson, D. (1995). From rhetoric to action: Implementing STS education through action research. *Journal of Research in Science Teaching, 32*(5), 463–485.

Pedretti, E., & Nazir, J. (2011). Currents in STSE education: Mapping a complex field, 40 years on. *Science Education, 95*(4), 601–626. doi:10.1002/sce.20435.

Petrina, S. (2000a). The political ecology of design and technology education: An inquiry into methods. *International Journal of Technology and Design Education, 10*(3), 207–237.

Petrina, S. (2000b). The politics of technological literacy. *International Journal of Technology and Design Education, 10*(2), 181–206.

Pilloton, E. (2009). *Design revolution: 100 products that empower people*. New York: Metropolis Books.

Pitt, J., & Heinemeyer, C. (2015). Introducing ideas of a circular economy. In K. Stables & S. Keirl (Eds.), *Environment, ethics and cultures: Design and technology education's contribution to sustainable global futures* (pp. 245–260). Rotterdam: Sense Publishers.

Riley, C. (2003). The cultural influence of brands: In defense of advertising. In S. Heller & V. Vienne (Eds.), *Citizen designer: Perspectives on design responsibility* (pp. 70–81). New York: Allworth Press.

Rose, R. (1991). What is lesson-drawing? *Journal of Public Policy, 11*(1), 3–30.

Saltman, K. J. (2005). *The Edison schools: Corporate schooling and the assault on public education. [Ebook]*. Florence, KY: Routledge.

Sandlin, J., & McLaren, P. (2010). Exploring consumption's pedagogy and envisioning a critical pedagogy of consumption – living and learning in the shadow of the shopocalypse. In J. Sandlin & P. McLaren (Eds.), *Critical pedagogies of consumption: Living and learning in the shadow of the shopocalypse* (pp. 1–19). London: Routledge.

Schnaiberg, A. (1980). *The environment, from surplus to scarcity*. New York: Oxford University Press.

Schor, J. B. (2004). *Born to buy*. New York: Schribner.

Seabrook, J. (1985). *Landscapes of Poverty*. Oxford: Basil Blackwell.

Shaull, R. (2000). Foreword to Pedagogy of the oppressed. In P. Freire (Ed.), *Pedagogy of the oppressed*, pp. 29–34). (trans: Ramos M. B.). New York, NY: Continuum International.

Slade, G. (2006). *Made to break: Technology and obsolescence in America*. Cambridge, MA: Harvard University Press.

Solomon, J., & Aikenhead, G. S. (Eds.). (1994). *STS education: International perspectives on reform*. New York: Teachers College Press.

Sperling, E., Wilkinson, T., & Bencze, L. (2014). We got involved and we got to fix it!: Action-oriented school science. In L. Bencze & S. Alsop (Eds.), *Activist science and technology education* (pp. 365–380). Dordrecht: Springer Press.

Spring, J. (2003). *Educating the consumer citizen*. Mahwah: Lawrence Erlbaum.

Steinberg, S. R. (Ed.). (2011). *Kinderculture: The corporate construction of childhood*. Boulder: Westview Press.

Taylor, E. W. (2008). Transformative learning theory. *New Directions for Adult and Continuing Education, 119*, 5–15. doi:10.1002/ace.301.

Vasquez, V. M. (2014). *Negotiating critical literacies with young children* (10th ed.). New York: Routledge.

Vienne, V. (2003). In a continuous state of becoming: Design responsibility tomorrow. In S. Heller & V. Vienne (Eds.), *Citizen designer: Perspectives on design responsibility* (pp. 244–246). New York: Allworth Press.

Wettergren, Å. (2009). Fun and laughter: Culture jamming and the emotional regime of late capitalism. *Social Movement Studies, 8*(1), 1–15. doi:10.1080/14742830802591119.

Wilkinson, T., & Bencze, L. J. (2015). With head, hand, & heart: Children address ethical issues of design in technology education. In K. Stables & S. Keirl (Eds.), *Environment, ethics and cultures: Design and technology education's contribution to sustainable global futures* (pp. 231–243). Rotterdam, NL: Sense Publishers.

Willis, S. (1991). *A primer for daily life*. [Ebook]. London: Routledge.

Worldmapper. (n.d.). *The world as you've never seen it before*. Retrieved from http://www.worldmapper.org/

Wright, C. (2012). Postcolonial cosmopolitanisms: Towards a global citizenship education based on 'divisive univeralism.' In V. Andreotti & L. M. T. *M. Souza* (Eds.), *Postcolonial perspectives on global citizenship education* [EBook] (pp. 47–67). New York: Routledge.

Yager, R. E. (Ed.). (1996). *Science/technology/society as reform in science education.* Albany: SUNY Press.

Zeidler, D. L., Sadler, T. D., Simmons, M., & Howe, E. (2005). Beyond STS: A research-based framework for socioscientific issues education. *Science Education, 89*(3), 357–377.

Ziman, J. (1980). *Teaching and learning about science and society.* Cambridge: Cambridge University Press.

Zoras, B., & Bencze, L. (2014). Utilizing social media to increase student-led activism on STSE issues. In L. Bencze & S. Alsop (Eds.), *Activist science and technology education* (pp. 365–380). Dordrecht: Springer.

Zuk, B., & Dalton, R. (1998). *Humour and cultural perspectives. Connections '98.* In Proceedings of a faculty conference (4th, Victoria, British Columbia, Canada, May 1998). Retrieved from: http://files.eric.ed.gov/fulltext/ED442613.pdf

Hyper Design Thinking: Critique, Praxis and Reflection

Belinda von Mengersen

Abstract A practice of critique, integrated with design as a 'disciplinary habit of mind' (Klebesadel and Kornetsky 2009, p. 99), sustains and hyperextends students' capacity for design thinking and metacognition. A forward-thinking, design-focused curriculum in design and technology education demands the evolution of such critical dispositions. Reflective thinking and writing practices unite creative and critical analysis with design process, enabling deeper engagement with praxis, metacognition and critique. This chapter observes how these critical, creative and reflective dialogic design-based thinking and writing practices, already employed in design and visual arts education, can augment design and technology curricula. Reflective practice and writing are able to enhance cyclic, critical and design thinking within design and technology curricula through the praxis-based application of critique. Practical methods to stimulate modes of design thinking and communication include critical, creative and reflective thinking and writing. Application of these dialogic methods occurs through opportunities for low-risk exploration through oral and written discourse within a critical and cylic design process. The integration of creative, critical and reflective thinking practices within a design process leads to the sustained reflexive habits and evolving critical dispositions crucial to design and technology education.

Keywords Critical thinking • Design thinking • Creative thinking • Reflective writing • Metacognition • Praxis

1 Introduction

Education researchers Klebesadel and Kornetsky (2009) describe 'critique' as 'a disciplinary habit of mind' (p. 99) within design and visual arts (D&A) education. Similarly, design and technology (D&T) educators engaging with the concept of critique aim for their student groups to develop critical dispositions, together

B. von Mengersen (✉)
National School of Arts, Australian Catholic University, Sydney, NSW, Australia
e-mail: belinda.vonmengersen@acu.edu.au

PJ. Williams, K. Stables (eds.), *Critique in Design and Technology Education*, Contemporary Issues in Technology Education, DOI 10.1007/978-981-10-3106-9_16

with practices of sustained critical engagement. This chapter suggests that the philosophical concept of critique within this book can be practically applied within curriculum through praxis, reflection and augmentation of existing pedagogical practices. The integration of both the theory and practice (praxis) of critique offers an opportunity to enhance existing pedagogies and enable a renewed focus on the development of critical behaviours and dispositions within D&T education environments.

2 Critique in Design and Technology Education

Critique is interpreted here as a philosophically driven approach to design thinking where engagement is enabled through questioning and ongoing enquiry into design. Critical thinking is activated through systematic reflection and analysis: the construction of a series of questions. Critical attributes in D&T include but are not limited to what is traditionally termed 'critical thinking' in a broader educational context. In D&T, the individual, self-directed and cyclic nature of any design process requires the seeking of evidence from a wide range of student activities: close observation, design thinking, critical analysis and the evaluation of contextual issues and constraints. Alongside physical design development, there must also be a coherent discourse and communication of this conceptual design thinking, reflection and analysis and, finally, an ability to apply and adapt these evolving understandings in a practical sense through the construction of prototypes. Reflection is, therefore, an effective practical tool that can augment design thinking.

The term 'critique' in a D&T education context is used to define two subtely different modes of reflective design thinking: firstly, as a verb (Oxford English online dictionary, 2015) meaning to review something critically, to undertake a very close review or where a systematic, analytical assessment may be undertaken during a design process by the designer and, secondly, as a noun (OED, 2015) meaning to review the work of another designer or as is more common in design and visual arts (D&A) education as a form of review of design or artwork for assessment or examination purposes. In D&A, regular dialogic reflection (verbal or written) in relation to design development has been shown to enable critical thinking and enhance metacognition. In terms of the role reflection plays in metacognition within D&T education, Keirl (2005) suggests that 'critiquing aids selection of thinking styles' (p. 10). Thus, sophisticated critiquing is a form of metacognition. It is *both* reflective and deconstructive (echoing Sullivan 2010, pp. 107–108). In D&T the practice of critique (including reflection and praxis) could augment design thinking and support an emerging critical disposition. Similarly, Klebesadel and Kornetsky (2009) employ critique in D&A education as 'a formative mode of feedback' (p. 101), with metacognition as the goal. Here lies further potential for the application of critique in D&T education, as a reflective and formative mode of feedback within design. Design is a highly individualised and non-linear sequence, so tracing evidence of critical engagement can be complex. The concurrent or cyclical nature of this process means that 'critical thinking' and any written, oral or diagrammatic communication could optimally occur in parallel to design development. Situating critique within D&T offers educators the opportunity to encourage the development of attributes including critical thinking, reflective thinking and communication skills that are manifested through design thinking.

3 Design Thinking and Metacognition

In D&T education students are encouraged to develop their design thinking skills through design development. Ideally, the evolution of these skills includes an emerging critical capacity. It is this analytical aspect of design thinking that can be enriched by a theoretical understanding and practical application of critique. The concept of critique could be prioritised within design thinking (or the conceptual side of design development as a core D&T pedagogy) and that this can be done effectively through augmentation of existing pedagogical approaches to design development (including D&T folios, e-portfolios, project work, other documentations including evaluations of design solutions and prototype development). Critique represents the vital nexus between 'making' and 'knowing' in D&T education, and this can be supported and evidenced through simultaneous prototype development and language-based reflection, communication and critique of design.

The term 'design thinking' as a noun can be defined as 'the act or practice of using your mind to consider design' (Ambrose 2010, p. 1). Further, that 'design is an iterative process and design thinking is present in each stage of the journey . . . ' (Ambrose 2010, p. 6). This iterative process includes the following aspects: stages of thinking, research, idea generation, refinement, prototyping and implementation. Whilst Ambrose (2010) and others have described 'stages' of design, in the context of critique, it is essential that the so-called stages are seen as indicative and non-linear. There can be no strict order by which design thinking or a design problem is solved because there is no 'the' design process only 'a' design process that is infinitely variable in accordance to the individual designer and their unique approach to design thinking. Design is an inherently reflective and creative process that encourages metacognition or inspires students to ask questions, take risks in their design experimentation and reflect upon their unique approach. Advanced design thinking assimilates a highly individual analysis of a student's unique design thinking or design development approach, another way to describe metacognition. In the best sense, metacognition organically encourages criticality. Metacognition within D&T education can be activated by a practice of conscious critique, praxis and reflection.

4 Critique: Making or Knowing?

Whilst reflection is incorporated to some extent in D&T education, its potential as a metacognitive tool capable of enhancing the learner's critical thinking, critique, praxis and reflection has perhaps not yet been fully realised. Traditionally, subject choices like D&T and D&A are considered practical rather than theoretical and less dependent on skills that may be perceived as academic, such as writing. As Owen-Jackson (2013) asserts: 'subjects focused on knowledge are perceived as "academic" and those focused on skills as "practical"' (p. 64). Consequently, some students are drawn to these areas in avoidance of formal academic writing assessment tasks, as Childers et al. (1998), Orr et al. (2004) and Owen-Jackson (2013) have identified. Orr et al. (2004) also go on to say that in their field (design and visual arts education research), 'the role of writing [has been] questioned' (p. 75). There is a viewpoint that through the artefact, the student communicates without the use of verbal language, using a visual language that is 'wordless'. However, the reality – shared by D&A and D&T – is that the curricula require students to 'create, design and write' (Orr et al. 2004, p. 75). Furthermore, of crucial importance is the knowledge that curricula 'are assessed via the textual and the visual' (Orr and Blythman 2002, p. 1).

In a recent publication, *Debates in Design and Technology Education*, Martin and Owen-Jackson (Owen-Jackson 2013) pose the question: 'Is design and technology about making or knowing?' (p. 64). Exploring the dichotomy within an England-specific context, Martin et al. (Owen-Jackson 2013) describe how it continues to exist within D&T education. They contend that such a separation is counter-productive, instead suggesting, 'it is the combined effect of both [making and knowing] working together that supports successful designing and making' (Owen-Jackson 2013, p. 71). They conclude by reinforcing the inherently 'dynamic' nature of the subject (Owen-Jackson 2013, p. 71). John Wood (2008), design theorist and educator, echoes this sentiment by proposing that 'designers can reunite "saying" and "showing"' (p. 304). The notion of critique, as explored in this book, relates to both theory (knowing) and practice (making). It offers an approach which is both philosophical and practically applicable, mirroring the very nature of design thinking and thus circumventing the perceived dichotomy between making and knowing that so concerned Martin and Owen-Jackson (2013).

5 Defining Critique in Design and Technology Education

Critique is not limited to or defined by one aspect of learning in D&T. Instead, it affords a broad-reaching, philosophically informed means for educators to embed more opportunities for the emergence of critical dispositions within all aspects of the learning. One affordance offered by an engagement with critique is a shift away from the traditional dichotomy between 'making' and 'knowing' or 'skills' and 'knowledge'. For the philosophy of critique, it necessarily applies to both: it is about an exchange between the two – or the 'praxis'. It can, therefore, facilitate a shift towards a more sound interpretation of the nexus between theory and practice in D&T education.

6 Modes of Reflection in D&T Education

Whilst the concept of reflection is well known to educators as it is found in their own practice – discussed in chapter "Critiquing Teaching: Developing Critique Through Critical Reflection and Reflexive Practice" by Susan McLaren – it is less commonly identified and applied as a highly beneficial element in *student* learning. Moon (2013) defines the difference between these two modes as (a) reflection in learning and (b) reflection in professional development and considers 'reflection' to be 'a technique for aiding and reinforcing learning, used in education and professional development' (p. vii). This discussion examines the benefits of requiring students to engage in a practice of written or verbal reflection, and it will also specifically address the notion of 'critique': the rich learning and linkages which occur when both teacher and student are simultaneously undertaking reflective practice. For the purposes of this discussion, the terms 'reflective practice for educators', 'reflective practice for students' and 'critique' have been used to distinguish between the three key concepts. Thus, the term 'reflection' is used in this chapter as an umbrella term for both (a) reflective verbal (oral) or (b) written language-based communication methods (Sullivan 2010, p. 100) and also techniques like critique, or 'crit' in a D&A context, language-based reflection (Sullivan 2010, pp. 110–111), non-linear (Wood 2004) and reflective writing methods. It suggests that many of these critic-reflective verbal or written communication methods can support and augment design thinking and could therefore be adapted to D&T education.

7 Praxis in Education

The term praxis, in an educational context, has been defined as:

> An active, continuous process of critical action and reflection upon accepted knowledge, experiences, and perceptions of reality in order to transform reality. (Collins & O'Brien 2011, p. 363)

In their definitions Collins and O'Brien (2011) cite Freire who considered the notion to be cyclical when used in a practical pedagogy, where: 'The process involves a cycle of reflection and action based on that reflection, followed by further reflection' (pp. 363–364).

Critique is related to praxis through its definition of a 'nexus' between theory and practice in D&A and by its philosophical and reflective nature, including the ethical attributes alluded to by Freire and others. Freire described how reflective praxis activates a more 'critical consciousness' which he called the process of 'conscientisation'. He suggests that this occurs:

> when adults are ... critically discussing an issue, acting on that discussion, and then reflecting on that action before moving to act again. (Freire 1970, cited by Collins and O'Brien 2011, p. 96)

The concept of praxis is innately related to practices of critique and reflection within design thinking. Praxis can supplement cognition of critique within a D&T context by illuminating the operation of a dynamic intersection between theory and practice. Praxis is supported by the cyclical reflection that mirrors design thinking, suggesting that reflection is the crucial conduit for both praxis and critique. The application of these terms, and their capacity to inform and complement each other, offers D&T educators an energetic critical methodology innately suited to design (and design thinking): a practical and adaptable support structure for thinking and learning more comprehensively.

8 The Link Between Praxis and Reflection

In D&A education, reflection in the shape of reflective thinking and reflective writing and verbal communication in the form of 'crits' have been utilised to encourage and support praxis. Reflective thinking and writing have been used to support student understanding of praxis or the interrelationship between theory and practice within their own design thinking. In D&A education praxis is described as the dynamic relationship between concept (theory or ideas behind making) and process (practice or the act of making and working with physical materials). The perceived dichotomy between 'making' and 'knowing' present in D&T is echoed in the discipline of D&A education.

> rather than constructing a false and arbitrary dichotomy between knowing and doing, knowledge and action, theory and practice, they rather sought to find a braiding and to further explore issues of reflective practice. (Orr et al. 2010, p. 199)

One of the most significant insights that has emerged through this research into the formative role of reflective writing in D&A education[1] is the auxiliary role of reflective practice in both the maintenance of praxis and the emergence

[1]This research was initiated at Goldsmiths college, University of London, UK, through a research project called 'Writing Pad' www.writing-pad.ac.uk/; this led to the development of Writing Pad (2007). *Journal of Writing in Creative Practice* (Intellect, UK). The Writing Pad project included an extensive list of international partner institutions. Another significant research has been published by LTSN Subject Centre for Art, Design & Communication (2002) (*Art, Design & Communication in Higher Education (Online) Art, Design & Communication in Higher Education*) (Intellect, UK), including a special guest edited a two-part edition (2004) *Textual and Visual Interfaces in Art and Design Education* and an *International Centre for Learning and Teaching in Art and Design* (CLTAD), University of the Arts, London conference including (2010) *Creative Parternships: helping creative writing and visual practice students to make*

of a critical disposition. Reflective practice in D&A occurs through a wide range of different modes including oral and written modes and individual and/or group communication. Reflection in its many different modes is an active process that can hold up a mirror to or actively enable critique to occur alongside design thinking within a process of design. Here lies its potential within D&T education in support of critique. Lockheart (2010) describes this potential in relation to reflective writing, in which:

> one purpose ... is discovery: learning whilst doing, as opposed to writing up when the learning is complete... highlight (ing) that this not only develops writing, but also reading across different modes of doing and thinking. Indeed their article refuses the separation of text and artifact, and suggests that imaginative multimodal approaches to learning are the only way to continue to serve the truly reflective practitioner. (pp. 194–195)

The Australian Curriculum, Assessment and Reporting Authority (ACARA 2012) defines a multimodal text as a 'combination of two or more communication modes (for example, print, image and spoken text, as in film or computer presentations' (p. 13). In a D&A education context, multimodal refers to the use of reflective writing alongside or within a D&A sketchbook: another 'signature' pedagogy that operates alongside critique or 'crit' sessions within D&A (Sims and Shreeve 2012, pp. 62–63). In a D&T education context, this 'multimodal' approach could be applied through a wide variety of different analogue or digital communication methods including interactive platforms like e-portfolios (pioneering work in e-portoflios has been done by the Technology Education Research Unit (TERU), Goldsmiths college, University of London, UK, through their 'unpickled' and 'e-scape' research and development programmes and extensive development work in the use and design of e-portfolios New Zealand (for an overview, see Kimbell 2012, Kimbell and Stables 2008, Stables and Kimbell 2000, Williams 2012; Williams and Newhouse 2013, Edwards 2015).[2] It is this understanding of how these practices of reflection and communication that focus on praxis and operate in relation to design and design thinking that offer clear examples of how D&T educators might augment their pedagogies in similar ways.

9 Critique Within a Design and Visual Arts Context

It is not solely in a D&T context that the terminology being discussed is useful. In design and visual arts (D&A) education, praxis has become a normative way to describe the complex interrelationship between theory and practice. Significantly,

---

links between their creative processes and their personal, vocational and academic development (DOI:10.1386/jwcp.3.3.285_1).

[2]For more details on e-scape, see Williams (2012) Eds. Special Issue on e-scape in Design Technology Association (1990). *International Journal of Technology and Design Education (Online)International Journal of Technology and Design Education*, May 2012, Volume 22, Issue 2.

it has been used where the practice of design or visual arts is defined as a form of research (Barrett 2010; Sullivan 2010; Carter 2004). Arguably, all design is a form of research in the sense that it requires speculation, testing, analysis, evaluation and reflection. If we consider 'design' to be the primary focus of D&T education and then consider the number of different practical and theoretical tasks that need to be undertaken to ensure a well-developed design concept and prototype, it is clear that a constant exchange between theory and practice is required. Critique offers us a way to understand and influence that exchange and to support praxis. Some examples of approaches developed within D&A that could be applied to or adapted for D&T are considered within this chapter.

10 Critical Thinking as a *Defining Concept* of the University

Critical thinking has been described as a 'defining' (Barnett 1997) concept of the university, yet the skills for the comprehension and application of critical thinking may easily be assumed by educators within their pedagogical practices and curriculum design in all fields. Adaptation of critique seems to offer D&T educators the opportunity to teach and reinforce the importance of critical thinking, thus enabling the development of essential critical dispositions in students. And this focus on critical thinking is not limited to higher education. For instance, the Australian Curriculum, Assessment and Reporting Authority (ACARA 2012), has recently published the new Australian national curriculum for design and technologies, 'The Shape of the Australian Curriculum: Technologies'. This policy document clearly outlines critical thinking as a key dispositional attribute within D&T education (ACARA 2012) at both junior and senior levels. Here, creative and critical thinking is placed side by side and their roles are described together: 'Critical and creative thinking underpin learning in Technologies . . . ' (ACARA 2012, pp. 26–27)[3]. What is significant here is the brevity of the description, presupposing comprehension of the term by educators. Education researcher Jennifer Moon (2008) reminds us that critical thinking, however, is 'not often explicitly taken into consideration in pedagogy', and yet it is usually very evident in the 'rhetoric of education, particularly higher education' (p. vii). This, at least within D&T education, is why critique offers us an opportunity to augment existing pedagogies, together with a specific, design-driven critical thinking vocabulary and practice.

[3]For the full quote, please refer to http://www.acara.edu.au/verve/_resources/Shape_of_the_Australian_Curriculum_-_Technologies_-_August_2012.pdf, p.26–27.

Table 1 List of graduate attributes related to critical thinking

Australian HE institution	Hierarchical placement	Graduate attribute
University of Technology, Sydney	Listed as the (1) first graduate attribute under 'intellectual (practice oriented)'	(1) *Critical* and independent thinking
University of New South Wales	Listed as (a) first graduate attribute	(a) Scholars capable of independent and collaborative enquiry, rigorous in their analysis, *critique* and reflection, and able to innovate by applying their knowledge and skills to the solution of novel as well as routine problems
Australian Catholic University	Listed as no. (4) nb Ethical graduate attributes are listed above which are uncommon in the Australian HE context due to the specific ethical focus of this national institution	(4) Think *critically* and reflectively

When Moon (2008) conducted an extensive literature review on the significance of critical thinking in higher education, she concluded that:

> Critical thinking is considered to be central to higher levels of education or a fundamental goal of learning (citing Kuhn 1999; Kelley and Shemberg 1995) (p. 6).

Moon (2008, p. 7) suggests higher education in particular has assumed the term with 'missionary zeal' as a graduate 'objective' and cites many examples from higher education institutions in the UK and the USA. Such enthusiasm is quickly evidenced in an Australian higher education context through the hierarchical arrangement of graduate attributes (Table 1).

The University of Technology (Sydney, Australia) also lists critical thinking and creativity in their grade descriptors when defining what constitutes a 'high distinction' for assessment purposes. The two pertinent points in this descriptor that distinguish a high distinction from other grades that may be awarded are:

(a) 'By means of criticism' [or evidence of a disposition of critique].
(b) 'This grade may also be given to recognise particular originality or creativity'[4].

It is interesting to note in this context that the term critical thinking is also often situated within education policy documents alongside another ambiguous term: 'creativity', just as it is in the new ACARA D&T curriculum (2012) cited above. A detailed examination of the relationship between critique, critical thinking and creativity lies beyond the scope of this chapter but would be well worth investigating further, especially in light of the success of more creative reflective writing methods

[4]For the full descriptor, please refer to http://www.gsu.uts.edu.au/rules/student/schedule-2.html.

used in D&A (von Mengersen 2013, 2015). Padget (2013, pp. 2–3) has explored the cognitive relationship between creativity and critical thinking, describing it as 'symbyotic' and noting how creativity, critical thinking and reflection actively cross reference in an engaged learning environment. The brief analysis of evidence from an Australian D&T education context, above, supports the findings of educational researchers like Moon (2008) and Barnett (1997), highlighting the inconsistency between aspirational educational rhetoric and the reality of assumption, the lack of explicit pedagogy and the need for greater clarity (and discipline specificity) of an enigmatic yet vital term. Here, within D&T education, there is perhaps an opportunity to differentiate between 'critical thinking' and 'critique', to redefine these terms from a discipline-specific perspective and to reconfigure practical learning activities that support the role of critique in design thinking.

11 Critique in the Australian Design and Technology Education Context

In Australia, critique emerged during the rewriting of the South Australian design and technology curriculum in 2001 – discussed in chapter "Critiquing as Design and Technology Curriculum Journey: History, Theory, Politics and Potential" by Steve Keirl. At that time it became an integral third strand in a non-hierarchical list within the policy including critiquing, designing and making. The relationship is considered further:

> These three strands *are interdependent and none of them is predominant*. Read alongside each other *they do not constitute a sequential process*. They interrelate to support rich understandings. A quality Design and Technology education weaves the three into a dynamic and holistic learning experience *for all students*. (Keirl 2001, South Australia Department of Education, Training and Employment (DETE), as cited in Chapter 7 by Kierl)

The inclusion of critique can be seen here as supporting a more holistic, dynamic and inclusive learning experience. Critique is about reflection as a component of critical design thinking, involving thinking across theory and practice (praxis). Ideally, the inclusion of all three strands in this non-hierarchical way allows for different learning styles and the design of flexible curriculum through critiquing. It should be delivered and assessed through alternative verbal or written modes to support inclusive and individual learning for students at any stage or age (Broughman and Hunt 2013). In D&T, critique is used in this way to offer valuable formative feedback and 'low-risk writing opportunities' (Broughman and Hunt 2013, p. 188) for students undergoing project development. This enables the building of skills, vocabulary and comprehension: critique as evident in the complex exchange between designing, thinking and making.

12 Critique as a 'Signature' Pedagogical Practice in Design and Art

Klebesadel and Kornetsky (2009) describe critique, or the 'crit', as a 'signature pedagogy in Art and Design' (p. iii), which operates as 'a formative mode of feedback' (p. 101). As a signature pedagogical practice, critique is seen as 'essential to developing a self-critical habit of mind, inculcating current value systems in art and design and enabling students to position themselves within professional practice' (Sims and Shreeve 2012, p. 61). Within D&A education contexts, critique is often used to describe a show-and-tell session where students and educators meet together to discuss the evolution of design ideas, show prototypes or illustrations and participate in a dialogic 'critique' through analysis and constructive criticism. 'Crit' sessions vary depending upon the emphasis of the educator and may focus on 'evaluating works in progress or completed works' (Sims and Shreeve 2012, p. 61) and on teacher feedback or peer feedback and discussion. These sessions are understood to be formative feedback opportunities which enhance metadesign (Wood 2008) thinking and metacognition through oral communication and subsequent reflection. This is an example of how a metadesign (Wood 2008) approach and a 're-languaging' (Wood 2013) of design towards a more communicative or 'dialogic' (Bain 2012) approach could augment existing design pedagogies within D&T education.

13 Metadesign and Re-languaging

Lockheart and Wood (Lockheart and Raein 2012), founders and editors of the Intellect journal *Writing in Creative Practice*, conclude, after 10 years of research in D&A education, that language and literacy remain the keys to reflection and reflective writing practices that augment design thinking. They have described their ongoing interest in the 'designerly' (Cross 1982) use of language and the role of writing for designers and how it can inform students about their own practice through creative and critical thinking. They assert the importance of 'languaging' (Lockheart and Raein 2012, p. 285) in the evolution of a design research process. Wood has introduced two new terms into the vernacular of design thinking that call for a more critical, dialogic, reflective, ethical, creative and multidimensional approach: 'metadesign' (2008) and the 're-languaging' [of design] (2013). These terms are useful for D&T education in relation to critique because they present design and design thinking as a non-linear approach informed by critique, praxis and reflection.

Wood has developed an approach to design practice called 'metadesign' (Wood 2008, 2011, 2013;) which encourages a critico-ethical approach to design and looks at how design operates within complex systems. According to Wood (2004, p. 175), metadesign 'requires' ethical attributes. Wood (2008) considers metadesign to focus on 'an outcome-centred mode of reasoning' which 'advocates a comprehensive superset of practices... in which "design as planning" would be replaced by "design as seeding process"' (p. 307). Wood aimed to (a) encourage designers and design students to take a more reflective, ethical and sustainable approach to design thinking, (b) to encourage designers and design students to ask more questions and be more critical of the socio-economic contexts in which they are working and (c) to thus encourage metadesign and metacognition (including critique and reflexivity).[5] Von Mengersen (2013, 2015) has identified timing and vocabulary as two key factors that enhance the use of reflection as a pedagogical practice within D&A education. Therefore, in seeking to equip students with the skills for critique, praxis and reflection, it seems necessary to provide not only formative assessment opportunities but also the appropriate metalanguage for expression and understanding. Lockheart and Wood's focus on 'languaging' (Lockheart and Raein 2012, p. 285) that encompasses key terminologies that students can use to demonstrate higher-order thinking in relation to their design process, like Wood's 're-languaging' (2013).

With this second term 're-languaging', Wood (2013) suggests a re-examination of the very nature of designing:

> the re-designing of design to be more relational and combinatorial, so that its primary focus is more associated with the co-creative relations between things, rather than on individual products or services... metadesign is intended to help designers to re-think the habits, assumptions and discourses that seem 'normal'. (p. 59)

In D&T education, just as in other disciplines, students' skills and vocabulary for the practice of critique, praxis and reflection can easily be assumed, embedded as they are in tasks like the selection and application of cognitive organisers. Reflective communication practices (verbal or written) for critique developed in D&A education suggest that the practice of reflective forms of communication including critique should be regular, cyclic, formative and intrinsic to every stage of design thinking and, further, that specific vocabulary for reflection, praxis and critique needs to be developed. It is evident that both fields (D&T and D&A) are concurrently seeking to move beyond the dichotomy of 'making' or 'knowing' towards refined literacy terminologies and communicative community-based or 'dialogic' (Bain 2012) modes of design development and assessment. A discussion of assessment practices lies beyond the scope of this chapter but is another important aspect that could be analysed in relation to these findings.

Many design researchers are calling for a more 'dialogic' (Bain 2012), 'autodidactic' (Wood 2013), 'critico-ethical' (Wood 2008), 'combinative' and 'co-creative' (Wood 2013) approach or a language-driven, community-of-practice approach to design thinking and practice. Authentic reflective practices are vital, documenting a shift in student perspective and awareness – Harfield (2012) describes this as 'transformative learning' and Atkinson (2012) as 'the intangible designerly thinking' or 'tacit design intelligence' where creative problem solving occurs;

[5]For more information, refer to the Metadesigners network: http://metadesigners.org/HomePage

Bain (2012) suggests it opens up a more 'dialogic' assessment practice which allows students to take more risks in design development. Significantly, all of these skills in terms of the opening up of student perception in design thinking have also have been linked to what these researchers call the capacity for 'lifelong learning' or 'self-learning' (Harfield, 2012) or 'autonomous learning' (Bain 2012).

Critical literacy in this context thus appears to be crucial to the meaningfulness of critique, reflection and praxis for D&T education. Keirl (2005) suggests that critique is expressly linked to 'more powerful meaning-making opportunities for students' learnings about technologies' (p. 1). If it is the students' vocabularies (through verbal or written modes) that make their cognition and meaning-making visible to us, the educators, then arguably, we enrich and augment the pedagogy in this area of D&T education through critique. Padget describes his 'belief in the primacy of language in the learning process and how this links with creative learning and teaching and critical thinking' (Padget 2013, p. xi) – therefore, highlighting the point that without precise, specific metalanguage with which to articulate their learning, students' efforts cannot be either fully expressed or adequately measured. In reality, and for the purpose of summative assessment particularly, our students must demonstrate and communicate what they know through at least one language-based mode alongside artefacts or prototypes. Also, see the discussion in chapter "Modelling as a Form of Critique" by Niall Seery of 'Modelling as a Form of Critique', in particular for the discussion of cognitive and related physical manifestations which support an external and internal dialectic.

Writing is an assumed mode of communication in many D&T assessment models, including written exams and digital or analogue design portfolios; therefore, a focus on vocabulary development for critique, reflection and praxis is logical. For Wood (2013), 're-languaging' is an active form of critique that focuses on a dialogic community of practice enabling the evolution of individual and group thinking and thus metacognition, through reflection and language usage. Wood has focused on developing a 'culture' of critique that augments existing design paradigms and pedagogies, expanding well beyond many of the commonly stated 'stages' of design thinking. Wood's 'metadesign' (2008) and 're-languaging' (2013) methodologies offer scaffolds for redesigning design thinking and learning within D&T education.

14 Critique, Praxis and Reflection in Design and Visual Arts (D&A) Education

Design (and visual arts) education shares a history of using critique, particularly in relation to the term *praxis*. Sullivan (2010) describes how the actions of create-critique work in association within the visual arts. He analyses different versions of the theory/practice relationship or what he terms 'dimensions of practice between theory' (Sullivan 2010, p. 106). This theory relates to what he has called the create-critique dynamic, in which:

> visual arts practice and critical components are linked as theoretical issues and investigated through creating and critiquing; [and, where] theoretical interests are investigated through a cycle of processes involving issues and context. (Sullivan 2010, p. 106)

Pointedly, Sullivan (2010) then goes on to describe how this create-critique dynamic is dependent upon 'language-based methods' (p. 107). He describes three different 'domains of practice around inquiry' including:

(a) Discursive – in which visual forms are developed
(b) Dialectical – the use of language-based methods to assess the adequacy of arguments, claims and actions (in the studio, visual forms of language such as metaphor and analogy are used to challenge and change things)
(c) Deconstruction – methods that critique areas of emphasis and omission in systems and structures (in studio contexts, visual and verbal methods are used to embody meanings that explain how things are and how they might be)

Orr and Blythman (2002) proposed that the practice of reflective writing parallels the practice of designing and suggested that reflective writing (or other modes of language-based communication) can operate effectively within the practice-theory loop (praxis). In this context, they suggest that writing for reflection can enable more effective learning when: (a) it is being used regularly to build up skills and confidence, (b) it is linked to both formative and summative assessment tasks (primary and secondary reflective practice) and Francis (2009, p. 36) adds (c) when it is clearly linked to [a students'] personal endeavour. In D&A education many researchers and educators have experimented with more creative verbal and written modes of reflection that can operate alongside or in parallel to the 'crit'. Here, students are encouraged to speculate upon what questions may arise during their 'crit' and then later reflect upon those anticipated and unanticipated questions that did arise. Instead of formal writing methods, these educators use informal ones including creative and non-linear or multimodal writing techniques, developed at a formative stage. Others have developed visual or diagrammatic methods. Creative, reflective writing models can enhance learning journals and other reflective writing practices. Potentially, they may support reflective writing practices within technology by offering more approachable, logical and arguably less-formal methods for *thinking* through writing.

One of the crucial factors in both D&T and D&A education is the capacity to think speculatively so that reflection here is not limited solely to a reflection of that which has past but towards a future thinking and speculative nature: a critique of possibilities. Sims and Shreeve (2012, p. 57) describe how in D&A education the conversations between student and educator 'often deal with indeterminate and unknown creative outcomes' or what (cited by Sims and Shreeve 2012) described as 'pedagogies of ambiguity' and (cited by Sims and Shreeve 2012) as uncertainty. Reflection, praxis and critique can augment D&T education in support of this necessary creative, questioning and speculative mode of design thinking. Akin to Keirl's (2005) description, 'critiquing is about questioning rather than answering' (p. 8) and also related to Atkin's term 'anticipatory thinking' (2012). Indeed, it is this aspect of critique that may be one of the most important in terms of D&T education and metacognition, whereby the emergence of a critical disposition signals the capacity for both autonomous learning (Bain 2012), another poignant aspirational graduate attribute that often sits alongside 'critical thinking'.

15 The Tetrahedron: Non-linear Writing Models for Design

In the 'tetrahedron' model of design, Wood (2004) describes a non-linear method of creative writing that he suggests is 'a productive way to explore and guide the practice of design' which can be 'applied to help designers become more self-reflexive' (p. 175). The method does not subscribe to normative academic writing conventions. Instead, it encourages writing as a critical practice that is more 'purpose defined, outcome-centred, reader-empathic, and self-reflexive' (Wood 2004, p. 175). In relation to critique, Wood (2004, p. 175) offers us a timely reminder of the 'autodidactic' and thus metacognitive potential of the writing process. D&A educators have described the reflective, metacognitive and critical benefits of verbal and/or written communication methods in their discipline. It is these communication methods that can be adapted for use in D&T education curricula, in particular where the focus is on self-directed design development. Wood (2004) describes a practice of non-linear writing, effective for use alongside design, to support critical design thinking, suggesting that this approach supports praxis and 'self-reflexivity'. Wood (2004) is optimistic about what he describes as the most important contextual issue in his research: 'the familiar tension between what we clumsily polarize as "practice" and "theory"'(p. 179). He outlines historical reasons for the development of this dichotomy, particularly in a university context. He suggests that because this dichotomy still exists, 'the cultures of doing or making and thinking or writing have yet to be integrated in an optimal way' (Wood 2004, p. 179). The terms that Wood has created such as metadesign and the tetrahedron model of writing for design and re-languaging suggest methods for shifting beyond this dichotomy towards increasing praxis and reflection in design thinking.

This optimism and these methods also offer opportunity for the development of critique within D&T, as Wood (2004) reflects: 'fortunately... positions are merging slowly into a more reflective discourse of praxis. Art and design education has been important in pointing the way to a healthy fusion of these two systems of thought and action' (p. 179). Evidence of this fusion, particularly in terms of the relationship between critique, praxis and reflection, can be seen in D&A education through the work of Wood (2004), Moon (2006), Francis (2009) and others who have developed a variety of non-linear design thinking exercises based on dialogic language, creative writing and multimodal methods which encourage students to find methods that suit their own learning. Moon's (2006, pp. 26–35) research looks at how these tasks can be used to encourage more consistent use of learning journals among students of all age groups and in a range of disciplines. She describes how students' learning occurs through the regular use of learning journals alongside project work, suggesting that such work offers many benefits including slowing the pace of learning [increasing time for reflection], increasing the sense of ownership of learning, acknowledging the role of emotion in learning, giving learners an experience of dealing with ill-structured material of learning [where design is a pedagogy of ambiguity and/or uncertainty], encouraging metacognition and enhancing learning through the [autodidactic] process of writing.

Francis (2009) and Moon (2006) have explored tasks that focus on non-linear creative design development and critical design thinking – and it is these that could also be used to augment design development pedagogies within D&T education. The following (see Table 2) have been identified as useful non-linear writing tasks within a studio-writing research project (von Mengersen, 2010–2015). These have

Table 2 List of creative design thinking and dialogic writing tasks

Task	Reference/s	Summary of approach
Use questions	Moon (2006), p. 142; Johns (1994)	'Questions help learners to get started in reflecting or to deepen their reflection' (Moon 2006, p. 142)
		Develop a set of programme-specific learning journal (or reflective) questions involving the writer (Johns 1994, p. 71–5)
Generate questions	Moon (2006), p. 142 citing Hahnemann (1986)	'An intermediate stage between the use of preposed questions and unstructured writing is to ask learners to develop their own questions' (Moon 2006, p. 142)
		Pat Francis calls these reflectionnaires (Francis 2009, p. 51)
Concept mapping or graphic representation of ideas	Moon (2006)	'A concept map encapsulates an idea and the themes radiate from the main idea and subdivide hierarchically' (Moon 2006, p. 143)
Writing lists	Francis (2009)	Lists are a 'way of limbering up and helping develop associations' (Francis 2009, p. 105)
		Non-hierarchical lists including a spiral (Francis 2009, p. 96), the plait (Francis 2009, p. 97), the daisy metaphor (Francis 2009, p. 98–99)
		Vocabulary extension including mnemonics (Francis 2009, p. 127), creating words (Francis 2009, p. 130–131), repetition (Francis 2009, p. 135), nouns and verbs (Francis 2009, p. 138–139)
Free-flowing or stream-of-consciousness writing	Moon (2006) citing Elbow (1973)	This process can be used as a less-formal warm-up (Moon 2006, p. 143)
Take a sentence	Moon (2006) citing Hahnemann (1986)	'Hahnemann (1986) asks her students to "take one sentence . . . and write on its meaning"' (Moon 2006, p. 144)

(continued)

Table 2 (continued)

Task	Reference/s	Summary of approach
Draw or map a research process	Moon (2006)/visualising research (Gray and Malins 2004)	'Draw your project' (Moon 2006, p. 151)
		'Undertaking a contextual review: mapping the terrain', visual models of mapping research (Gray and Malins 2004, p. 48–64)
Draw an image	Moon (2006, p.150)	'Progoff uses the drawing of images to facilitate reflection or to summarise a session of reflection' (Moon 2006, p. 150)
Write a poem	Moon (2006)/create textual ontology (Hall 2012)	'The writing of poetry can enable the emotional content of a topic to be more freely expressed' (Moon 2006, p. 157)
		'A poetics of textual practice may . . . encourage student designers to put more meaning into their writing by making challenges to form' (Hall 2012, p. 365–366)

been integrated into a fully online class for cross-disciplinary design and visual arts students. These tasks have been tested in this context specifically in terms of how they assist individual design development and design thinking where students are working on self-directed projects. All of these have been found to increase the students' capacity for critique, praxis and reflection and have been tested over a period of 6 years within a D&A higher education context.

These new terms – metadesign and the tetrahedron model of writing for design and re-languaging (Wood) – all rely upon creative, non-linear design thinking and 'languaging' methods. Moon, Wood, von Mengersen and others suggest that it is these non-linear dialogic methods that teach students to expose the connection between critique, praxis and reflection in design thinking and practice. The inherent feeling of optimism suggested by Wood (and others) in response to their success with D&A education students offers a clear proposal as to how D&T educational pedagogies could be augmented by thoughtful inclusion of critique, praxis and reflection. In this way we might begin to bridge the gap in a dichotomy between 'making' and 'knowing' and move towards an emerging disposition of critique which includes more meaningful and dialogic design thinking and critical comprehension, communicated through multimodal languaging methods to mirror the uniquely non-linear nature of design practice.

16 Conclusion

The nature of critique is ongoing inquiry, where the construction of dialogic questions is primary. This may even be the very definition of what determines a critical disposition within D&T education. This kind of philosophical approach to critique can augment existing D&T pedagogical practices and underpin the integration of praxis and reflection. Praxis offers a more evolved approach to the dated dichotomy between 'making' and 'knowing' that has traditionally existed within D&T education. Reflection can be used to describe a wide range of language-based verbal or written modes of communication. It seems logical for critiquc to evolve into a 'signature' pedagogical practice which offers formative feedback opportunities, with the broader aim of encouraging higher-order thinking and lifelong learning, and above all seeks the emergence of a disposition of inquiry. Dialogic reflection can augment design thinking and practice by enabling praxis and critique, thus encouraging the emergence of metacognitive processes through expanding vocabulary, technological literacy and evolving critical insight. This chapter suggests that some of the key factors enabling this application and evolution of critique to augment design thinking and existing D&T pedagogies are praxis and reflection. To reinforce these, it also suggests that the timing of formative feedback opportunities, a focus on conscious vocabulary development for critique (equipping students with the language literacy and terminology for accurate, more meaningful engagement and self-awareness), and the use of more creative and less declarative modes of communication appear to be significant. A focus on critique aims to embed a culture of sustained questioning that ideally should reside at the heart of any learning environment or endeavour.

References

ACARA. (2012). *Australian curriculum and reporting authority. The shape of the australian curriculum: Technologies*. Retrieved from http://www.acara.edu.au/verve/_resources/Shape_of_the_Australian_Curriculum_-_Technologies_-_August_2012.pdf

Atkinson, S. (2012). What constitutes good learning in technology education: How can we ensure that technology education graduates can provide it? *Explorations of best practice in Technology, Design and Engineering Education, 1*, 1–12.

Ambrose, G. (2010). *Design thinking*. Lausanne: AVA Publishing.

Bain, J. (2012). Negotiating the vacuum: construction and applying assessment criteria to focus design learning. *Explorations of best practice in Technology, Design and Engineering Education, 1*, 13–24.

Barnett, R. (1997). *Higher education: A critical business*. Buckingham/Bristol: The Society for Research into Higher Education & Open University Press.

Barrett, E. (2010). *Practice as research: Approaches to creative arts enquiry* (Paperback ed.). London: I.b. Tauris.

Broughman, C. & Hunt, L. (2013). Inclusive teaching, Chapter 11. In Hunt, L. (2013). *University teaching in focus a learning-centred approach* (pp. 182–198). London: Routledge.

Carter, P. (2004). *Material thinking: The theory and practice of creative research.* Carlton: Melbourne University Press.

Childers, P., Hobson, E., & Mullin, J. (1998). *ARTiculating: Teaching writing in a visual world.* Portsmouth: Institute of Education Sciences.

Collins, J. W., & O'Brien, N. P. (2011). *The Greenwood dictionary of education* (2nd ed.). Santa Barbara: ABC-CLIO.

Critique. (v, n). (2015). *OED online*. London: Oxford University Press. Retreived 4 March 2016, from http://www.oed.com/

Cross, N. (1982). Designerly ways of knowing. *Design Studies, 3*(4), 221–227.

Edwards, R. (2015). E-portfolio environment design principles in practice: A case study of a collaborative project in technology teacher education. *Australasian Journal of Technology Education, 2*(1).

Elbow, P. (1973). Appendix essay. The doubting game and the believing game: An analysis of the intellectual process. In *Writing without teachers* (pp. 147–191). Oxford: Oxford University Press.

Francis, P. (2009). *Inspiring writing in art and design.* Bristol: Intellect.

Freire, P. (1970). *Pedagogy of the oppressed.* New York: Seabury Press.

Gray, C., & Malins, J. (2004). *Visualizing research: A guide to the research process in art and design.* Surrey: Ashgate.

Hahnemann, B. K. (1986). Journal writing: A key to promoting critical thinking in nursing students. *Journal of Nursing Education, 25*(5), 213–215.

Hall, S. (2012). Designing writing/designing reading: Textual ontologies and poetic practice. *Journal of Writing in Creative Practice, 5*(3), 365–385. doi:10.1386/jwcp.5.3.365_1.

Johns, C. (1994). Nuances of recollection. In *Journal of Clinical Nursing 3*, 71–5.

Keirl, S. (2001). *Design and technology and the five 'essential learnings' of a new curriculum framework.* Doctoral dissertation, Loughborough University.

Keirl, S. (2005). *Critiquing, designing and making in the middle years in design and technology education – A commentary on the interplay.* In Proceedings of the national biennial conference of the Australian curriculum studies association: Blurring the boundaries, Sharpening the Focus.

Kimbell, R. (2012). The origins and underpinning principles of e-scape. *International Journal of Technology and Design Education, 22*(2), 123–134.

Kimbell, Richard, & Stables, Kay. (2008). *Researching design learning: Issues and findings from two decades of research and development.* (Science & technology education library). Guildford/Secaucus: Springer London Springer distributor.

Klebesadel, H. & Kornetsky, L. (2009). Critique as signature pedagogy in the arts, chapter 6. In Chick, N. (2009). *Exploring signature pedagogies: Approaches to teaching disciplinary habits of mind* (pp. 99–120) Sterling: Stylus Publishing.

Lockheart, J. (2010). Challenging the curriculum: Exploring the discipline boundaries in art, design and media. *Journal Of Writing In Creative Practice, 3*(3), 193–196. doi:10.1386/jwcp.3.3.193_2.

Lockheart, J., & Raein, M. (2012). No one expects the design inquisition: Searching for a metaphorical solution for thinking, researching and writing through design. *Journal Of Writing In Creative Practice, 5*(2), 275–289. doi:10.1386/jwcp.5.2.275_1.

Moon, J. (2006). *Learning journals: A handbook for reflective practice and professional development* (2nd ed.). Abingdon/Oxford/New York: Routledge.

Moon, J. (2008). *Critical thinking: An exploration of theory and practice.* London/New York: Routledge.

Moon, J. (2013). *Reflection in Learning and professional development theory and practice.* Hoboken: Taylor and Francis.

Orr, S., & Blythman, M. (2002). The process of design is almost like writing an essay. *Writing Center Journal, 22*(2), 39–54.

Orr, S., Blythman, M., & Mullin, J. (2004). Textual and visual interfaces in art and design education (editorial). *Art, Design and Communication in Higher Education, 3*(2), 75–79. doi:10.1386/adch.3.1.75/0.
Orr, S., Dorey Richmond, J., & Richmond, D. (2010). Reflect on this! *Journal of Writing in Creative Practice, 3*(3), 197–210. doi:10.1386/jwcp.3.3.197_1.
Owen-Jackson, G. (Ed.). (2013). *Debates in design and technology education.* Oxon/New York: Routledge.
Padget, S. (2013). *Creativity and critical thinking*. Abingdon: Routledge.
Sims, E. & Shreeve, A. (2012). Signature pedagogies in art and design, chapter 5. In Chick, N. (2012). *Exploring more signature pedagogies: Approaches to teaching disciplinary habits of mind* (pp. 55–67). Sterling: Stylus Publishing.
Stables, K., & Kimbell, R. (2000). The unpickled portfolio: Pioneering performance assessment in design and technology. In R. Kimbell (Ed.), *Design and technology international millennium conference* (pp. 195–203). Wellesbourne: The D&T Association.
Sullivan, G. (2010). *Art practice as research: Inquiry in visual arts* (2nd ed.). Thousand Oaks: Sage Publications.
Williams, P. J. (2012). Investigating the feasibility of using digital representations of work for performance assessment in engineering. *International Journal of Technology and Design Education, 22*(2), 187–203.
Williams, P. J., & Newhouse, C. P. (Eds.). (2013). *Digital representations of student performance for assessment.* Rotterdam: Springer Science & Business Media.
Wood, J. (2004). The tetrahedron can encourage designers to formalize more responsible strategies. *Art, Design & Communication In Higher Education, 3*(3), 175–192. doi:10.1386/adch.3.3.175/1.
Wood, J. (2008). Auspicious reasoning: Can metadesign become a mode of governance? *Journal Of Writing In Creative Practice, 1*(3), 301–316. doi:10.1386/jwcp.1.3.301/1.
Wood, J. (2011). Languaging change from within; Can we metadesign biodiversity? *Journal of Science and Innovation, 1*(4), 27–31.
Wood, J. (2013). Re-languaging the creative: Designing as a comprehensive act of combination. *Journal Of Writing In Creative Practice, 6*(1), 59–70. doi:10.1386/jwcp.6.1.59_1.
von Mengersen, B. (2013). Activating creative forms of reflective writing for sustainable self-directed learning in the lab/workshop/design-studio. In *PATT27, Technology education for the future: A play on sustainability* (pp. 346–354). Christchurch: Technology Environmental Science and Mathematics Education Research Centre, University of Waikato.
von Mengersen, B. (2015). Reflective writing for design and technology: Shifting the focus from justification to critique. In *PATT29, Plurality of approaches in design and technology education* (pp. 441–448). Marseille: Presses Universitaire de Provence.

The manufacturer's authorised representative in the EU is Springer
Nature Customer Service Centre GmbH, Europaplatz 3, 69115 Heidelberg,
Germany. If you have any concerns regarding our products, please
contact ProductSafety@springernature.com

Printed and bound by CPI Group (UK) Ltd, Croydon, CR0 4YY
22/09/2026
02221058-0016